Methods of Non-α-Amino Acid Synthesis

Michael B. Smith
University of Connecticut
Storrs, Connecticut

Marcel Dekker, Inc. New York • Basel • Hong Kong

Library of Congress Cataloging-in-Publication Data

Smith, Michael.
Methods of non-α-amino acid synthesis / Michael B. Smith.
p. cm.
Includes bibliographical references.
ISBN 0-8247-9604-7
1. Amino acids–Synthesis. I. Title.
QD431.S685 1995
547.7'50459–dc20 95-11621
CIP

The publisher offers discounts on this book when ordered in bulk quantities. For more information, write to Special Sales/Professional Marketing at the address below.

This book is printed on acid-free paper.

MARCEL DEKKER, INC.
270 Madison Avenue, New York, New York 10016

Current printing (last digit):
10 9 8 7 6 5 4 3 2 1

PRINTED IN THE UNITED STATES OF AMERICA

Preface

Amino carboxylic acids (or just amino acids) are, as a class of organic molecules, among the most important and useful compounds known. The chemistry of amino acids has been studied for well over a century. The importance of α-amino acids [$RCH(NH_2)COOH$] in mammalian biology and as synthetic intermediates is well established and there have been many reviews, several monographs and thousands of individual research papers in this area. The synthesis of α-amino acids has also been the subject of many books and reviews. In addition to α-amino acids, there are many other amino carboxylic acids where the amino group is not on the carbon immediately adjacent to the carboxyl group (the α position) but rather is attached to another carbon in the chain (e.g., the β, γ, δ carbon, etc.). These *non-α-amino acids* are also components of biologically important molecules, are important in the pharmaceutical industry, and are useful in materials science, particularly for polymer synthesis.

This monograph gives a representative overview of the synthesis of non-α-amino acids beginning with the year 1962. With a few exceptions, citations prior to that date are leading references taken from other citations. The work prior to 1962 is well represented, however, in the synthetic approaches incorporated into this book since the "early work" is the basis of most of those approaches. Every attempt was made to give the most common and useful methods for the synthesis of non-α-amino acids. This monograph will focus attention on acyclic amino acids of C_3-C_{10} and includes aminoalkanoic carboxylic acids, aminoalkenoic acids and aminoalkynoic acids. The synthesis of amino carboxylic acids attached to or incorporated in rings of three to ten carbons is also presented, including amino cycloalkanoic and amino cycloalkenoic acids. Although saturated heterocyclic derivatives and aryl substituted amino acids are discussed, aromatic amino carboxylic acids and heteroaromatic amino carboxylic acids are not discussed, except where they are synthetic precursors to or related to aliphatic amino carboxylic acids.

The primary goal of this book is to summarize synthetic approaches to non-α-amino acids, particularly those amino acids that are key synthetic intermediates or important compounds in their own right. Achiral amino carboxylic acids are discussed throughout and special attention is focused on both chiral nonracemic and chiral racemic amino acids in chapter five, emphasizing the diastereoselectivity and/or enantioselectivity of each synthetic process. Having such synthetic information collected in one place will, it is hoped, facilitate current research and stimulate new research in this important area.

Saturated amino acids of all types are discussed, as well as unsaturated, alkyl and aryl substituted, and amino acids that bear a heteroatom functional group. In general, four structural types of alkenyl amino acids are considered. In the first type, the double bond can be conjugated to the carbonyl group and the amine moiety can be attached directly to the double bond. Secondly, the amino group can be attached to the saturated carbon chain. The third and fourth types focus on the double bond, which can be in or out of conjugation with the carboxyl, and the amine group can be attached anywhere on the chain. Substituents can appear anywhere on the chain. Less formalism is required for discussing structural variations in alkynyl amino acids. The triple bond is both conjugated and unconjugated and the amino group can appear at almost any point in the carbon. Only one derivative has the amino attached directly to the triple bond, however, and it is 3-amino 2-propynoic acid.

Non-α-amino acids have been important for over a century and their synthesis and chemical transformations constitute a rich and varied chapter in organic chemistry. Despite this, virtually no compilations exist that discuss these compounds. This stands in stark contrast to α-amino acids, for which thousands of articles have appeared, as well as numerous books, monographs and reviews. Since part of my own research involves the used of non-α-amino acids in the development of synthetic methods, I was particularly interested in this type of reference book. This was the impetus for preparing this monograph, which is a compilation of synthetic methods that can be used to prepare non-α-amino acids.

This monograph is a comprehensive review of synthetic methods used to prepare non-α-amino acids. The source material surveyed in this book formally begins with literature after 1962, but there are many citations and examples taken from literature published prior to 1962. Virtually all of these early references were taken from literature citations in later articles. I am confident that the "older" literature is well represented, particularly in its survey of the synthesis of and the synthetic methods applied to non-α-amino acids.

This monograph is divided into seven chapters. In the first chapter, methods for synthesizing the fundamental structural types found in acyclic amino acids are presented. In this chapter, the synthetic methods rely on substituent refunctionalization. In chapter two, refunctionalization again plays a prominent role. The major difference is that cyclic precursors are used to prepare acyclic amino acids. Chapter three discusses those methods that rely on conjugate addition reactions. Chapter four brings together those synthetic strategies that rely on condensation reactions involving a nucleophilic species with a carbonyl compound, or a carbonyl surrogate. This technique is particularly useful for incorporating heteroatom substituents such as hydroxyl, oxo, or amino into an amino acid. Chapter five collects a variety of synthetic methods together under the umbrella of stereoselectivity. Those reactions and strategies that lead to good diastereoselectivity and/or enantioselectivity during the synthesis are discussed. Chapter six again collects several different strategic approaches for the synthesis of amino acids. The unifying theme of this chapter is the preparation of biologically important amino acids, with particular emphasis given to GABOB and carnitine as well as statine and its derivatives. Chapter seven concludes the book and discusses amino acids that have a carboxyl group attached to a ring. Those derivatives for which the amino group is attached to the ring and those in which the nitrogen is part of the ring (heterocyclic derivatives) are discussed.

This monograph will be particularly useful to those of us engaged in synthesis and synthetic manipulations involving amino acids. Its utility is not limited to that audience, however. Those involved in developing synthetic methods and those pursuing the total synthesis of natural products will also find the book useful because it contains a useful review of strategies and reaction types found in nitrogen-containing molecules. This monograph will be a useful reference to quickly find what synthetic methods have been used, and in what systems. With such a background, it is my hope that new areas of research in this important area will be discovered more quickly. I believe this work will be a valuable contribution to the ever-growing field of non-α-amino acids.

Where there are errors, I take complete responsibility. Every effort has been made to keep the manuscript error-free but if errors are found please contact me and I will try to answer any questions that may be raised.

This manuscript was prepared with a Macintosh IIci computer using Microsoft Word, version 5.1. All drawings in this book were prepared using CSC ChemDraw Plus, from Cambridge Scientific Computing, Inc., Cambridge, MA.

I want to thank Dr. Maurits Dekker, who made an intriguing suggestion several years ago concerning one of my papers. This suggestion for a book led to the preparation of this manuscript and I thank Dr. Dekker for inspiring this work. I want to thank Mr. Joe Stubenrauch and Ms. Anita Lekhwani of Marcel Dekker Inc., for editing this manuscript and providing the care and knowledge required to convert it into a book. I particularly want to thank the independent reviewers of this manuscript, with whom I must share credit for the final form of this monograph. Their reviews were insightful, very helpful, and I thank them very much. Most of all I want to thank my wife Sarah and my son Steven for their patience and understanding during the research for, and preparation of this manuscript.

Michael B. Smith

Contents

Abbreviations

Ac	Acetyl	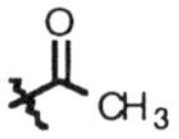
acac	Acetylacetonate	
AIBN	*azo-bis*-isobutyronitrile	
aq.	Aqueous	
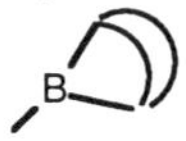	9-Borabicyclo[3.3.1]nonylboran-9-yl	
9-BBN	9-Borabicyclo[3.3.1]nonane	
BINAP	*2R,3S*-2,2'-*bis*-(diphenylphosphino)-1,1'-binaphthyl	
Bn	Benzyl	
Bz	Benzoyl	
BOC	*t*-Butoxycarbonyl	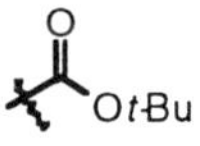
bpy (Bipy)	2,2'-Bipyridyl	
Bu	*n*-Butyl	$-CH_2CH_2CH_2CH_3$
CAM	Carboxamidomethyl	
CAN	Ceric ammonium nitrate	$(NH)_2Ce(NO_3)_6$
c-	Cyclo-	
cat.	Catalytic	
Cbz	Carbobenzyloxy	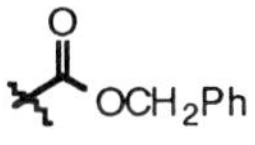
Chirald	2S,3R-(+)-4-Dimethylamino 1,2-diphenyl-3-methylbutan-2-ol	

COD	1,5-Cyclooctadienyl	
COT	1,3,5-Cyclooctatrienyl	
Cp	Cyclopentadienyl	
CSA	10-Camphorsulfonic acid	
CTAB	Cetyltrimethylammonium bromide	$C_{16}H_{33}NMe_3^+ Br^-$
Cy (*c*-C_6H_{11})	Cyclohexyl	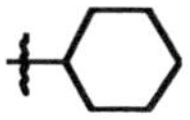
°C	Temperature in Degrees Centigrade	
DABCO	1,4-Diazabicyclo[2.2.2]octane	
dba	Dibenzylidene acetone	
DBE	1,2-Dibromoethane $BrCH_2CH_2Br$	
DBN	1,8-Diazabicyclo[5.4.0]undec-7-ene	
DBU	1,5-Diazabicyclo[4.3.0]non-5-ene	
DCC	1,3-Dicyclohexylcarbodiimide	*c*-C_6H_{13}-N=C=N-*c*-C_6H_{13}
DCE	1,2-Dichloroethane	$ClCH_2CH_2Cl$
DDQ	2,3-Dichloro-5,6-dicyano-1,4-benzoquinone	
%de	% Diastereomeric excess	
DEA	Diethylamine	$HN(CH_2CH_3)_2$
DEAD	Diethyl azodicarboxylate	EtO_2C-N=NCO_2Et
Dibal-H	Diisobutylaluminum hydride	$(Me_2CHCH_2)_2AlH$
Diphos (dppe)	1,2-*bis*-(Diphenylphosphino)ethane	$Ph_2PCH_2CH_2PPh_2$
Diphos-4 (dppb)	1,4-*bis*-(Diphenylphosphino)butane	$Ph_2P(CH_2)_4PPh_2$
DMAP	4-Dimethylaminopyridine	
DME	Dimethoxyethane	$MeOCH_2CH_2OMe$
DMF	*N,N'*-Dimethylformamide	
dppb	1,4-*bis*-(Diphenylphosphino)butane	$Ph_2P(CH_2)_4PPh_2$
dppe	1,2-*bis*-(Diphenylphosphino)ethane	$Ph_2PCH_2CH_2PPh_2$
dppf	*bis*-(Diphenylphosphino)ferrocene	
dppp	1,3-*bis*-(Diphenylphosphino)propane	$Ph_2P(CH_2)_3PPh_2$
dvb	Divinylbenzene	
e^-	Electrolytic reactions	
%ee	% Enantiomeric excess	
EE	1-Ethoxyethyl	EtO(Me)CH-
Et	Ethyl	$-CH_2CH_3$
EDA	Ethylenediamine	$H_2NCH_2CH_2NH_2$
EDTA	Ethylenediaminetetraacetic acid	
FMN	Flavin mononucleotide	
fod	*tris*-(6,6,7,7,8,8,8)-Heptafluoro-2,2-dimethyl-3,5-octanedionate	
Fp	Cyclopentadienyl-*bis*-carbonyl iron	
FVP	Flash vacuum pyrolysis	

h	hour (hours)	
hν	Irradiation with light	
1,5-HD	1,5-Hexadienyl	
HMPA	Hexamethylphosphoramide	$(Me_2N)_3P{=}O$
HMPT	Hexamethylphorous triamide	$(Me_2N)_3P$
iPr	Isopropyl	$-CH(CH_3)_2$
LICA (LIPCA)	Lithium cyclohexylisopropylamide	
LDA	Lithium diisopropylamide	$LiN(iPr)_2$
LHMDS	Lithium hexamethyl disilazide	$LiN(SiMe_3)_2$
LTMP	Lithium 2,2,6,6-tetramethylpiperidide	
mCPBA	*meta*-Chloroperoxybenzoic acid	
Me	Methyl	$-CH_3$
MEM	β-Methoxyethoxymethyl	$MeOCH_2CH_2OCH_2-$
Mes	Mesityl	2,4,6-tri-Me-C_6H_2-
MOM	Methoxymethyl	$MeOCH_2-$
Ms	Methanesulfonyl	CH_3SO_2-
MS	Molecular sieves (3Å or 4Å)	
MTM	Methylthiomethyl	CH_3SCH_2-
NAD	Nicotinamide adenine dinucleotide	
NADP	Sodium triphosphopyridine nucleotide	
Naphth	Naphthyl ($C_{10}H_7$)	
NBD	Norbornadiene	
NBS	*N*-Bromosuccinimide	
NCS	*N*-Chlorosuccinimide	
NIS	*N*-Iodosuccinimide	
Ni(R)	Raney nickel	
NMO	*N*-Methylmorpholine-*N*-oxide	
Oxone	2 $KHSO_5 \cdot KHSO_4 \cdot K_2SO_4$	
	Polymeric backbone	
PCC	Pyridinium chlorochromate	
PDC	Pyridinium dichromate	
PEG	Polyethylene glycol	
Ph	Phenyl	
PhH	Benzene	
PhMe	Toluene	
Phth	Phthaloyl	
Pip	Piperidyl	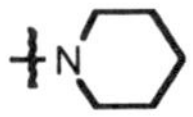
Pr	*n*-Propyl	$-CH_2CH_2CH_3$
Py	Pyridine	
quant.	Quantitative yield	

Red-Al	$[(MeOCH_2CH_2O)_2AlH_2]Na$	
sBu	*sec*-Butyl	$CH_3CH_2CH(CH_3)$
sBuLi	*sec*-Butyllithium	$CH_3CH_2CH(Li)CH_3$
sec	seconds	
Siamyl	Diisoamyl	$(CH_3)_2CHCH(CH_3)$-
TASF	*tris*-(Diethylamino)sulfonium difluorotrimethyl silicate	
TBAF	Tetrabutylammonium fluoride	*n*-$Bu_4N^+ F^-$
TBDMS	*t*-Butyldimethylsilyl	*t*-$BuMe_2Si$
TBHP (*t*-BuOOH)	*t*-Butylhydroperoxide	Me_3COOH
t-Bu	*tert*-Butyl	$-C(CH_3)_3$
TEBA	Triethylbenzylammonium	$Bn(Et)_3N^+$
TEMPO	Tetramethylpiperdinyloxy free radical	
TFA	Trifluoroacetic acid	CF_3COOH
TFAA	Trifluoroacetic anhydride	$(CF_3CO)_2O$
Tf (OTf)	Triflate	$-SO_2CF_3$ [$-OSO_2CF_3$]
THF	Tetrahydrofuran	
THP	Tetrahydropyranyl	
TMEDA	Tetramethylethylenediamine	$Me_2NCH_2CH_2NMe_2$
TMS	Trimethylsilyl	$-Si(CH_3)_3$
TMP	2,2,6,6-Tetramethylpiperidine	
Tol	Tolyl	$-C_6H_4(4-CH_3)$
Tr	Trityl	$-CPh_3$
TRIS	Triisopropylphenylsulfonyl	
Ts(Tos)	Tosyl (*p*-Toluenesulfonyl)	4-MeC_6H_4-
·)))	Sonication	
X_c	Chiral auxiliary	

1

Functional Group Exchange Strategies

1.1. Functional Group Transformations that Generate Amines

Even the most simple non-α-amino carboxylic acids are important compounds. The initial focus is on the ω-aminoalkanoic acids (the amino group is on the carbon furthest removed from the carboxyl group), since they describe the types of amino acids to be discussed. Perhaps the simplest non-α-amino carboxylic acid is 2-amino propanoic acid, which has the common name of β-alanine (*1.1*). 4-Aminobutanoic acid (*1.2*) is also known as γ-aminobutyric acid and given the abbreviation GABA.

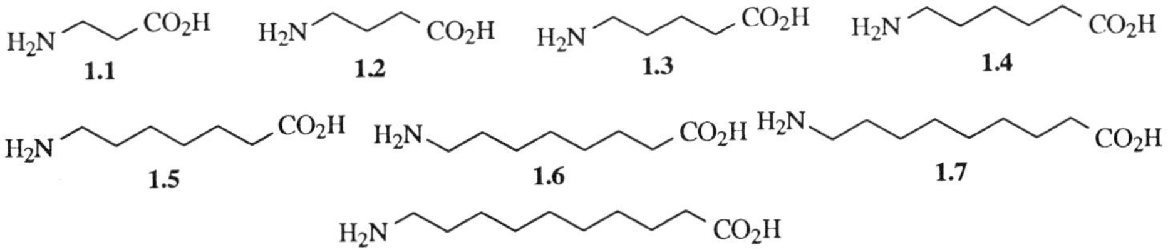

1.1 **1.2** **1.3** **1.4** **1.5** **1.6** **1.7** **1.8**

5-Aminopentanoic acid (*1.3*) is known as δ-aminovaleric acid and given the abbreviation DAVA. Both 6-aminohexanoic acid (*1.4*), also known as aminocaproic acid, and 7-aminoheptanoic acid (*1.5*) are well known. 8-Aminooctanoic acid (*1.6*),

9-amino-nonanoic acid (***1.7***) and 10-aminodecanoic acid (***1.8***), which is also known as aminocapric acid, complete the first eight amino carboxylic acids in this series.

These particular non-α-amino carboxylic acids can be generally termed ω-amino carboxylic acids. There are several synthetic routes to this structurally simple yet important class of compounds. These routes include functional group transformations that generate either the amino group or the carboxyl group. In some cases, both groups are incorporated in a single synthetic step. The synthetic routes to these "simple" amino acids can generally be used to prepare substituted and functionalized amino acids.

1.1.A. Reaction with Ammonia and Amines

This section will include functional group transformations that convert one group into an amine, using ammonia or an amine as a nucleophile. The fundamental idea behind this approach is to functionalize a halide (or another molecule bearing a suitable leaving group) by direct incorporation of an amino group.

1.1.A.i. Ammonia

Both ammonia or ammonium hydroxide (often these are used in a reaction bomb) react with an alkyl halide or sulfonate ester to give an amine. If the molecule also has a carboxyl substituent or an acid surrogate, the product is an amino acid or can be converted to one. 1-Bromononanoic acid (***1.9***), for example, was converted to 9-aminononanoic acid (***1.7***) by this procedure.[1] There are many examples of this type

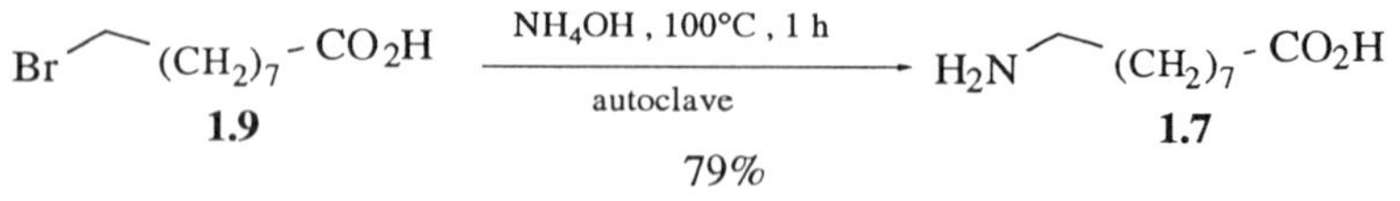

of reaction, involving many types of acid derivatives. Reaction of bromo-ester ***1.10*** with aqueous ammonia for nine days, for example, gave ethyl 2,2-diethyl-5-aminopentanoate (***1.11***).[2] As in this example, the yield of direct displacement reactions of a halide by ammonia can be low. The yield of the amino acid product can sometimes be improved if an amine surrogate is used for the displacement (see section 1.1.B).

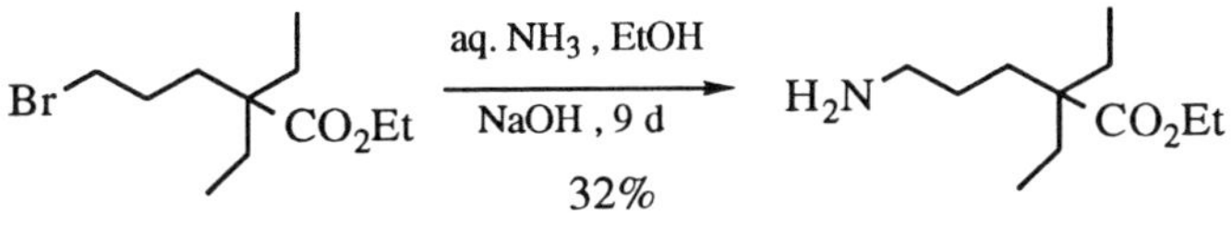

1.10 **1.11**

Both alkyl and alkenyl amino acids can be prepared by this approach. A common method for introducing the halide into an alkene-bearing molecule is illustrated by the reaction of *E*-pent-2-enoic acid with N-bromosuccinimide to form ***1.12***.[3] Subsequent treatment with ammonia led to displacement of the bromine moiety to give 4-aminopent-2-enoic acid (***1.13***). An alternative method reacted ***1.12*** with sodium azide and then reduced the azide with zinc and acetic acid[3] (see section 1.1.B.iv). Allylic halogenation in systems such as ***1.12*** are well known.

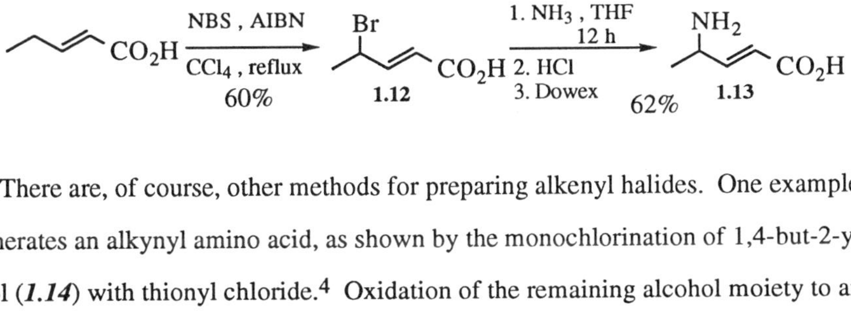

1.12 **1.13**

There are, of course, other methods for preparing alkenyl halides. One example generates an alkynyl amino acid, as shown by the monochlorination of 1,4-but-2-yne diol (***1.14***) with thionyl chloride.[4] Oxidation of the remaining alcohol moiety to an

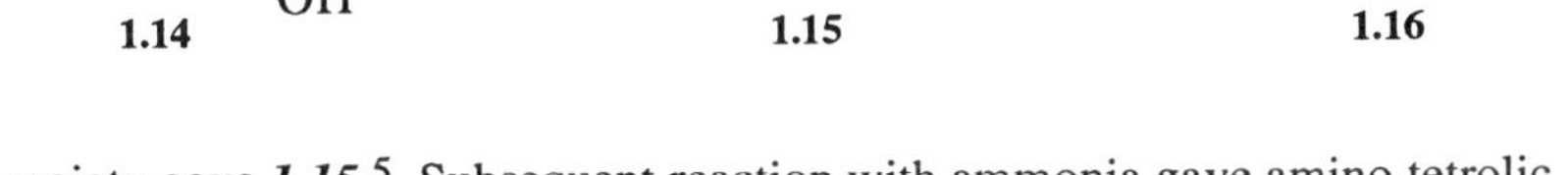

1.14 **1.15** **1.16**

acid moiety gave ***1.15***.[5] Subsequent reaction with ammonia gave amino tetrolic acid (***1.16***, 4-aminobut-2-ynoic acid). A variety of amines also reacted with ***1.15***, leading to substituted alkynylamino acids.[5,6]

Table 1.1. Synthesis of amino acids by direct displacement of halides with ammonia.

No.	Transformation	Reference
1.	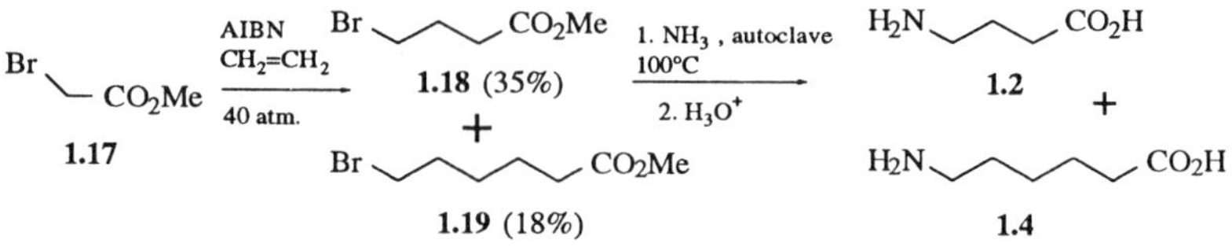	8
2.	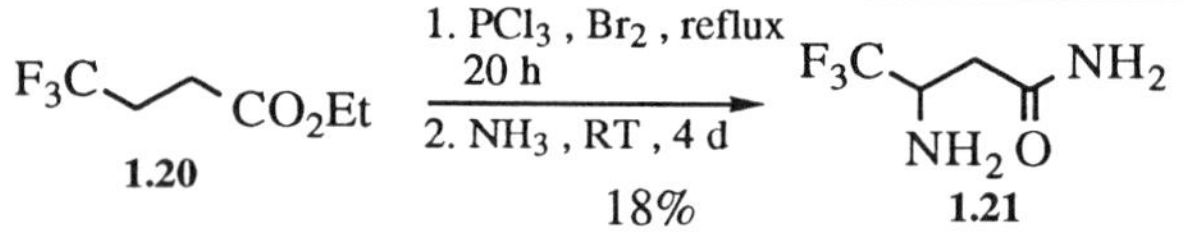	9
3.	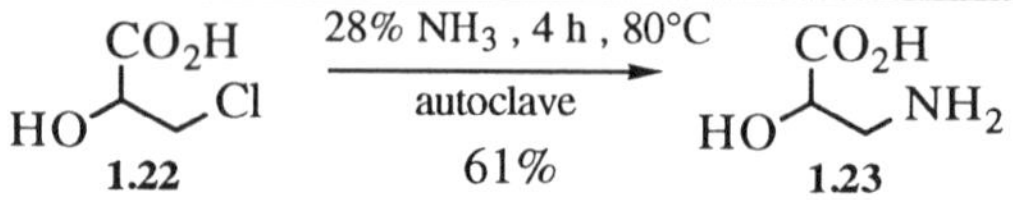	10
4.	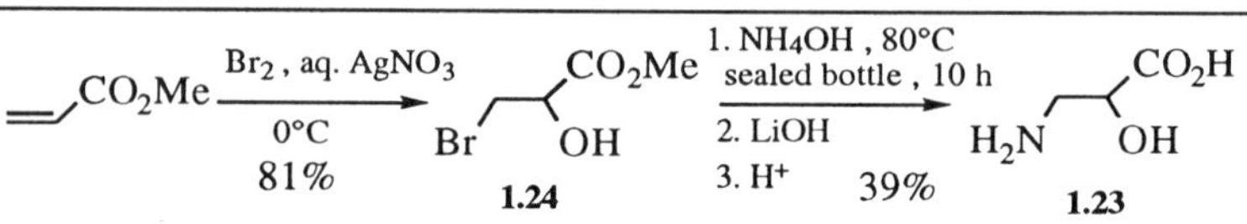	13
5.	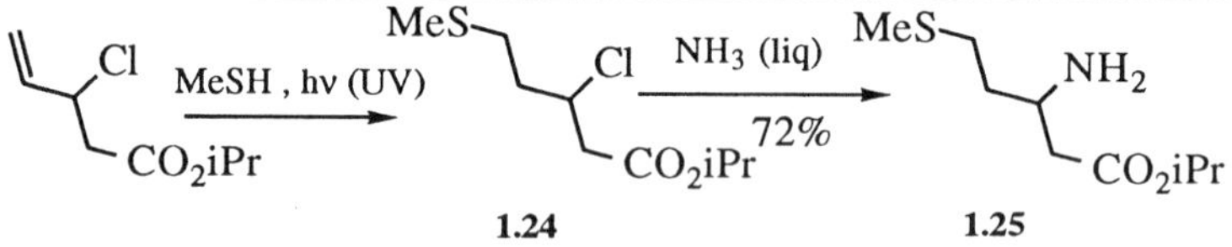	14
6.	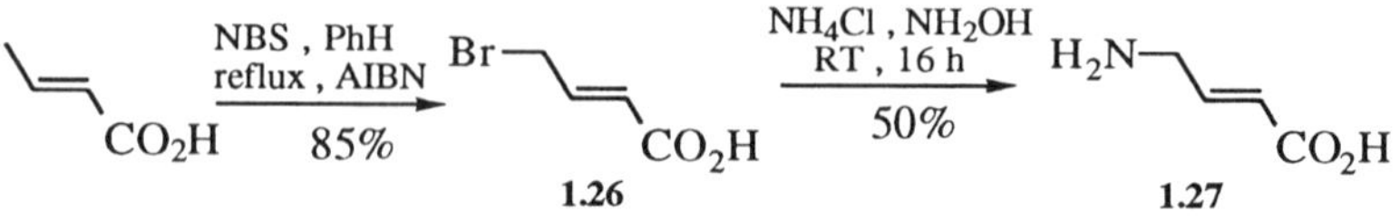	15
7.	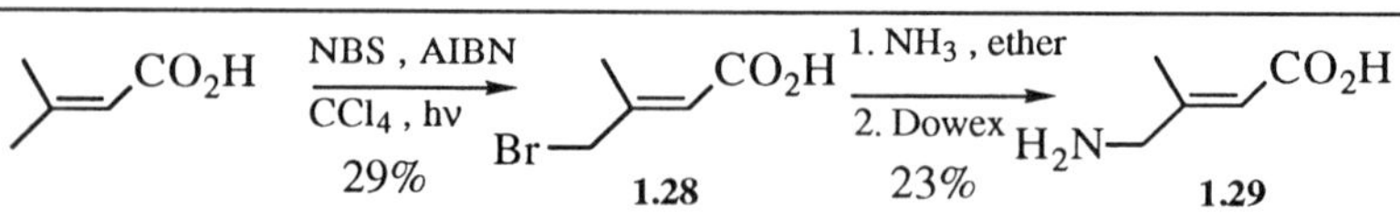	16

8.	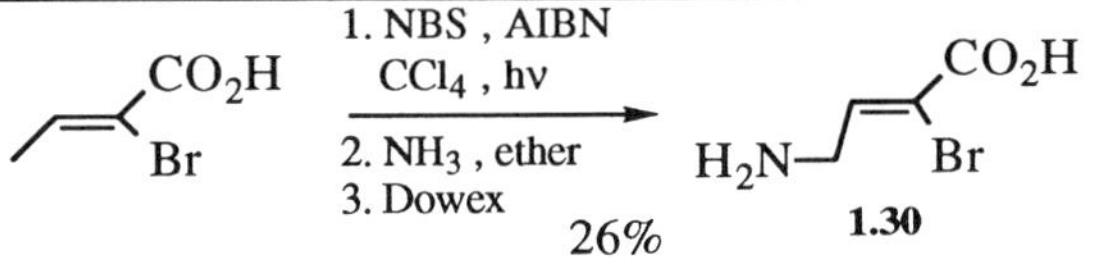	16a
9.	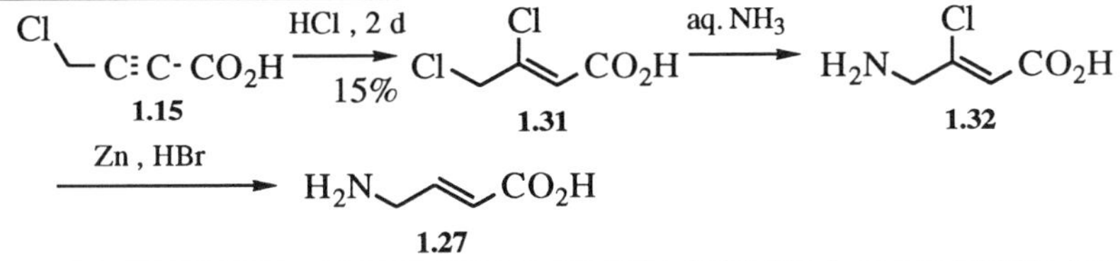	6

There are many reports in the literature that use this strategy and generate unsubstituted, alkyl-substituted and functionalized amino acids. Many of them are virtually identical from a strategic viewpoint, but a variety of substituents and functional groups can be incorporated. Several examples are shown in Table 1.1.

Examples 1 to 5 in Table 1.1 illustrate simple displacement of a bromide (reactions 1 and 4) or a chloride (reactions 2, 3, and 5). Reaction 1 illustrates a radical chain extension reaction[7] that reacts ethylene and methyl bromoacetate (***1.17***), with AIBN (*azobis*-isobutyronitrile) as a catalyst, to produce the requisite bromo-ester. The reaction is not very selective, however, since it produced a mixture of ***1.18*** and ***1.19***. The reaction was not very selective and the observed products arose from addition of one and two equivalents of ethylene, respectively, to ***1.17***. Subsequent treatment with ammonia, the key reaction for this section, led to a mixture of 4-aminobutanoic acid and 6-aminohexanoic acid, ***1.2*** and ***1.4***.[8] In reaction 2, ethyl 4,4,4-trifluoropropanoate (***1.20***) reacted with PCl_3 and bromine to give ethyl 3-bromo-4,4,4-trifluorobutanoate, which reacted with ammonia to give 3-amino-4,4,4-trifluorobutanamide, ***1.21***.[9] Reaction 3 shows the reaction of β-halo lactic acid (3-chloro-2-hydroxypropanoic acid, ***1.22***) with 28% aqueous ammonia, in an autoclave, to give isoserine (3-amino-2-hydroxypropanoic acid, ***1.23***).[10] Both halo-acids[11] and halo-esters[12] have been converted into hydroxy amino acids: typical products are 3-

hydroxy-4-aminobutanoic acid[10,12a] and 2-hydroxy-4-aminobutanoic acid.[12b] Reaction 4 involves a reaction of methyl acrylate with bromine and silver nitrate to give ***1.24***.[13] This intermediate is obviously closely related to ***1.22***. Subsequent reaction with ammonia led to isoserine (***1.23***). Reaction 5 illustrates incorporation of a methylthio moiety into an amino acid. Photolysis of isopropyl 3-chloropent-4-enoate in the presence of methanethiol led to ***1.24***, and reaction with ammonia led to displacement of the chloride and formation of isopropyl 3-amino-5-methylthio-pentanoate, ***1.25***.[14]

Reaction 6 shows another displacement of an allylic bromide, analogous to formation of ***1.12*** above. Reaction of crotonic acid with N-bromosuccinimide gave ***1.26***[15] and subsequent reaction with aqueous ammonia gave 4-aminobut-2-enoic acid (***1.27***). Several other alkenyl acids were used in this particular study as synthetic precursors. Another example (reaction 7) began with the bromination of 3-methylbut-2-enoic acid using N-bromosuccinimide to give ***1.29***. Subsequent reaction with ammonia displaced the bromide to give ***1.80***.[16,3] A brominated amino acid was prepared in reaction 8 by treatment of 2-bromobut-2-enoic acid with N-bromosuccinimide and then ammonia to give 4-amino-2-bromobut-2-enoic acid, ***1.30***[16a] (the analogous 4-amino-2-chlorobut-2-enoic acid was formed by reaction with N-chlorosuccinimide, but in only 9% yield). In reaction 9, the starting material is alkynyl acid ***1.15***, which reacted with HCl to give ***1.31***.[6] Displacement at the primary chloride by ammonia gave 4-amino-3-chlorobut-2-enoic acid, ***1.32***. Subsequent reduction of the vinyl chloride with zinc and HBr gave 4-aminobut-2-enoic acid (***1.27***).

1.1.A.ii. Reaction with Amines

This reaction is identical to the ammonia reactions shown in the first section except that an amine is used as the nucleophile. Both N-alkyl and N,N-dialkylamino

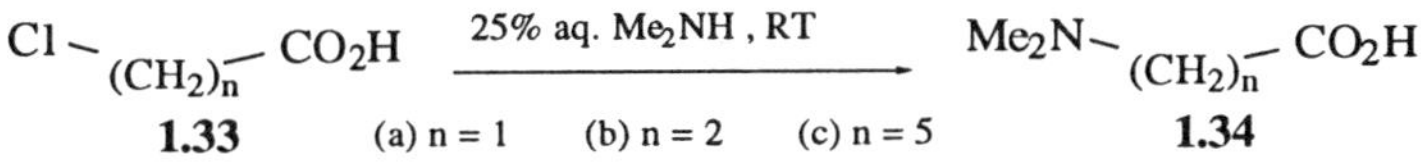

acids can be produced by this approach. A simple example is the reaction of 2-chloroethanoic acid (***1.33a***), 3-chloropropanoic acid (***1.33b***) or 6-chlorohexanoic acid (***1.33c***) with a 25% aqueous solution of dimethylamine. The products of these reactions are 69% of ***1.34a***, 54% of ***1.34b*** and 53% of ***1.34c*** [2-(N,N-dimethylamino)ethanoic acid; 3-(N,N-dimethylamino)propanoic acid; 6-(N,N-dimethylamino)hexanoic acid, respectively].[17] In these examples, the yields are better than the analogous reaction with ammonia.

Allylic halides can be useful synthetic precursors, as seen for reactions with ammonia. The reaction of ***1.35*** with diethylamine to give ethyl 4-(N,N-diethylamino)but-2-enoate, ***1.36***.[18] In this case the yield of ***1.36*** is moderate rather than high, but generally better than the identical reaction with ammonia (*vide supra*). A

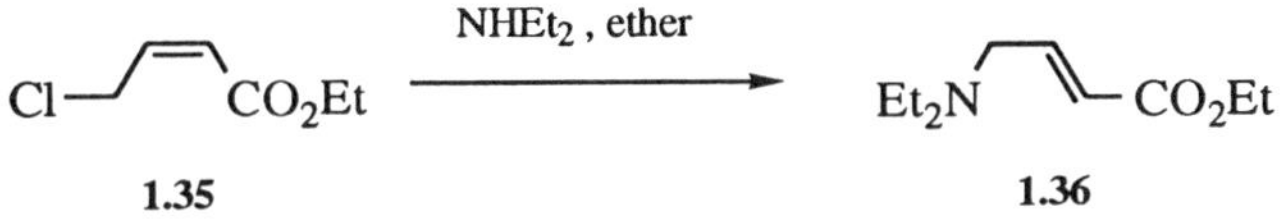

last example is given to show that the halide can be incorporated at a position removed from the alkene and the acid moiety. When ***1.37*** was treated with dimethylamine, 8-(N,N-dimethylamino)oct-2-enoic acid (***1.38***) was obtained.[19] Similarly, 6-(N,N-dimethylamino)hex-2-enoic acid was prepared in 58% yield and 7-(N,N-dimethylamino)hept-2-enoic acid was prepared in 61% yield using this method.[19] A variety of amines can be used including *S*-phenethylamine, which was reacted with ethyl 4-bromobut-2-enoate to give ethyl 4-(N-phenethylamino)but-2-enoate.[20] The use of chiral auxiliaries such as this to induce formation of a chiral, nonracemic product is discussed in chapter five.

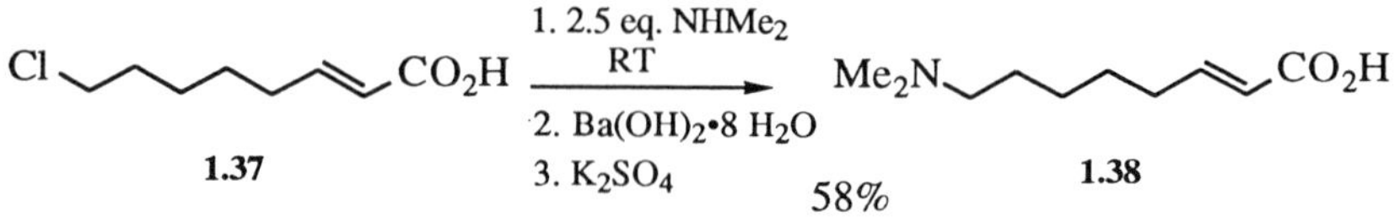

1.37 1.38

1.1.B. Reaction with Amine Surrogates

The reaction of ammonia or amines with halides, even allylic halides, often gave poor yields of the amine product, as seen in section 1.1.A. Such poor yields can often be improved by a two-step sequence that uses an amine surrogate (a reagent that incorporates a nitrogen functional group into a molecule which is later converted to an amino group). The first step is the displacement reaction (usually with an alkyl halide) that introduces the functional group and the second step converts the functional group (the surrogate) into an amine. Common surrogates are phthalimide, nitro (NO_2), cyanide (C≡N), or azide (N_3), which replace a halide or sulfonate ester leaving group and are then hydrolyzed or reduced to give an aminomethyl moiety[21] ($-CH_2NH_2$) or an amino group ($-NH_2$).

1.1.B.i. Phthalimide: The Gabriel Synthesis

The reaction of halides with phthalimide to generate an N-alkyl phthalimide and the subsequent conversion to an amine is well known. This process is referred to as

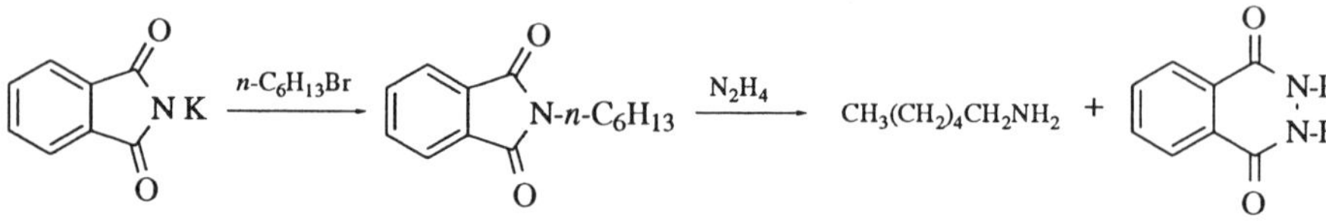

1.39 1.40 1.41

the Gabriel Synthesis.[22] One of the more important methods for converting this N-alkyl phthalimide to the targeted amine is by treatment with hydrazine (the Ing-Manske procedure).[23] A typical example is the reaction of potassium phthalimide

(*1.39*) with 1-bromohexane to give ***1.40***. Subsequent treatment with hydrazine gave hexylamine and ***1.41***.[6] An earlier method for converting the phthalimide to an amine treated the phthalimide derivative first with base and then with a carboxylic acid.[6]

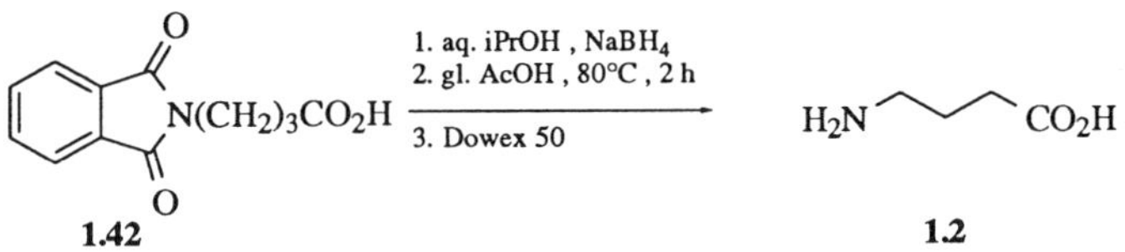

1.42 **1.2**

This classical reaction is also an excellent method for the synthesis of amino carboxylic acids, as illustrated by the conversion of ***1.42*** to 4-aminobutanoic acid, ***1.2***.[24] In this work Ganem introduced another mild method for the deprotection of phthalimides, as shown.

Other functional groups can be tolerated in this procedure, even normally labile groups such as a chlorine that is α- to a cyano group. Reaction of 2,3-dichloropropanenitrile and phthalimide, for example, gave ***1.43***.[25] In this case the cyano group was an acid surrogate since its hydrolysis gave 3-amino-2-chloropropanoic acid, ***1.44***.[25,26]

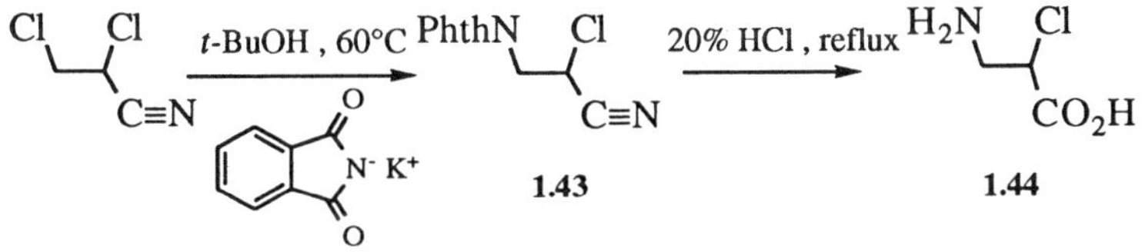

1.43 **1.44**

Amino acid derivatives containing a vinyl chloride moiety can be prepared by adding HCl to an alkynyl amino acid precursor. The reaction of 4-chloro-2-butyn-1-ol (***1.45***) with potassium phthalimide was followed by Jones oxidation[27] to give phthalimidoyl acid ***1.46***.[6] In this particular case, HCl was added to convert the phthalimide moiety to the requisite amine (-NH_2), but HCl also added to the alkyne moiety to give a vinyl chloride, as seen in ***1.32***. The yield of this latter process was very poor, and treatment of ***1.46*** with HBr or HI gave no vinyl halide at all.[6]

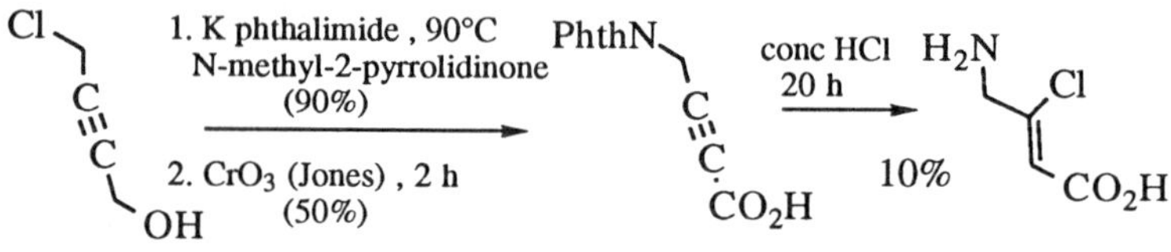

1.45 **1.46** **1.32**

Electrophilic agents other than halide can be used in reactions with phthalimide. One example involved opening the epoxide moiety in ***1.47*** with phthalimide, catalyzed by palladium (0), to give methyl 4-(N-phthalimido)-5-hydroxyhex-2-enoate, ***1.48***.[28] This example is also an example of the powerful palladium (0) chemistry that has been applied to many systems over the last few years.

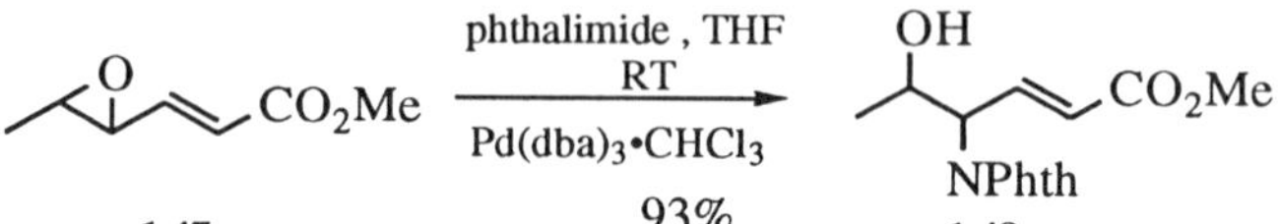

1.47 **1.48**

1.1.B.ii. Cyanide and Nitriles

Cyanide is always an excellent source of nitrogen for amino acid derivatives and acid hydrolysis of nitriles allows them to function as carboxyl surrogates as well (see section 1.3.A.i). The examples in this section will focus on using the cyano moiety as an amine surrogate.

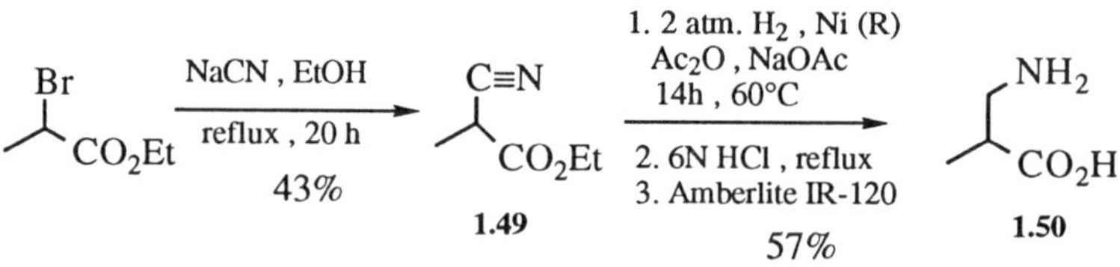

1.49 **1.50**

The most fundamental reaction involves displacement of a halide by cyanide (:NC:$^{\ominus}$), as in the reaction of ethyl 2-bromopropanoate and sodium cyanide to give

1.49.[29] Catalytic hydrogenation, in this case using a Raney nickel catalyst, converted the cyano group into an aminomethyl group (-CH_2NH_2). This was followed by acid hydrolysis of the ester to give 2-methyl-3-aminopropanoic acid, ***1.50***. Note that an ion exchange resin was required for isolation of the "free" amino acid. When the zwitterionic acid is the final product, ion exchange is often necessary for its purification.

An alternative preparation of a nitrile is illustrated by the reaction of the half-ester of 2,2-diethyl malonate (***1.51***) with $SOCl_2$ and then NH_3 to give the amide. Subsequent heating with phosphorus pentoxide led to dehydration and gave nitrile ***1.52***.[30] Catalytic hydrogenation reduced the nitrile to an aminomethyl group, and acid hydrolysis gave 2,2-diethyl-3-aminopropanoic acid (***1.53***). 2-Ethyl-2-cyclohexyl-3-aminopropanoic acid and 2-ethyl-2-benzyl-3-aminopropanoic acid were also prepared by this method.[30]

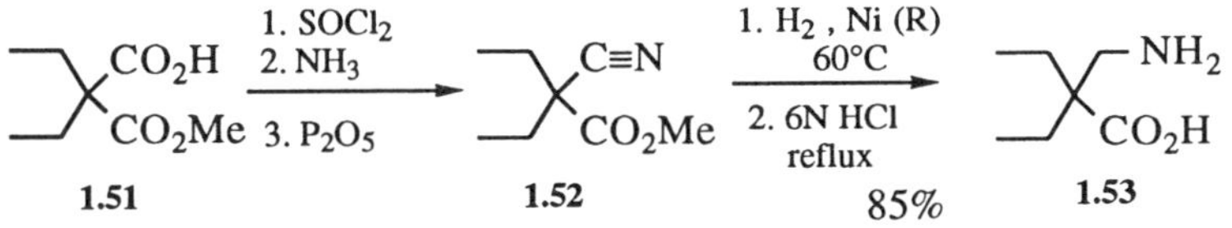

1.51 1.52 1.53

Another example illustrates that catalytic hydrogenation for reduction of a nitrile can tolerate labile functional groups. Raney nickel catalyzed hydrogenation of ***1.54*** gave methyl 5-amino-2-methoxypentanoate (***1.55***).[31] No elimination of the labile α-methoxy group occurred, despite the reaction temperature of 265°C. Methyl 6-amino-3-methoxyhexanoate was prepared in an identical manner from methyl 5-cyano-2-methoxypentanoate.

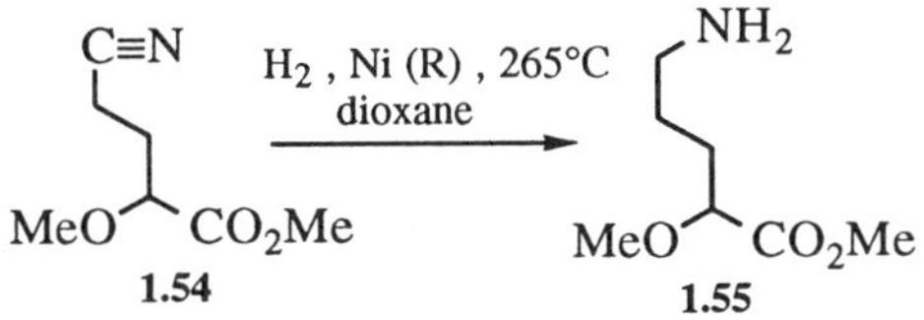

1.54 1.55

A different reduction strategy was used for the conversion of ***1.56*** to ***1.57***. Using a platinum anode and a lead cathode, one nitrile moiety in ***1.56*** was electrolytically reduced to an aminomethyl group. The other cyano group functioned as a carboxyl surrogate and acid hydrolysis of ***1.57*** gave 6-aminohex-4-enoic acid, ***1.58***.[32]

There is another important method by which a cyano group can be converted to an amino group. When a nitrile reacts with a nucleophilic organometallic such as a Grignard reagent, addition to the acyl carbon generates an iminium salt (***1.59***).

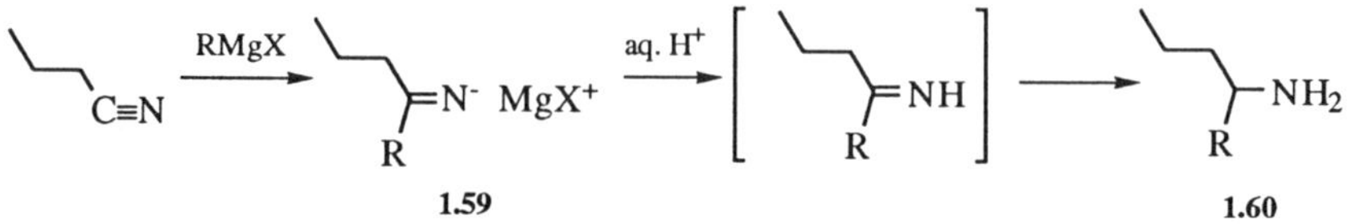

Normally, aqueous acid hydrolysis liberates the imine, which is converted to a carbonyl under the reaction conditions by loss of ammonia. If the intermediate imine is reduced *in situ*, an amine product such as ***1.60*** is generated. If the cyano group is conjugated to an ester (or another carbonyl group), the imine isomerizes to the enamine form (an enamine ester). This latter process is illustrated by the reaction of ethyl α-cyanoacetate and ethylmagnesium bromide, and the product was ethyl 3-aminopent-2-enoate, ***1.61***.[33]

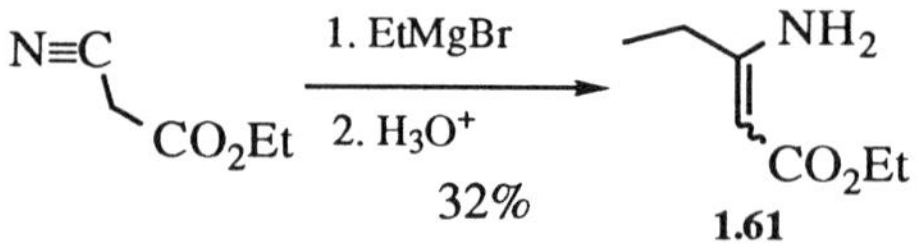

Ethyl 3-aminohex-2-enoate (in 13% yield) and ethyl 3-aminohept-2-enoate (in 16%

yield) were also prepared from the appropriate cyano ester. As can be seen, the yields were generally poor, except for the reaction of ethyl α-cyanoacetate with phenylmagnesium bromide, which gave ethyl 3-amino-3-phenylprop-2-enoate in 83% yield.[33]

A variation of this method prepared the Grignard of 3-bromotoluene and reacted it with the magnesium enolate of ethyl α-cyanoacetate to give ethyl 3-amino-3-(3-methylphenyl)prop-2-enoate, ***1.62***.[34] Ethyl 3-amino-3-phenylprop-2-enoate was prepared in 60% yield by this method, using bromobenzene as the source of the Grignard reagent. Other aryl derivatives were prepared as well. Although the

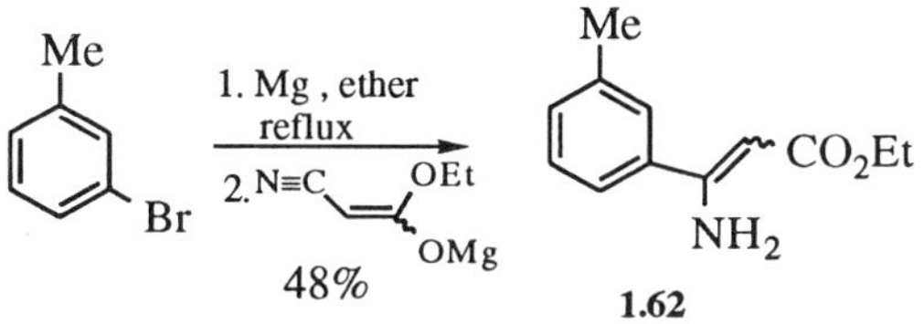

1.62

yields are poor to moderate when alkyl Grignard reagents react with cyano-esters, this method constitutes a fast and relatively easy method for the preparation of substituted 3-aminopropenoic acid derivatives. It is usually a reasonable method for the synthesis of 3-aryl-3-aminopropenoic acid derivatives, and the yields of product range from moderate to good.

1.1.B.iii. Nitro Compounds

Another very important amine surrogate is the nitro group. The reaction of halides and nitrite (NO_2^-) is not a good general route to these compounds, although silver nitrate has been used to prepare nitro compounds. Oxidation of an amine to a nitro group is possible, but strong oxidizing conditions are required. The most common use of a nitro group as a surrogate involves manipulation of an existing nitroalkane, which can be converted to the corresponding enolate anion. Such carbanions react with a variety of electrophilic reagents. An example is reaction of the enolate of nitro-methane, which displaced the chlorine in methyl 3-chloropropanoate to give nitro-ester ***1.63***.[35] Reduction of the nitro group in this example was accomplished by treatment with ammonium formate and palladium-on-carbon, giving ethyl 4-aminobutanoate (***1.64***). This method is limited only by the availability of the nitroalkane precursor and the relative reactivity and availability of the ω-halo ester. The yield of alkylated products such as ***1.63*** often decreases as the chain length of the starting nitroalkane increases, however. It is noted that $LiAlH_4$ will reduce nitro groups to an amine,[36] but in this sequence the ester (or acid) moiety would also be reduced to an alcohol. For this reason, nitro groups are reduced by catalytic hydrogenation or a selective reducing agent milder than hydride (such as ammonium formate).[37]

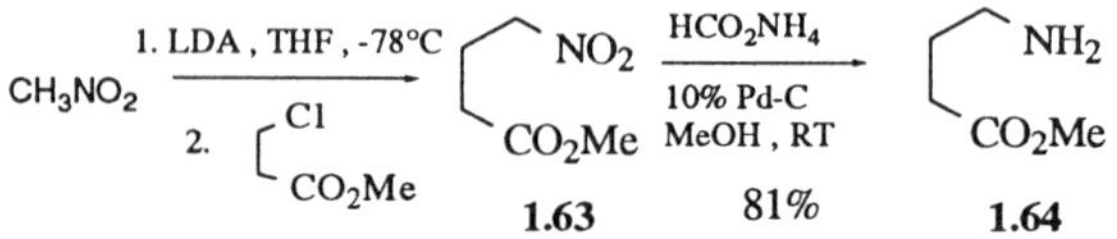

1.63 **1.64**

1.1.B.iv. Azides

Alkyl azides are an important source of amines, where azide ion (N_3^-) is the amine surrogate in nucleophilic displacement reactions. The azide group is readily

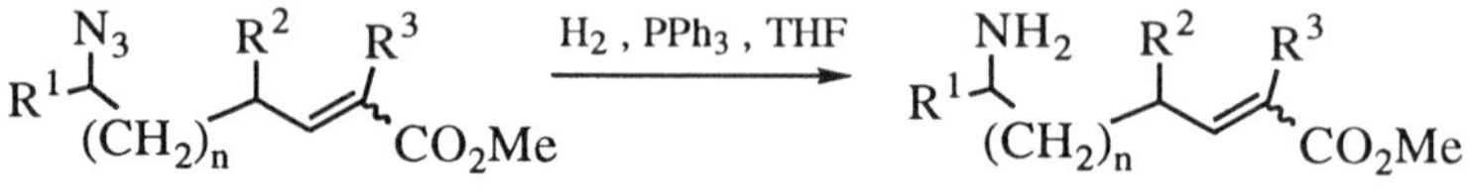

1.65
1.66 a-c

1.67

3.67	n	R^1	R^2	R^3	% yield
a	2	H	H	Me	86
b	2	H	Me	H	83
c	2	Me	H	H	82

converted to an amine ($-N_3 \rightarrow -NH_2$) via catalytic hydrogenation, as illustrated by the conversion of azide ***1.65*** (n = 1, $R^1 = R^2 = H$; R^3 = Me) to ***1.67*** using triphenylphosphine and hydrogen.[38] Reduction of ***1.66a-c*** produced 7-amino-2-methyl-hept-2-enoate (***1.67a***), 7-amino-4-methylhept-2-enoate (***1.67b***), or 7-aminooct-2-enoate (***1.67c***).

Probably the most common approach for inserting an azide moiety into a molecule is by nucleophilic displacement of a halide, as just mentioned. Reaction of oct-9-enoic acid (***1.68***) with HBr generated 9-bromodecanoic acid, and the bromine moiety was displaced by sodium azide to give ***1.69***.[39] Catalytic hydrogenation

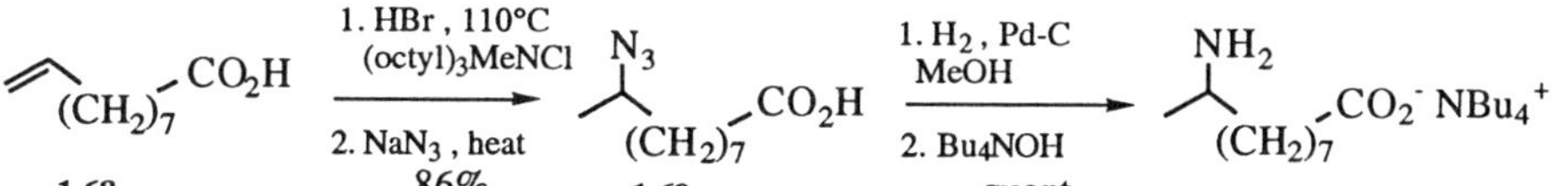

1.68 1.69 1.70

converted the azido moiety to an amine, and 9-aminodecanoic acid was isolated as its tetrabutylammonium salt, ***1.70***. A variation of the azide displacement incorporated the leaving group in a different manner. The carboxylic acid moiety in *S-2.71* was converted to an acid chloride by reaction with oxalyl chloride, allowing its reduction to a

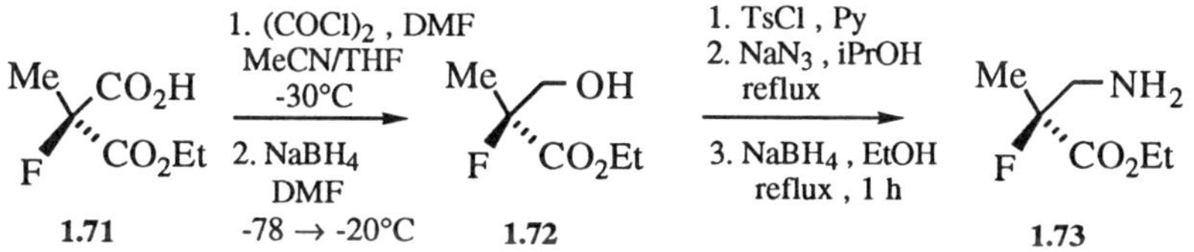

1.71 1.72 1.73

hydroxymethyl moiety in ***1.72***.[40] Reaction of the alcohol with tosyl chloride and pyridine gave a tosylate, which was displaced by azide to give the alkyl azide. In this case, $NaBH_4$ reduction was used to obtain the amine ($NaBH_4$ also reacts with esters[41] but in most cases that reaction is slow relative to more labile functional groups such as azide). The final product was ethyl 3-amino-2-fluoro-2-methylpropanoate, ***1.73***. This example also illustrates that a fluoro group can be incorporated and that the asymmetric center in ***1.71*** is preserved in ***1.73*** (see chapter 5).

Azide is a potent nucleophile and can react with electrophilic substrates other than halides and sulfonate esters. Reaction of azide with the epoxide moiety in carboethoxy oxirane (ethyl glycidate) was followed by catalytic hydrogenation of the azide and hydrolysis of the ester to give isoserine (3-amino-2-hydroxypropanoic acid, ***1.23***).[42] This synthetic approach to isoserine dates to 1879.[43]

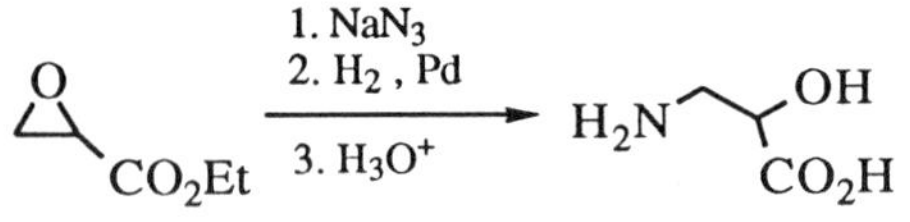

1.23

An alternative route incorporated the acid moiety via reaction of ***1.74*** with sodium cyanide to give ***1.75***, after catalytic hydrogenation of the alkyne.[44] This alkyne starting material (***1.74***) was itself prepared by coupling reaction of the lithium salt of but-3-yn-1-ol with 1-bromo-3-chloropropane, as shown. Hydrolysis of the nitrile group in ***1.75*** gave the acid, which was converted to the corresponding ester. This allow the hydroxyl group to be converted to a mesylate. Subsequent displacement with sodium azide gave an azide that was reduced to the amine by reaction with aqueous triphenylphosphine, giving methyl 8-aminooct-5-enoate, ***1.76***.[44]

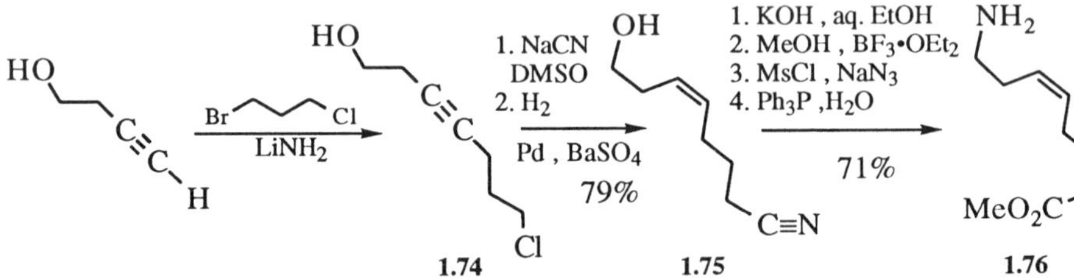

1.74 **1.75** **1.76**

There are several rather specific transformations that involve different reagents. Under acidic conditions, hydrazoic acid (HN_3; usually generated by reaction of sodium azide with an acid) reacted with 1,6-hex-3-ynedioic acid (***1.77***), to convert one carboxyl group to an amine. The final product was 5-aminohex-3-ynoic acid, ***1.78***.[45] This is not a synthetically useful approach in this case, since the yield of ***1.78*** was negligible. It is included simply to show that other forms of azide can be used.

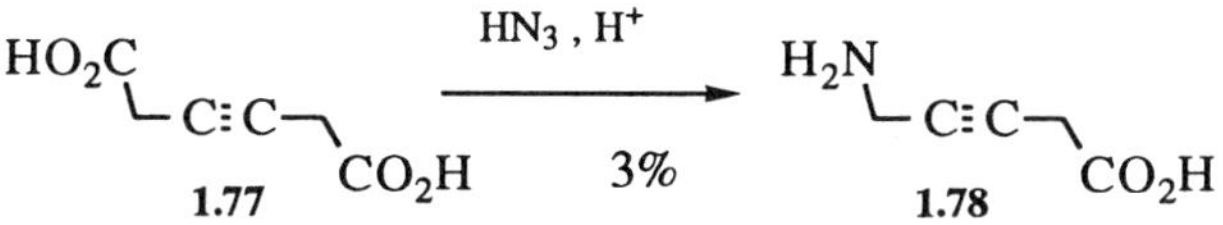

1.77 **1.78**

A last example is more useful. When the acid moiety in ***1.79*** reacted with the phosphoryl azide shown, at 90°C, rearrangement occurred to convert the $-CO_2H$ moiety into a $-NHCO_2H$ moiety. This amino acid [3-amino-2-phenylpropanoic acid] was trapped as the 4-methylbenzyloxy carbamate (***1.80***) shown.[46] For the synthesis of amino acids, this approach requires the preparation of half-esters of dicarboxylic acids. There are several methods known to do this, including selective enzymatic hydrolysis. Where the half-ester is available, this is a good synthetic approach.

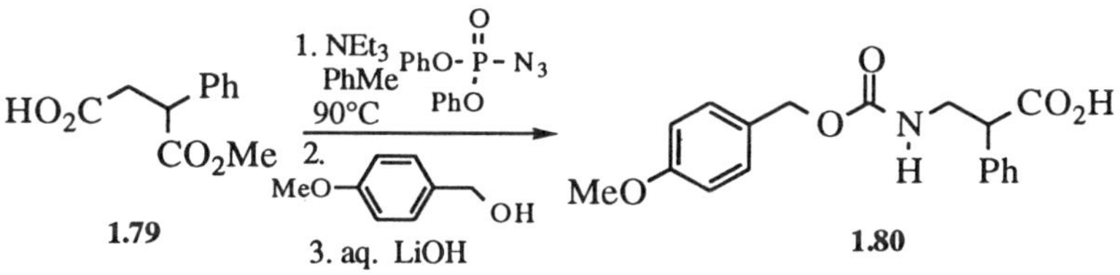

1.79 **1.80**

1.1.B.v. Isocyanates

There are other amine surrogates that can be used to prepare amino acids by reaction with halides or other electrophilic species. Most are rather specific, but one shows some generality. Effenberger showed that reaction of methyl 5-bromohexanoate (***1.19***) with potassium isocyanate gave ***1.81***. In this case, the isocyanate is the amine surrogate, and hydrolysis with concentrated HCl at 100°C gave 6-aminohexanoic acid, ***1.82***, as its HCl salt.[47] 4-Aminobutanoic acid, 3-aminopropanoic acid and 5-aminopentanoic acid were prepared in an identical manner.[47]

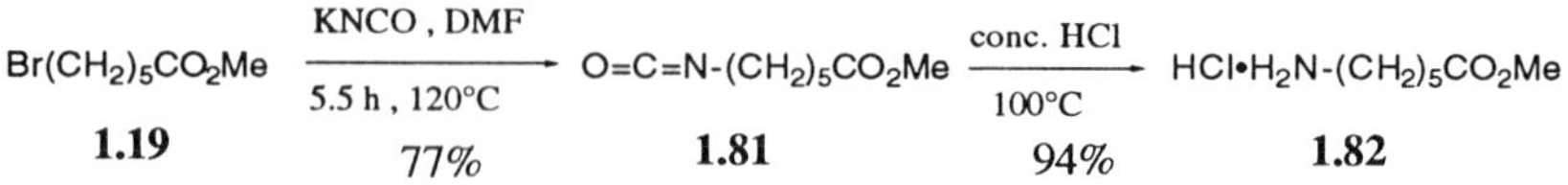

1.19 **1.81** **1.82**

1.1.C. Conversion of Carbonyls to Amines

Aldehydes and ketones are converted to amines by a handful of methods. The most common is reductive amination. When a keto-acid or keto-ester molecule is utilized and the carbonyl is not in conjugation, reductive amination leads to an amino acid or an amino-ester. When the carbonyl of an aldehyde or ketone is β- to the carboxyl moiety, reductive amination is unnecessary for preparing amino acids and alkenyl amino acids are possible (specifically, 3-aminopropenoic acid derivatives). There are several methods that can be applied to keto-acid derivatives, largely dependent on whether the aldehyde or ketone carbonyl is in or out of conjugation with the carboxyl moiety. The first part of this section will focus on reductive amination procedures, and the latter part of the section will focus on β-keto acids and esters. A few other specialized methods for converting aldehydes and ketones to amino acids will be mentioned at the end of this section.

1.1.C.i. Reductive Amination

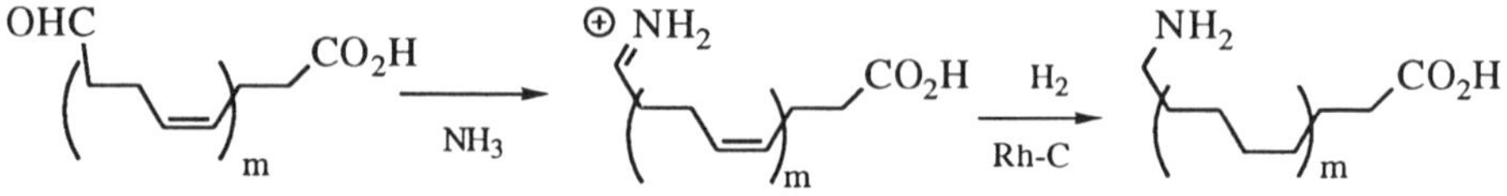

1.83 m = 1,2 **1.84** **1.6** m = 1 **1.85** m = 2

Aldehydes and ketones serve as useful precursors to amino acids when they are converted to a Schiff base (from aldehydes) or an imine (from ketones). Reduction of the imino C=N moiety, usually *in situ*, generates the amine. When ***1.83*** (m = 1, 2) was treated with ammonia, for example, the resulting iminium salt (***1.84***) was hydrogenated to give either 8-aminooctanoic acid (***1.6***) or 12-aminododecanoic acid (***1.85***) in the two examples shown.[48] This reactions works with ketones also, and the most common form of the reaction combines ammonia with a hydrogenation catalyst such as Raney nickel. An example is the reductive amination of keto-acid ***1.86*** to give 4,4-dimethyl-7-aminooctanoic acid (***1.87***) as the final product.[49]

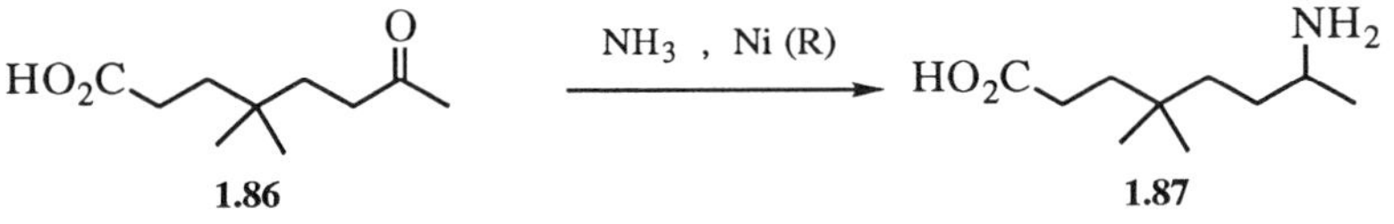

Reductive amination was applied to the product obtained by ozonolysis of undec-10-enoic acid (***1.88***). Hydrogenation of the resulting ozonide in the presence of ammonia and Raney nickel led to 10-aminodecanoic acid (***1.8***).[50] Similar treatment of the ether ester of ***1.88*** led to ethyl 10-aminodecanoate, in 38% yield.

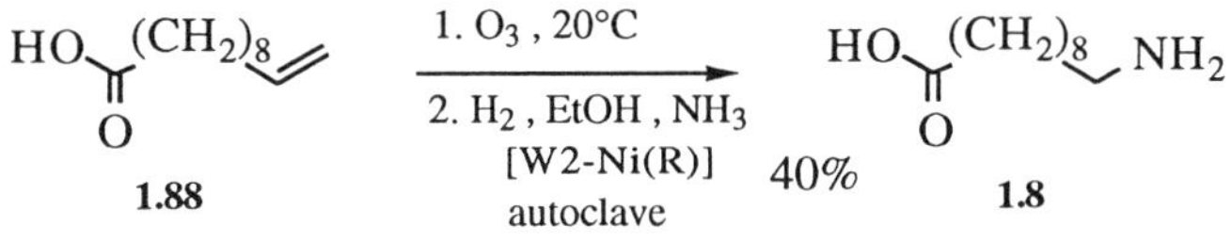

Reduction of the imine intermediate can also be done using zinc and acetic acid.

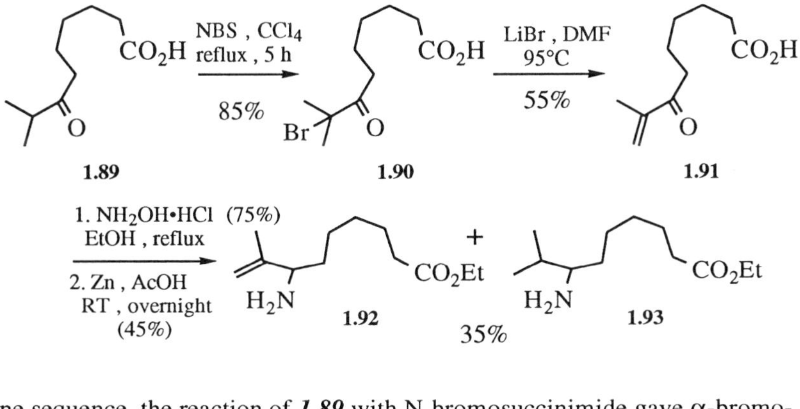

In one sequence, the reaction of ***1.89*** with N-bromosuccinimide gave α-bromo-ketone ***1.90*** and elimination, induced by lithium bromide in hot DMF, gave ***1.91***.[51]

Conversion of the ketone moiety to the oxime followed by reduction gave an 80:20 mixture of ***1.92:1.93*** (ethyl 7-amino-8-methylnon-8-enoate:ethyl 7-amino-8-methyl-nonanoate).

As seen in previous examples, the intermediate iminium salt produced by reductive amination need not be isolated, and reduction *in situ* is very common.[52] A mixture of sodium cyanoborohydride and ammonium acetate was used for the reductive amination of the aldehyde derived from acid hydrolysis of the dioxolane moiety in ***1.94***. In this example, the dioxolane moiety was hydrolyzed to release the aldehyde, which reacted with $NaBH_3CN/NH_4OAc$ to produce an amine. The amine was then protected as its benzenesulfonate. The final product was methyl 9-amino-nonanoate benzenesulfonamide, ***1.95*** (20% overall yield).[53]

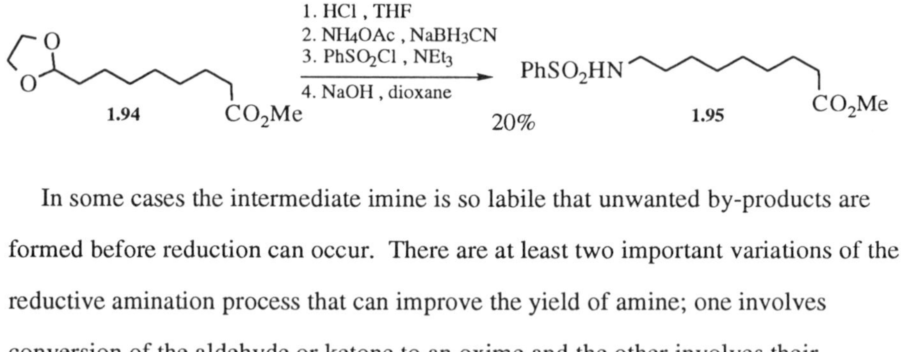

1.94 **1.95**

In some cases the intermediate imine is so labile that unwanted by-products are formed before reduction can occur. There are at least two important variations of the reductive amination process that can improve the yield of amine; one involves conversion of the aldehyde or ketone to an oxime and the other involves their conversion to a hydrazone. When keto-acid ***1.96*** was treated with hydrazine, oxime ***1.97*** was formed.[54] In this approach, a second step was required to convert the oxime to an amine. Catalytic hydrogenation is the most common procedure that is used, as in the reduction of ***1.98*** to 7-aminooctanoic acid, ***1.99***. In this particular example, the starting material for this sequence was diketone ***1.96***, which was cleaved with aqueous hydroxide to give 7-oxooctanoic acid, ***1.97***[54] (see chapter two, section 2.1).

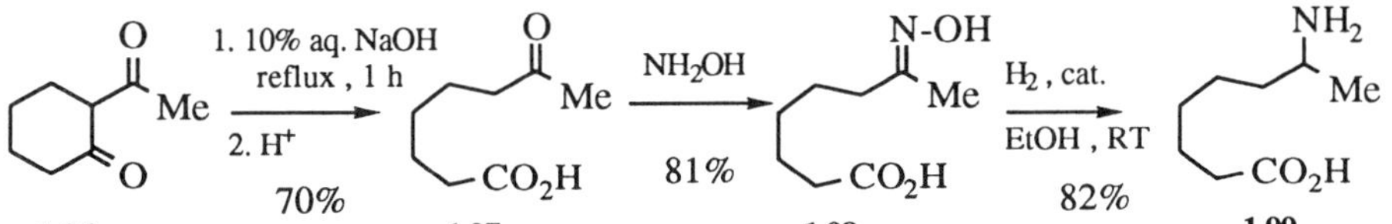

1.96 1.97 1.98 1.99

A similar "oxime-driven" methodology was applied to the aldehyde moiety in 4,4-dimethyl-glutaraldehyde (***1.100***),[55] which was converted to oxime ***1.101***. In this example, catalytic hydrogenation of the oxime gave an amine, but the amino acid spontaneously cyclized to give 4,4-dimethyl valerolactam, ***1.102***[56] (see chapter two, section 2.3). An acid hydrolysis step was required to liberate 5-amino-4,4-dimethyl-pentanoic acid, ***1.103***. In some cases, reduction in this type of system leads to a mono-amide as an intermediate product rather than a lactam.[57] Similar formation of

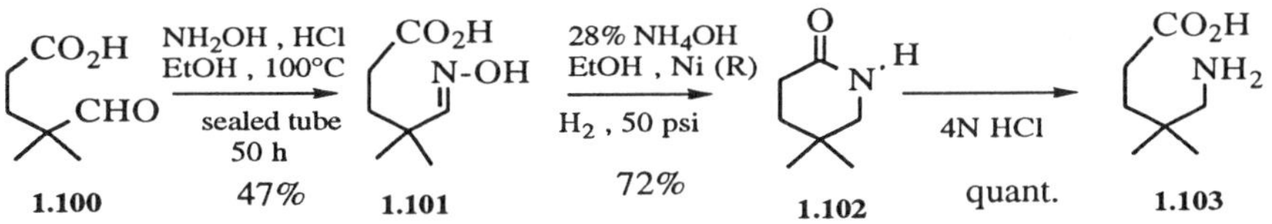

1.100 1.101 1.102 1.103

lactam ***1.105*** was observed when oxime ***1.301*** 104 hydrogenated.[58] Once again, acid hydrolysis was required to give the amino acid, which in this case was isolated as its ethyl ester, ethyl 4,8-dimethyl-5-aminononanoate (***1.106***).

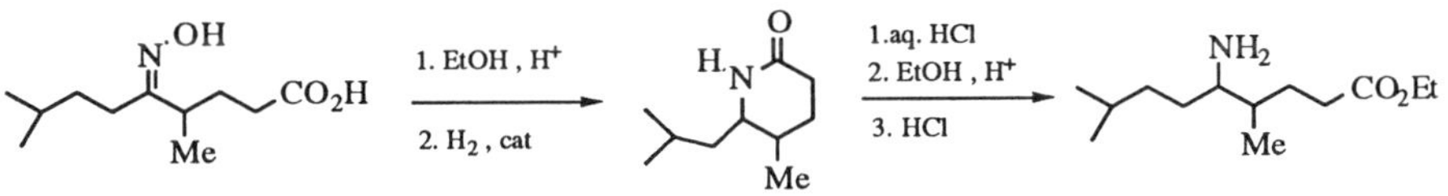

1.104 1.105 1.106

Other types of cyclic compounds can sometimes be generated in the oxime-forming reaction if the carbonyl is sufficiently close to the ester/acid moiety. Reaction of ***1.107*** (prepared by literature methods)[59] with hydroxylamine, for example, gave isoxazolidine ***1.108***[60] via reaction of the initially formed oxime with the proximal ester moiety. This heterocyclic ring was cleaved with sodium in isopropanol and then esterified to give methyl 3-amino-2,2-dimethylhept-6-enoate, ***1.109***. This example also illustrates an alternative method (dissolving metal

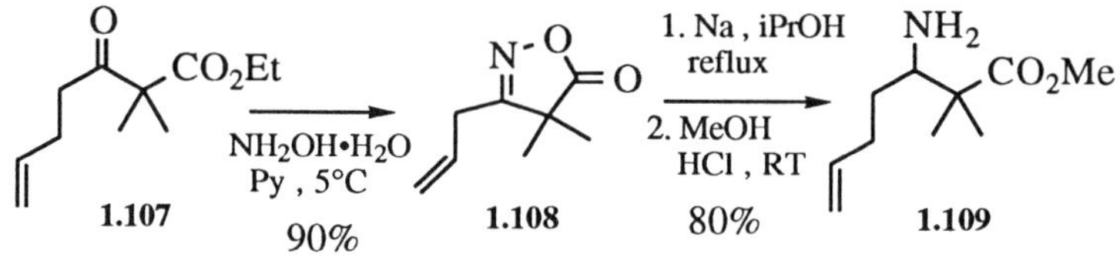

reduction) for converting a key intermediate to an amine, when compared to previous examples in this section. The traditional reagent for dissolving metal reductions is sodium metal in liquid ammonia. This latter reagent was used to convert oxime ***1.110***, prepared from 4-oxo-4-phenylbutanoic acid, to 4-phenyl-4-aminobutanoic acid, ***1.111***.[34] A related procedure converted 5-phenyl-4-oxopentanoic acid to 5-phenyl-4-aminopentanoic acid in 77% yield.[61]

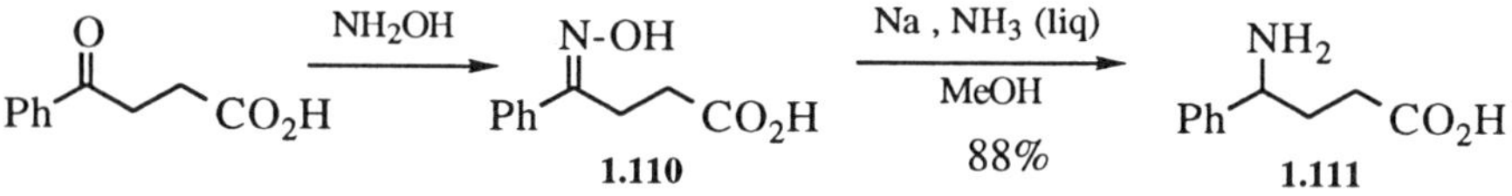

Just as aldehydes and ketones react with hydroxylamine to form an oxime, they react with hydrazine derivatives to give hydrazones. The C=N moiety in these

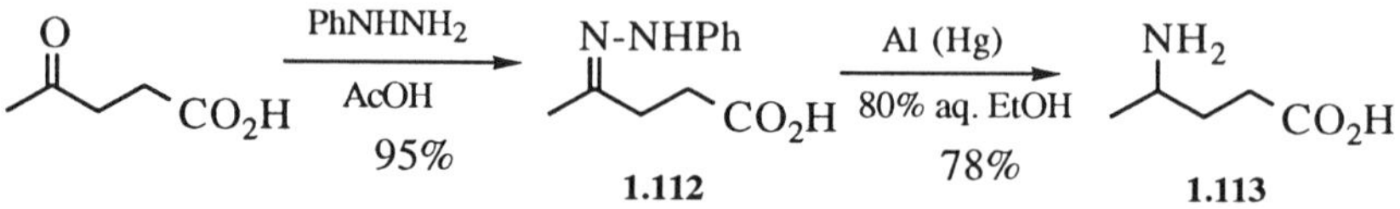

hydrazones can be reduced to give amines. When levulinic acid (4-oxo-pentanoic acid) was treated with phenylhydrazine, for example, hydrazone ***1.112*** was obtained. Reduction of the C=N moiety used a reagent new to this section, aluminum amalgam in ethanol, and it gave 4-aminopentanoic acid, ***1.113***.[62]

The last example presented in this section also involved a specialized hydrazone intermediate, a N-phthalimide hydrazone. Once the amine group in ***1.114*** (prepared by literature methods)[63] was blocked as the phthalimide, the acid moiety in ***1.114*** was subjected to the Arndt-Eistert sequence shown.[64] This reaction sequence extended the carbon chain by one and this new acid was converted to its *t*-butyl ester (***1.115***).[65] The N-phthalimido protecting group was removed, with reduction of the imine moiety, to give 5-methyl-7-aminooctanoic acid, ***1.116***.[65] When another

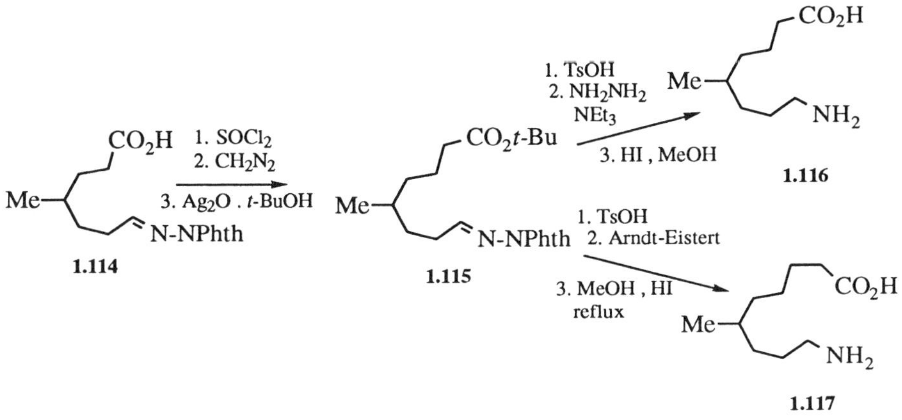

Arndt-Eistert sequence was applied to ester ***1.115***, the carbon chain was extended by an additional carbon, and amino acid 6-methyl-8-aminononanoic acid (***1.117***) was produced.[65]

1.1.C.ii. From β-Keto Acid Derivatives

The reaction of β-keto acid derivatives and ammonia or amines relies on initial reaction with an aldehyde or ketone moiety. The carbonyl must be β- to the carboxyl group, however, in order to obtain the enamino acid or enamino-ester derivative (an enamine is a "vinyl amine," first used synthetically by Stork,[66] and they are well known). These enamino derivatives are actually 3-aminopropenoic acid derivatives and can be used as such or reduced to 3-aminopropanoic acid derivatives.

The reactions of β-keto-esters just described are variations of a very old procedure. The treatment of ***1.118*** with ammonia was reported in 1887 and gave ***1.119***.[67] This approach has remained virtually unchanged for over a century and a wide range

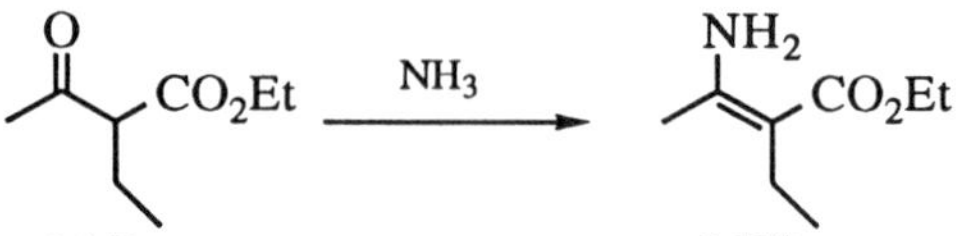

1.118 **1.119**

of β-keto-esters can be used. One modification used a mixture of ammonia and *p*-toluenesulfonic acid to induce the reaction.[68] The procedure is not limited to keto-esters, however, as shown by the reaction of 2,4-pentanedione with ammonia to give ***1.120*** (reported in an 1892 publication).[69]

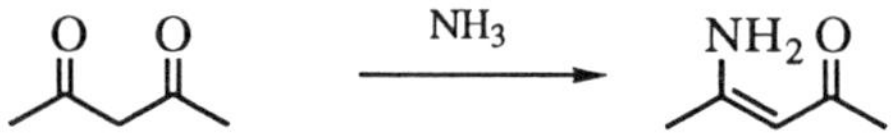

1.120

Both ammonia and amines react with the carbonyl group. The reaction of dimethylamine and ethyl acetoacetate, for example, gave ethyl 3-(N,N-dimethylamino)but-2-enoate, ***1.121***.[70] Similar reaction with ammonia and ethyl acetoacetate gave methyl 3-aminobut-2-enoate in 90% yield.[71]

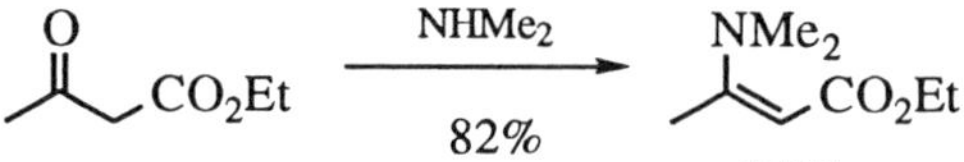

1.121

This reaction sequence leads to aminoalkenoic acids, but reduction of the double bond gives the corresponding aminoalkanoic acid, as mentioned above. The platinum catalyzed hydrogenation of ***1.122*** gave methyl 4-methyl-3-(N-acetylamino)pentanoate (***1.123***), for example.[72] The N-acetyl methyl esters of several 3-aminoalkanoic acids were prepared by hydrogenation of the alkenylamino acid precursor. The yield of saturated products can be low using catalytic hydrogenation, as seen for ***1.123***, and other methods for reducing the alkenyl moiety gave been developed.

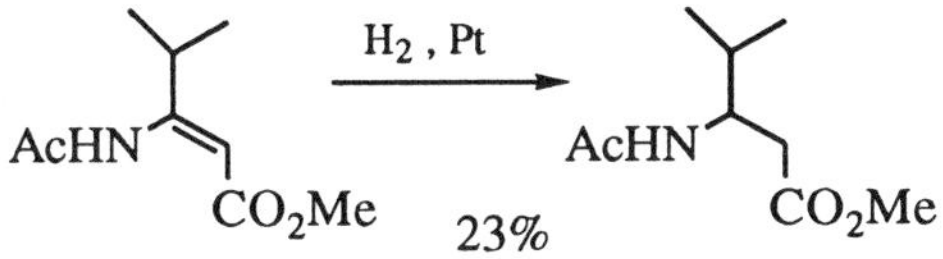

1.122 **1.123**

Sodium cyanoborohydride is an effective reagent for the reduction of β-amino-propenoic acid derivatives. Reaction of benzylamine with ethyl acetoacetate gave ***1.124*** and reduction with cyanoborohydride led to ethyl N-benzyl-3-aminobutanoate, ***1.125***.[73] When R-phenethylamine was used in place of benzylamine, the amino acid product showed asymmetric induction (2-28% ee, see chapter five) after the phen-ethyl group was removed by treatment with hydrogen and $Pd(OH)_2$ on carbon.[73]

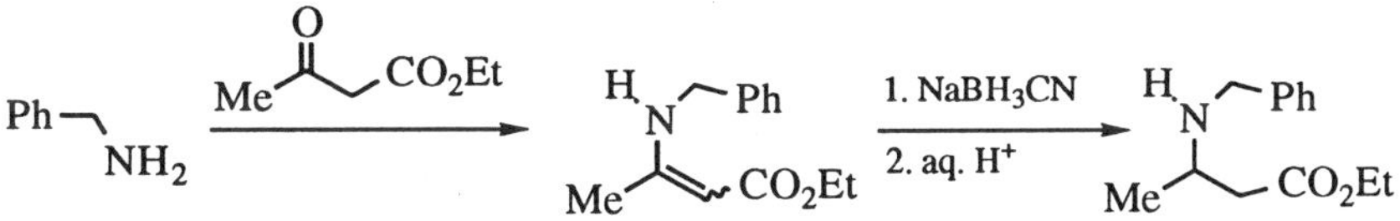

1.124 **1.125**

A variety of β-keto esters have been used in reactions with ammonia or amines. So many of them are the same or very similar that several examples have been collected into Table 1.2 to illustrate the scope of this approach.

This approach is used to prepare many substituted aminopropenoic acid derivatives from the appropriate β-keto-ester.[68,74,73] In reaction 1, ***1.126*** reacted with ammonia in ethanol to give ethyl 3-amino-4-methylpent-2-enoate, ***1.127***.[75] β-Keto-esters can also be condensed with nitriles to give the corresponding alkenyl

Table 1.2. Reaction of β-keto esters with ammonia and amines.

Reaction	Transformation	Reference
1.	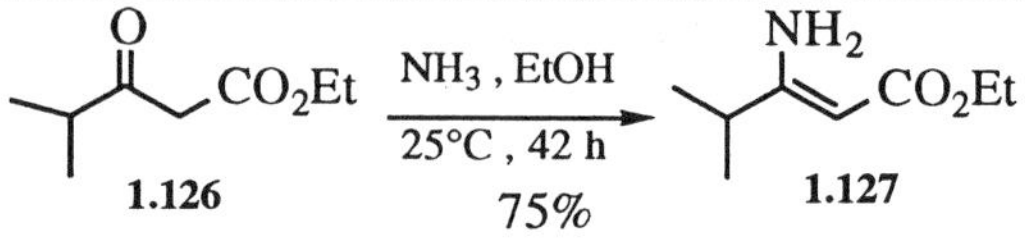	75
2.	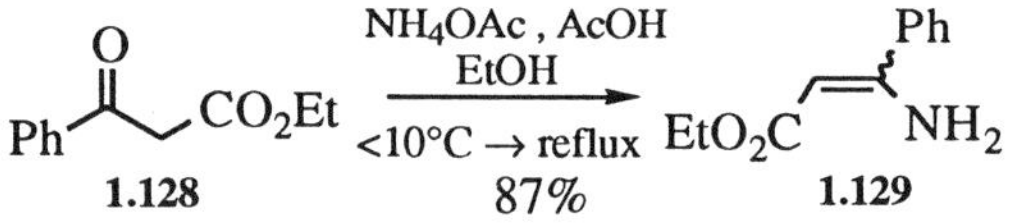	77
3.	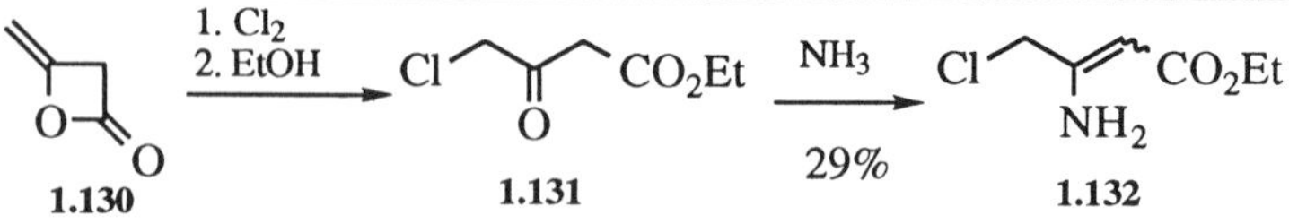	79
4.	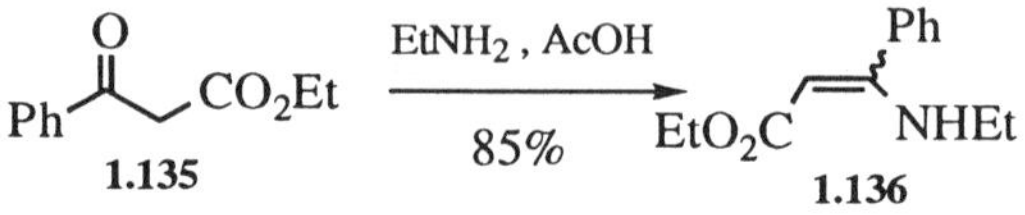	80
5.	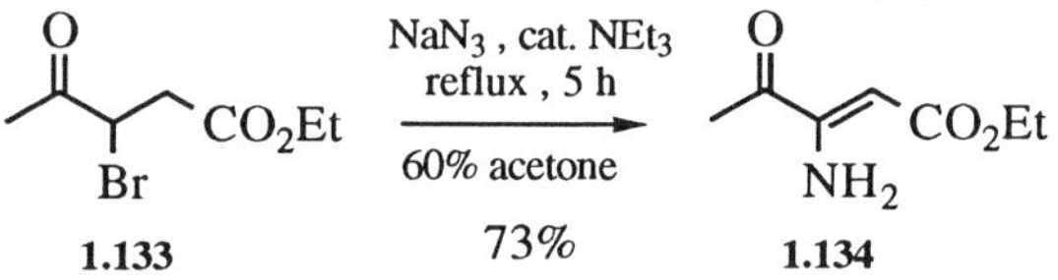	81
6.	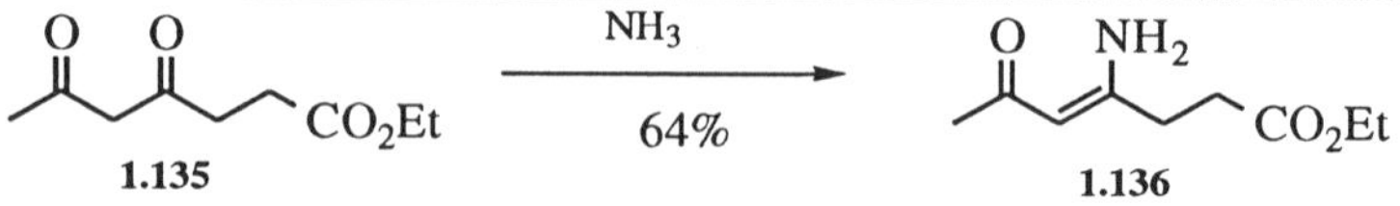	82
7.	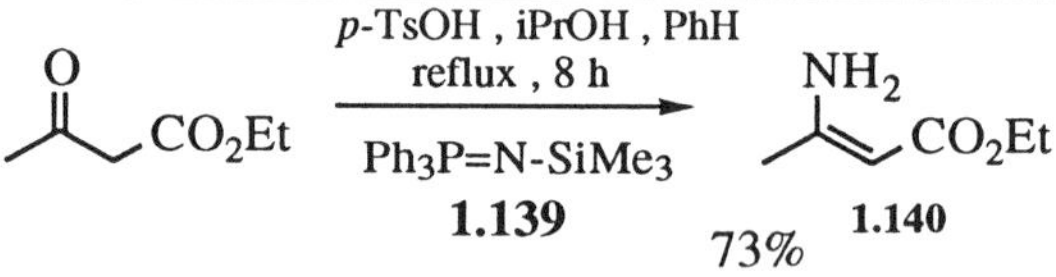	84

amino acid.[76] Aryl substituted alkenyl amino acids such as ethyl 3-amino-3-phenyl-prop-2-enoate (***1.129***) was prepared in reaction 2 by reacting keto-ester ***1.128*** with ammonium acetate in acetic acid.[77] A chlorine substituent is compatible with this reaction since the β-carbonyl is much more reactive with ammonia. This is illustrated in reaction 3 when β-lactone ***1.130*** was first converted to β-keto-ester ***1.131*** by a literature method.[78] Subsequent reaction with ammonia gave ethyl 3-amino-4-chlorobut-2-enoate, ***3.132***.[79]

Amines also react with β-keto-esters to give the corresponding amino acid derivative. When ***1.133*** was treated with ethylamine in reaction 4, ethyl 3-(N-ethylamino)-3-phenylprop-2-enoate (***1.134***) was obtained.[80] Several other 3-aryl derivatives (piperonyl, *p*-bromophenyl, *p*-methylphenyl, and *p*-nitrophenyl) were prepared by this method. The use of different amines (methylamine, isopropylamine, benzylamine, allylamine, aniline) led to several different N-substituted derivatives.[80]

It is also possible to use amine surrogates in this type of reaction. When sodium azide reacted with the bromide moiety in ***1.135*** (see reaction 5), an α-azido-ketone was formed. This azide was reduced *in situ* to give ethyl 3-amino-4-oxopent-2-enoate, ***1.136***.[81]

Another variation takes advantage of the fact that 1,3-diketones give enamino ketones such as ***1.120***, and if those β-diketones also contain an acid or ester moiety, they can be used to prepare amino acids. Diketo-ester ***1.137***, in reaction 6, reacted with ammonia to give exclusively ethyl 4-amino-6-oxohept-4-enoate, ***1.138***.[82] The final example (reaction 7) in Table 1.2 illustrates an interesting variation that uses another surrogate. In the presence of an acid catalyst, $Ph_3P{=}N{-}SiMe_3$ (***1.139***)[83] reacted with ethyl acetoacetate to give ethyl 3-aminobut-2-enoate, ***1.140***.[84]

This section is concluded with a different type of reaction that gives products similar to those observed above. In this study,[85] acetonitrile was first reacted with HCl and then with Meldrum's acid[86] (***1.141***) to give ***1.142*** in a Pinner synthesis.[87] Hydrolysis was followed by decarboxylation (*in situ*) to give ethyl 3-aminobut-2-enoate, ***1.143***.[85] This reaction used a nitrile as an amine surrogate (acetonitrile) and a variety of nitriles can be used. Using propanenitrile in this sequence led ethyl 3-aminopent-2-enoic acid in 36% overall yield and butanenitrile led to 3-aminohex-2-enoic acid in 37% yield.[85] Similarly, pentanenitrile, hexanenitrile, 4-methylpentane-nitrile, and 3-phenylpropanenitrile were used to generate alkenyl amino acids (3-aminohept-2-enoic acid, 3-aminooct-2-enoic acid, 3-amino-6-methylhept-2-enoic acid, and 3-amino-5-phenylpent-2-enoic acid, respectively).[85]

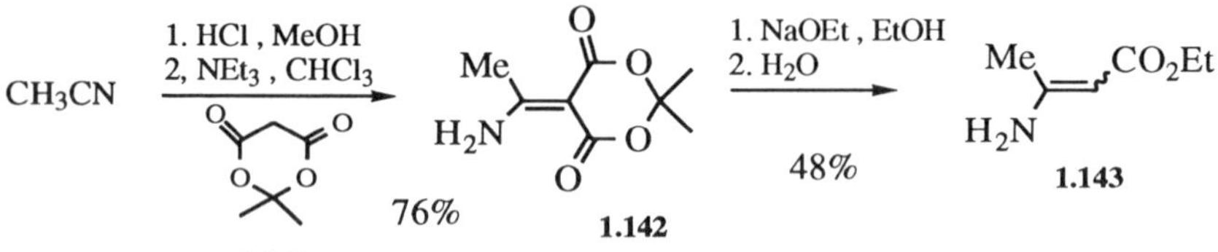

1.141 **1.142** **1.143**

1.1.C.iii. Specialized Methods

Apart from reductive amination and the reaction of β-keto acid derivatives with ammonia or amines, there are other carbonyl derivatives that can serve as precursors to amino acids. One sequence begins with the conversion of α,ω-dicarboxylic acids

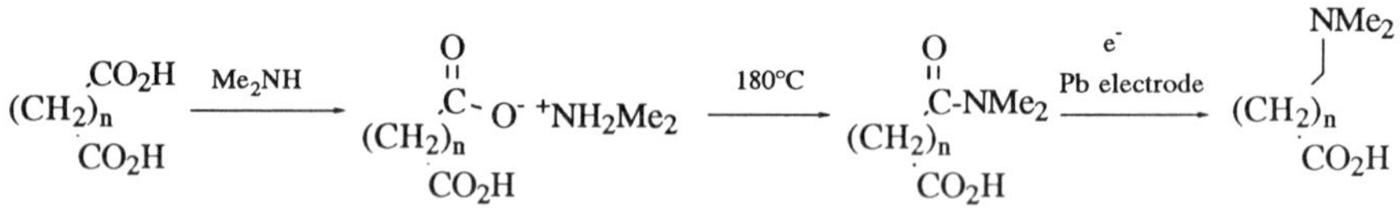

1.144 **1.145** **1.146** **1.147**

such as ***1.144*** to the corresponding ammonium salt, ***1.145***. Heating that salt to 180°C generated the amide (***1.146***), which was electrolytically reduced to the N,N-dimethylamino acid (***1.147***).[88] Several "straight-chain" amino acids were prepared by this route.

Another route converted ethyl ω-bromocaproate (ethyl 6-bromohexanoate, ***1.148***) to ethyl 7-bromo-8-oxo-nonanoate, ***1.149***.[89] The bromide was displaced by phthalimide and later unmasked to an amine. The ketone moiety in ***1.149*** was converted to an amine by reductive amination to give ethyl 7,8-diaminononanoate, ***1.150*** .

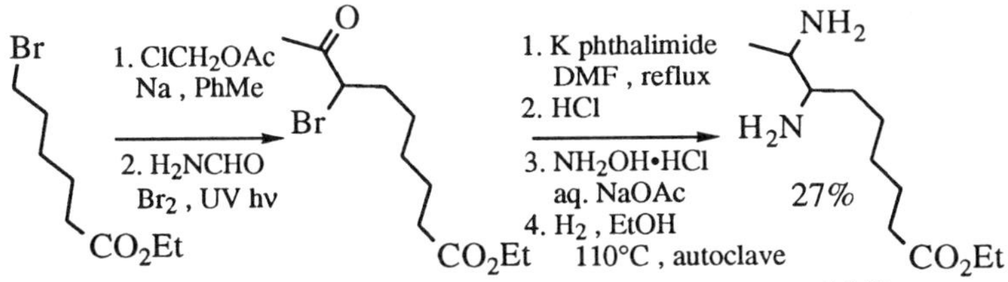

1.148 **1.149** **1.150**

A related method also involved an α-bromo-ketone. Succinic acid derivative (***1.151***) was brominated and then converted to the phthalimidoyl derivative (***1.152***).[90] Acid hydrolysis gave ethyl 5-amino-4-oxopentanoate, ***1.153***. Alkyl substituents can be incorporated into the backbone of oxo-amino acids by simple modification of the method just described.

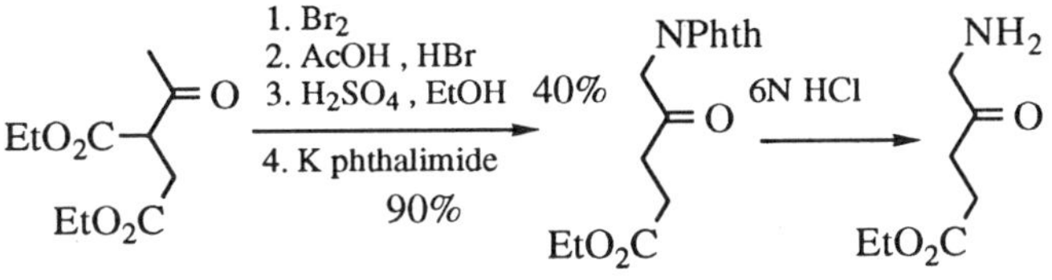

1.151 **1.152** **1.153**

A last approach involved an acid chloride, which served as a vehicle for introducing both an amine group and an oxo moiety. Succinic acid derivative ***1.154*** was first converted to cyano ketone ***1.155***.[91] Reduction of the cyano group in ***1.155*** with zinc and acid was followed by acid hydrolysis to give 5-amino-4-oxopentanoic acid, ***1.156*** (the free acid form of ***2.153***).

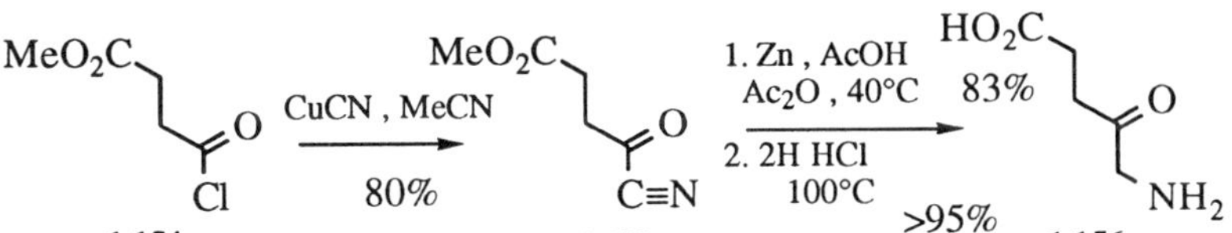

1.154 **1.155** **1.156**

1.2. Functionalization to Form Carboxyl Groups

1.2.A. Carboxyl Equivalents

Just as there are several methods for incorporating an amino group into a molecule, there are several strategies for incorporating a carboxyl moiety (-COOH) into the molecule. Certain acid chlorides (such as ethyl chloroformate) may be the best "direct" method for introducing a carboxyl group. There are also several reagents such as cyanide and carbon dioxide that can be used to insert a carboxyl into a putative amino acid.

1.2.A.i. Carboxyl Surrogates (Cyanide)

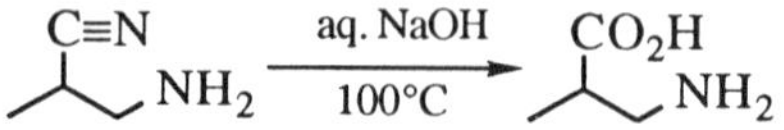

1.157 **1.50**

One of the most common methods for incorporating an acid moiety into a molecule involves the reaction of cyanide with a suitable leaving group to give a nitrile. Subsequent hydrolysis of that cyano group leads to an acid moiety. An

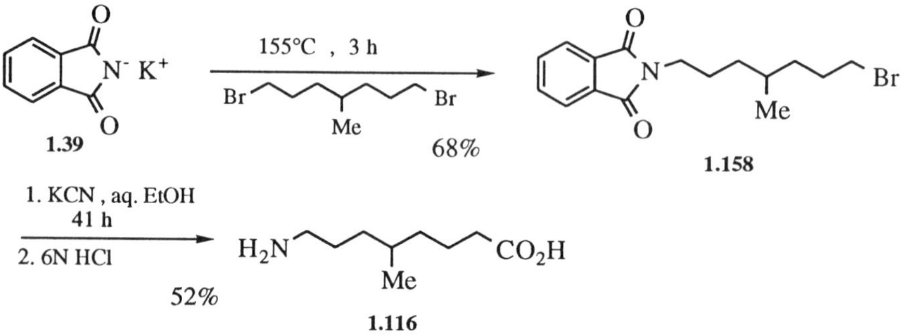

example is the basic hydrolysis of the cyano group in ***1.157*** to give 2-methyl-3-aminopropanoic acid, ***1.50***.[92] Note that this hydrolysis is sluggish since the reaction is done at 100°C. Another example begins with reaction of potassium phthalimide

(*1.39*) and 4-methyl-1,7-dibromoheptane to give ***1.158***. A subsequent reaction with cyanide led to 5-methyl-8-aminooctanoic acid (***1.116***) after acid hydrolysis.[93] In this case, the phthalimide group functions as an amine surrogate and cyanide as a carboxyl surrogate.

A variation in this approach is the reduction of conjugated nitriles such as ***1.159***, which gave 2-ethyl-3-aminopropanoic acid (***1.160***) after hydrolysis of the cyano group to COOH and removal of the phthalimido protecting group.[94] This route was also used to prepare 3-amino-2-methylpropanoic acid, in 65% yield.

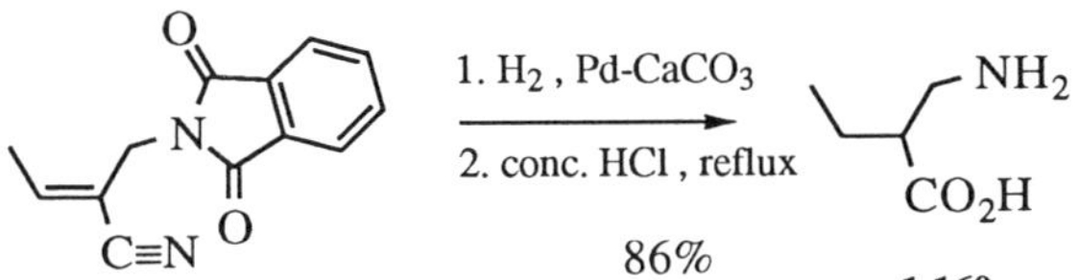

1.159 **1.160**

Nitriles can be used in reactions other than displacement. The Boc-protected aldehyde derived from alanine (known as N-Boc alinal, ***1.161***) was condensed with potassium cyanide to give a cyanohydrin. Hydrolysis of the cyano group to give an acid and removal of the N-Boc group gave 3-amino-2-hydroxybutanoic acid, ***1.162***.[95]

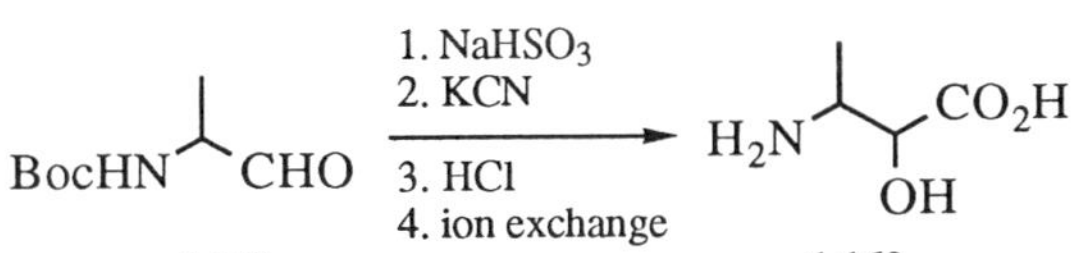

1.161 **1.162**

Specialized variations are also possible. Treatment of *bis*-cyano-ester ***1.163*** with an acid such as HCl, for example, gave ***1.164***.[96] Similar reaction of ***1.163*** with methanol, however, gave methyl 3-amino-2-cyano-3-methoxyprop-2-enoate, ***1.165***.[96]

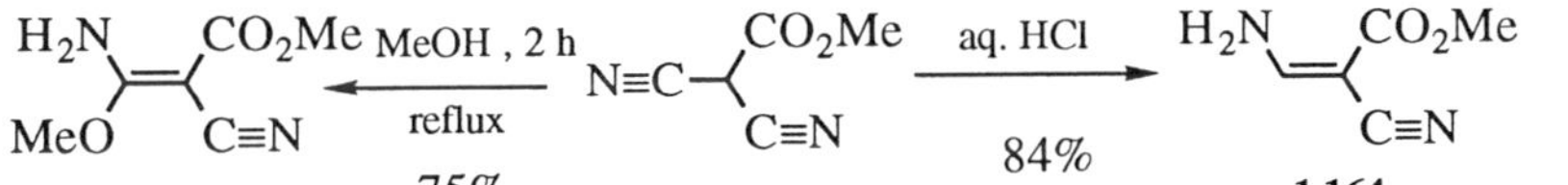

1.165 **1.163** **1.164**

1.2.A.ii. Carboxyl Surrogates (Carbon Dioxide)

A different approach uses the reaction of an organometallic with the electrophilic carbonyl of carbon dioxide, giving a carboxyl group directly. Formation of the Grignard reagent of ***1.166*** and subsequent reaction with carbon dioxide gave ethyl 4-(N,N-diethylamino)-but-2-enoate (***1.36***), but the yield was poor.[97] Poor yields of acid in the reaction of carbon dioxide and Grignard reagents are very common, except when aryl Grignard reagents are used.[98]

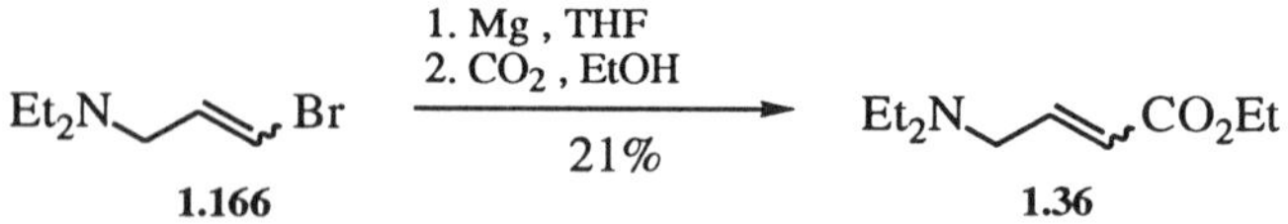

1.166 **1.36**

A variation used the amine moiety of an amino-alkyne such as ***1.167*** rather than a Grignard regent. Condensation of the corresponding alkyne anion with carbon dioxide gave 4-(N,N-dimethylamino)but-2-ynoic acid, ***1.168***.[99] Catalytic hydrogenation of the triple bond gave 4-(N,N-dimethylamino)but-2-enoic acid, ***1.169*** (also see ***1.27***, *vide supra*).[99] In addition to ***1.169***, 5-(N,N-dimethylamino)-pent-2-enoic acid, and 8-(N,N-dimethylamino)oct-2-enoic acid were prepared by this method using the appropriate amino-alkyne.[99]

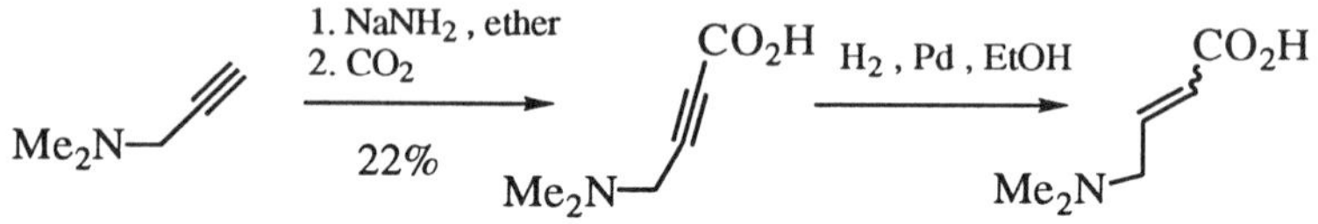

1.167 **1.168** **1.169**

The yield of product from the carbon dioxide reaction with the alkyne anion was very poor, as it was with the vinyl Grignard derived from ***1.166***. Extending the carbon chain between the triple bond and the dimethylamino group, as in the reaction of ***1.170*** with sodium amide and then carbon dioxide,[100,98] however, led to a greatly improved yield of acid. Amino acid ***1.171*** [5-(N,N-dimethylamino)pent-2-ynoic acid] was isolated in 70% yield.[100] It is clear from this example that good

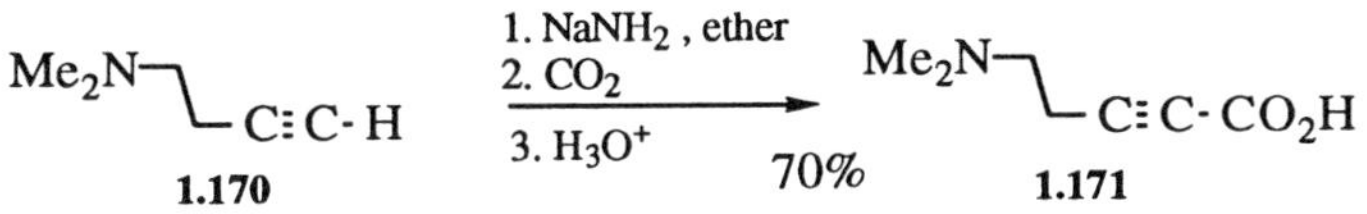

yields are possible in carbanion reactions with carbon dioxide. It should be noted that sodium amide or sodium metal is not the only base that is used to generate the alkyne anion. *n*-Butyl-lithium can also be used for this purpose.[101]

1.2.A.iii. Acid Chlorides

The reaction of a carbanion with ethyl chloroformate incorporates an ester group directly into a molecule. Amino-alkyne ***1.172*** was converted to an amino-ester via reaction with sodium metal to give the alkyne anion and then with ethyl chloroform-

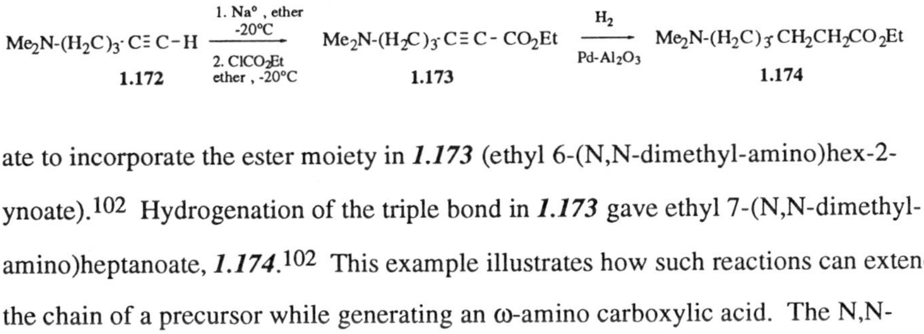

ate to incorporate the ester moiety in ***1.173*** (ethyl 6-(N,N-dimethyl-amino)hex-2-ynoate).[102] Hydrogenation of the triple bond in ***1.173*** gave ethyl 7-(N,N-dimethyl-amino)heptanoate, ***1.174***.[102] This example illustrates how such reactions can extend the chain of a precursor while generating an ω-amino carboxylic acid. The N,N-diisopropyl analog of ***1.173*** was also prepared by this method, in 44% yield.[102] Amino alkynes such as ***1.172*** are usually prepared by reaction of a chlorinated alkyne, such as 6-chloro-1-hexyne and an amine such as dimethylamine.[103] It is also possible to do the amination step <u>after</u> formation of, or incorporation of, the acid moiety, and this generates amino tetrolic acids (4-aminobut-2-ynoic acids).[104]

A related route reacted enamino-nitrile ***1.175*** with phosgene (another carboxyl surrogate) to give ***1.176***.[105] Methanolysis of the acid chloride gave methyl 3-amino-

2-cyanobut-2-enoate, ***1.177***. In addition to the *trans-* starting material shown (***1.175***), *cis*-enamino nitriles can also be used.[106]

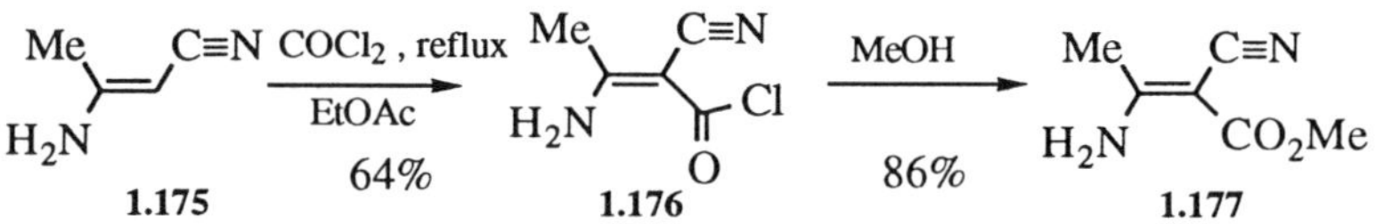

1.175 **1.176** **1.177**

1.2.B. Carboxyl-bearing Carbanions and Ylids

There are many carbanion and ylid derivatives that also contain an acid moiety, an ester or an acid surrogate. Ester and acid enolates, for example, react with a variety of electrophilic reagents to produce new acid or ester derivatives. If the electrophile contains an amine or an amine surrogate, amino acids are produced as the final product. In addition, ylids bearing an acid or ester moiety react with aldehydes or ketones bearing an amine moiety via a Wittig reaction or a Horner-Wadsworth-Emmons reaction to give alkenyl amino acids.

Malonate anions are convenient sources of ester enolates. They react with halides and a variety of other electrophiles. Acid hydrolysis of the ester followed by decarboxylation gives the mono-acid. An example of this approach used phthalic anhydride (***1.178***) as a starting material in a reaction with 2-aminoethanol to give ***1.179***.[107] Conversion of the alcohol moiety in ***1.179*** to its O-benzenesulfonate ester

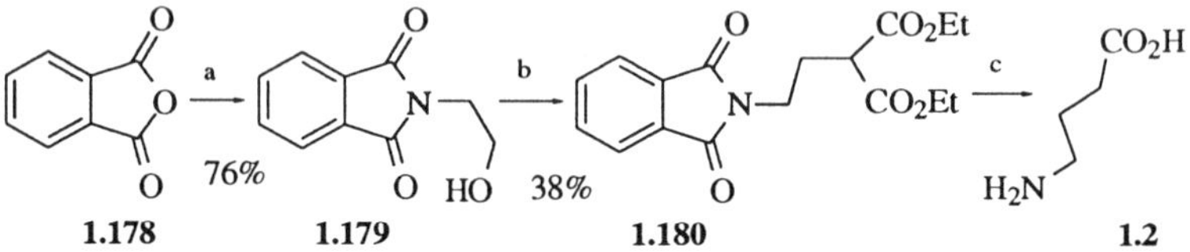

1.178 **1.179** **1.180** **1.2**

(a) $HOCH_2CH_2NH_2$, PhMe (b) 1. $PhSO_2Cl$, Py 2. $CH_2(CO_2Et)_2$, PhMe , Na°, reflux
(c) 1. aq. H_2SO_4 , reflux 2. aq. $BaCO_3$

was followed by displacement with sodium diethyl malonate to give adduct ***1.180***, in low yield. Treatment with aqueous sulfuric acid and neutralization with barium carbonate hydrolyzed the ester, cleaved the phthalimide group, and induced decarboxylation to give 4-aminobutanoic acid, ***1.2***.[107] A similar approach converted the ester moiety of ethyl 3-(N,N-di-*n*-propylamino)propanoate (***1.181***) to a chloride via reduction of the ester and treatment of the resulting alcohol with thionyl chloride. Subsequent displacement by the sodium enolate of diethyl malonate was followed by hydrolysis and decarboxylation to give 4-(N,N-di-*n*-propyl-amino)butanoic acid, ***1.182***.[108] In a similar manner, ethyl 2-(N,N-di-*n*-propylamino)ethanoate was converted to 3-(N,N-di*n*-propylamino)propanoic acid.

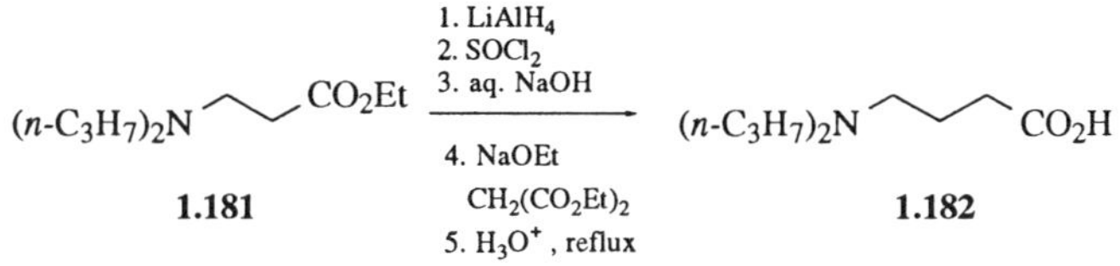

"Simple" ester enolates (those derived from mono-esters) can be used to prepare amino acids. The lithium enolate of *t*-butyl acetate reacted with the methoxy group in ***1.185***, for example, to give the methyl carbamate of *t*-butyl 3-aminobutanoate,

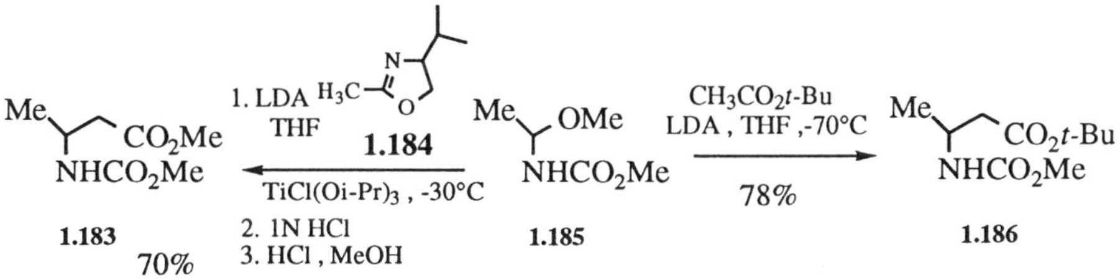

1.186.[109] Similar results were obtained with methyl acetate (70% yield of ***1.183***).[109] Another acid surrogate was used with ***1.185*** to produce an amino acid. Meyers developed an oxazolidine derivative (***1.184***)[110] that was readily converted to the carbanion by reaction with lithium diisopropylamide. In the presence of Ti (IV), this carbanion was used in a reaction with ***1.185*** to give the methyl carbamate of methyl

3-aminobutanoate, ***1.183***.[109]

A variation in this approach used the lithium enolate of diethyl malonate in a reaction with the allylic acetate moiety in ***1.187***. Malonate anion reacted with an intermediate π-allyl palladium species [formed from the allylic acetate moiety in ***1.187***, catalyzed by the palladium (0) species] to give ethyl 6-(N-Boc amino)-7-(4-benzyloxyphenyl)-2-carboethoxyhept-4-enoate, ***1.188***.[111]

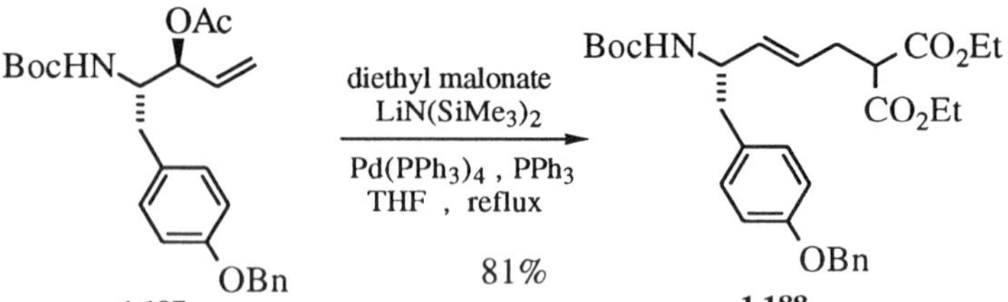

1.187 **1.188**

Ylids are also convenient sources of carbanions and they react with aldehydes and ketones to give alkenes. With a suitable reactant, alkenyl amino acids can be prepared using this approach. One example used a phosphonate ester ylid in a Horner-Wadsworth-Emmons reaction[112] with amino aldehyde ***1.189***. The product

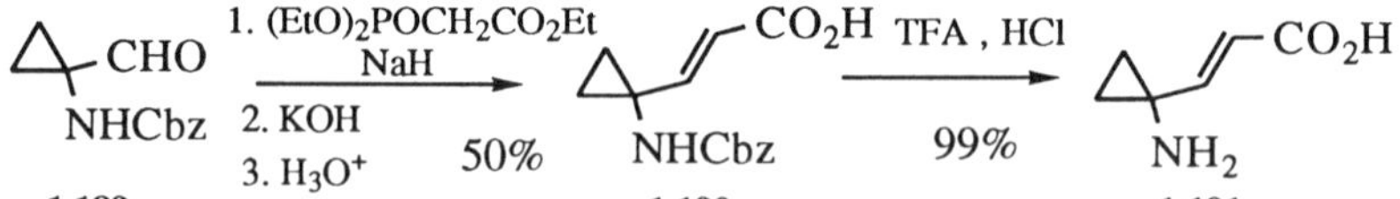

1.189 **1.190** **1.191**

after hydrolysis was conjugated acid ***1.190***.[113] Cleavage of the N-Cbz group gave 3-(1-cyclopropyl-1-amino)prop-2-enoic acid, ***1.191***. An identical strategy was used with aldehyde ***1.192***, which was generated *in situ* by ozonolysis of N-Boc allyl-amine. Horner-Wadsworth-Emmons olefination gave methyl 4-(N-Boc amino)but-2-enoate, ***1.193*** in good yield.[114] Treatment with trifluoroacetic acid removed the Boc group to give methyl 4-aminobut-2-enoate, ***1.194***.

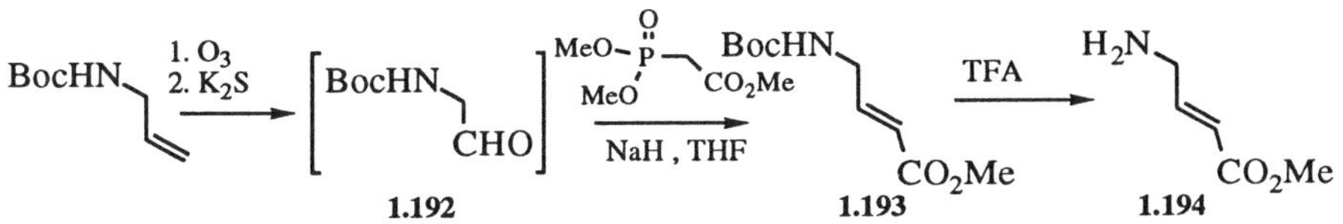

1.192 1.193 1.194

Other functionality can be incorporated into the ylid prior to reaction with an aldehyde or ketone. Aldehyde ***1.195***, for example was converted ***1.196*** by reaction with a fluorine-bearing phosphonate ylid.[115] The ester group was reduced to an aldehyde moiety (see ***1.197***) with diisobutylaluminum hydride. This allowed final conversion to 5-(N-Boc amino)-4-fluoro-6-phenylhex-3E-enoic acid (***1.198***), in four steps (3% overall yield; the first and second steps gave a combined yield of 14% and step five proceeded in 28% yield).[115]

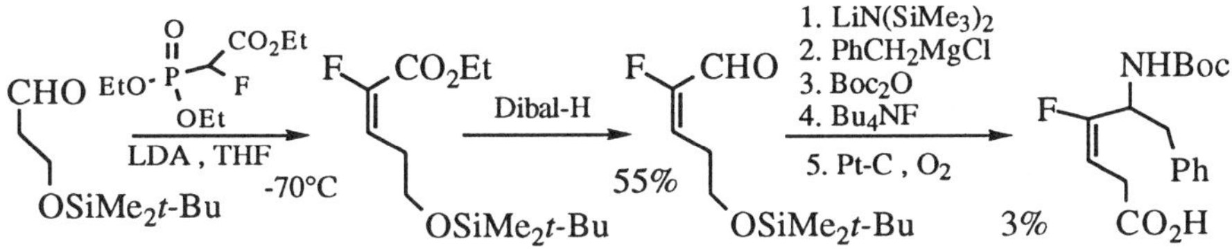

1.195 1.196 1.197 1.198

1.2.C. Oxidative Methods

There are several acid precursors that use an oxidation reaction to "unmask" an acid surrogate to give a carboxylic acid. Several of these approaches are outlined in this section. Oxidation of a primary alcohol to a carboxylic acid using chromium (VI) reagents[116] is perhaps the most common approach and is illustrated by oxidation of the primary alcohol moiety in phthalimidoyl alcohol ***1.199*** (prepared by literature methods).[6] Subsequent catalytic hydrogenation of the triple bond gave 4-(phthalimido)but-2Z-enoic acid, ***1.200***.[117] Treatment with diethylamine converted the phthalimide group to an amine moiety in 4-aminobut-2Z-enoic acid, ***1.201***.

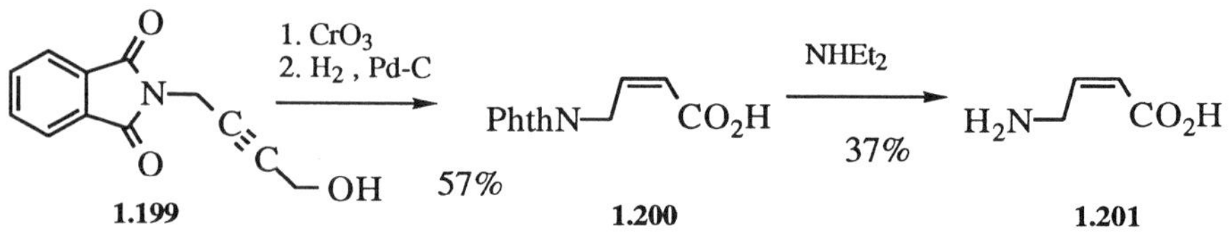

1.199 1.200 1.201

A related approach reacted amino-alkyne ***1.202*** with ethylmagnesium bromide and condensed it with formaldehyde to give ***1.203***. This allowed oxidation of the

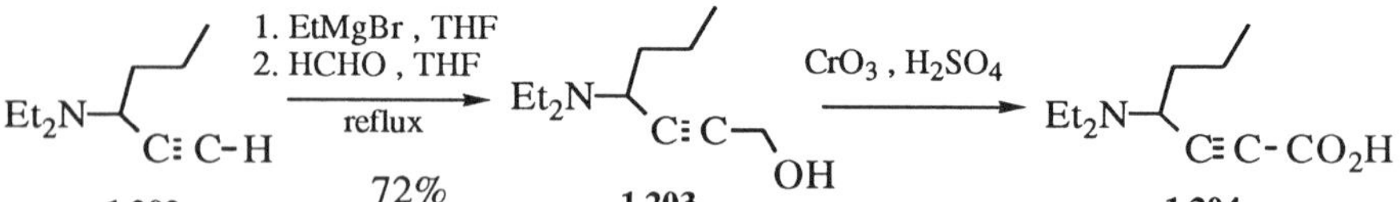

1.202 1.203 1.204

alcohol moiety to the acid moiety, giving 4-(N,N-diethylamino)hept-2-ynoic acid, ***1.204***.[118] A number of other alkynyl amino acids were prepared by this method, including 4-(N,N-diethylamino)pent-2-ynoic acid, 4-(N,N-diethylamino)but-2-ynoic acid, and 4-(N,N-diethylamino)-4-(2-furyl)but-2-ynoic acid.

One oxidative method used an aromatic ring as a carboxyl surrogate. Oxidation of ***1.205*** with ruthenium oxide, for example, gave 4-aminobutanoic acid (***1.2***).[119] Similar oxidation of tyramine gave an 86% yield of 3-aminopropanoic acid (***1.1***).

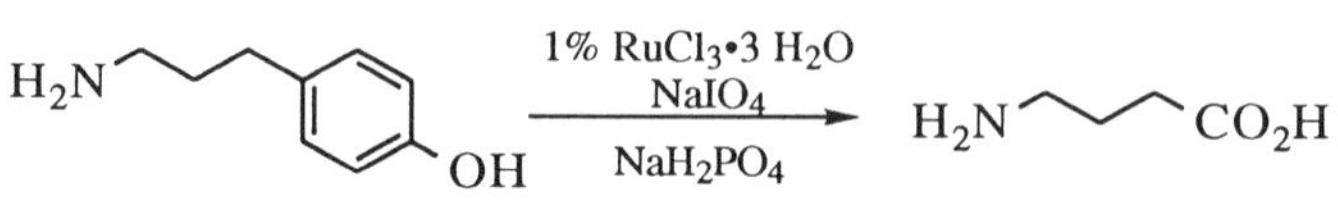

1.205 1.2

1.3. Functionalization of Amino Acids

A rather obvious approach to prepare functionalized amino acids is to begin with another amino acid. Most transformations involve conversion of chiral, nonracemic α-amino acids to chiral, nonracemic non-α-amino acids and these approaches will be discussed in chapter five. There are several procedures that generate functionality in an amino acid product in which the chiral center is racemized or lost altogether. There are many other sequences that use chiral, racemic amino acid precursors to prepare substituted non-α-amino acids. This section will focus on those methods that manipulate and/or add to the functionality of an existing amino acid to produce a new amino acid. As mentioned above, diastereoselective and enantioselective transformations of chiral, nonracemic amino acids will be discussed in chapter five, section 5.1.

1.3.A. Alanine and β-Alanine

The acyl carbon of readily available amino acids such as alanine can be converted to a ketone moiety by activation of the acid with carbonyl diimidazole (CDI) and then condensation with an enolate, such as the magnesium enolate of malonic acid, mono ethyl ester. In this particular example, N-Boc alanine was converted to ***1.206*** using this method[120] Catalytic hydrogenation of the ketone moiety gave the alcohol group in ***1.207***, and conversion to the chloride and base induced dehydrohalogenation gave ethyl 4-(N-Boc amino)pent-2-enoate (***1.208***).[121]

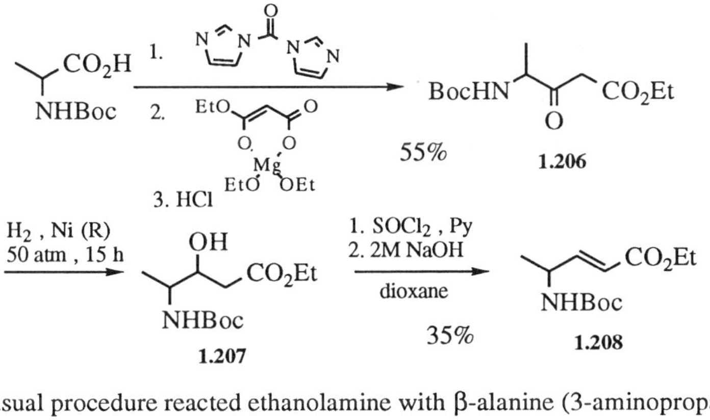

An unusual procedure reacted ethanolamine with β-alanine (3-aminopropanoic acid). Apparently, the reaction proceeds via formation of an enamine, which reacted with β-alanine to give ***1.209*** (2-aminomethyl-4-aminobutanoic acid).[122] The yield of this reaction sequence was quite good, but its scope is somewhat limited.

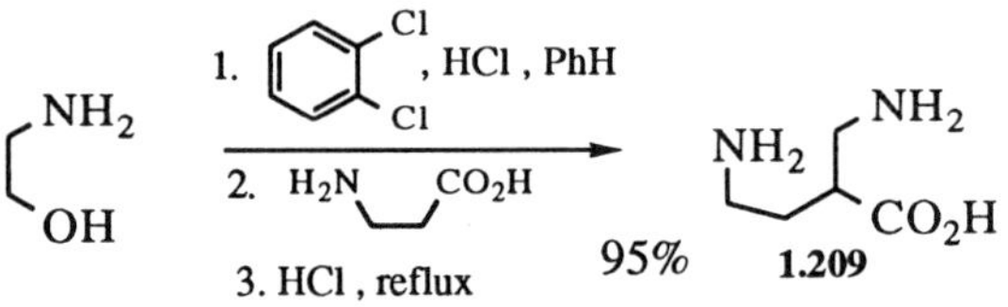

1.209

1.3.B. γ-Aminobutyric Acid (GABA) and Synthetic Derivatives

A more widely used amino acid is γ-aminobutyric acid (GABA, 4-aminobutanoic acid), along with its synthetic derivatives. This stems from the fact that GABA is a mammalian neural transmitter (see chapter six, section 6.1) and preparation of GABA and analogs has received quite a bit of synthetic attention.

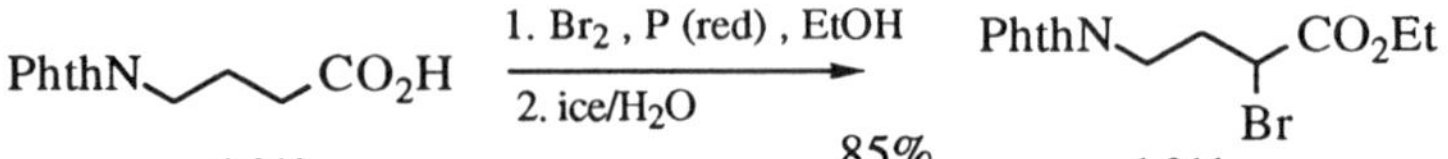

1.210 1.211

The major synthetic use of GABA is to prepare functionalized and/or substituted GABA derivatives. A simple example reacted N-phthaloyl-4-aminobutanoic acid (***1.210***) with ethanolic bromine and red phosphorus to give ethyl 2-bromo-4-phthal-imido butanoate, ***1.211***.[123] Clearly, the labile bromine moiety in ***1.211*** can be converted to a variety of other functionality via elimination or substitution reactions, making it an important synthetic intermediate. This is illustrated with 4-amino-

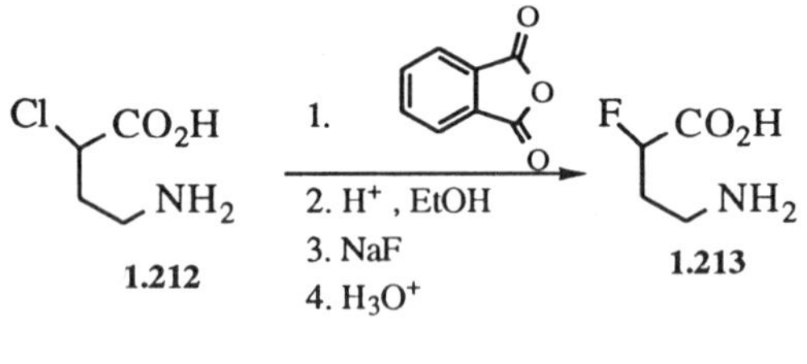

1.212 1.213

2-chlorobutanoic acid (***1.212***), which was converted to the 2-fluoro derivative, ***1.213***.[124] The amino moiety was protected, allowing the fluoride to be incorporated via displacement of the chlorine by fluoride ion. Prior to this work, this strategy had been reported for the conversion of phthalimidoylbutanoic acid (***1.210***) to α-bromoamide ***1.214***.[125] Subsequent reaction with KF led to displacement of bromine to give ***1.215*** (45% overall yield). Treatment with NOCl removed the phthalimidoyl group and acid hydrolysis gave 4-amino-2-fluorobutanoic acid, ***1.216***.

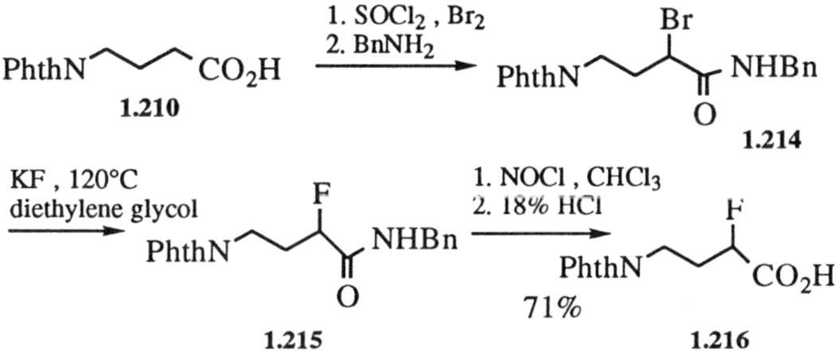

A synthetic derivative of GABA, tosylate ***1.217*** (prepared by literature methods),[126] was converted to the 3-methoxy derivative (***1.218***) via initial reaction with sodium methoxide in methanol.[127] Acid hydrolysis gave 3-methoxy-4-aminobutanoic acid, ***1.219***.

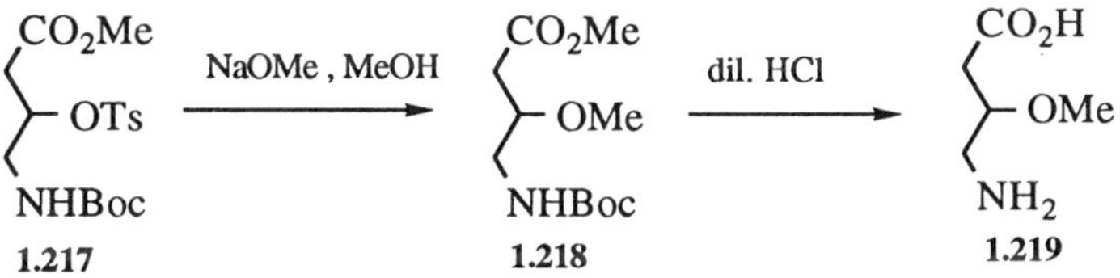

Even more complex synthetic derivatives such as N-Cbz baclofen[128] (***1.220***, see chapter six, section 6.1) can be used as synthetic precursors to new amino acids. α-Selenation of ***1.220*** via the enolate anion and oxidation to the selenoxide led to spontaneous *syn*-elimination, giving ***1.221*** in poor overall yield.[129] Removal of the carbobenzoxyl moiety with trifluoroacetic acid gave 4-amino-3-(4-chlorophenyl)but-2-enoic acid, ***1.222***.[129]

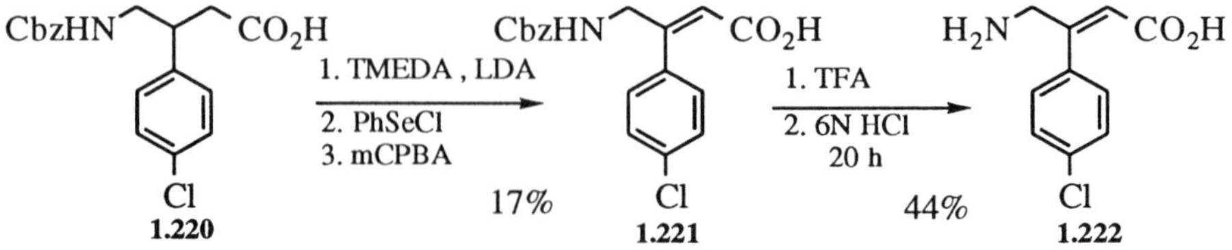

1.3.C. Glycine

A typical reaction that uses an amino acid derivative involves initial conversion to an enolate anion. This nucleophilic species is then reacted with an alkyl halide or a carbonyl derivative. An example that produces a new amino acid is the reaction of the ethyl ester of *n*-benzyl glycine with lithium diisopropylamide to give the enolate. Subsequent reaction with the mixed anhydride shown below proceeded with displacement of acetate to give ***1.223***.[130] Acid hydrolysis generated a β-keto amino acid, which decarboxylated under the reaction conditions to give 4-oxo-5-aminopentanoic acid (***1.156***, also known as 5-aminolevulinic acid).

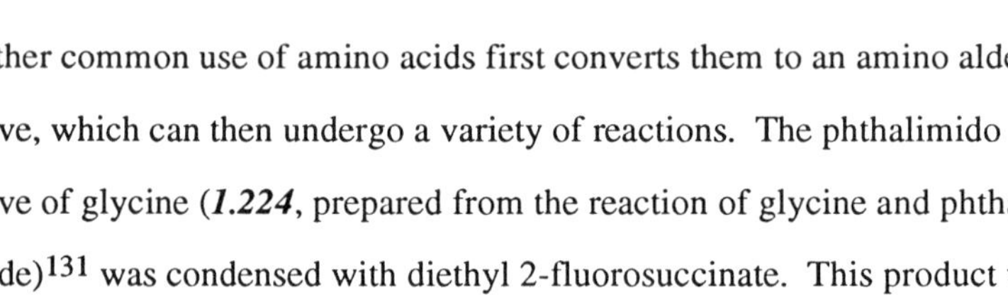

Another common use of amino acids first converts them to an amino aldehyde derivative, which can then undergo a variety of reactions. The phthalimido derivative of glycine (***1.224***, prepared from the reaction of glycine and phthalic anhydride)[131] was condensed with diethyl 2-fluorosuccinate. This product was hydrolyzed (and it proceeded with decarboxylation) to give 2-fluoro-4-phthalimidobut-2-enoic acid, ***1.225***, but in very poor yield.[132] Despite the poor conversion, this is a simple and fast route to these compounds.

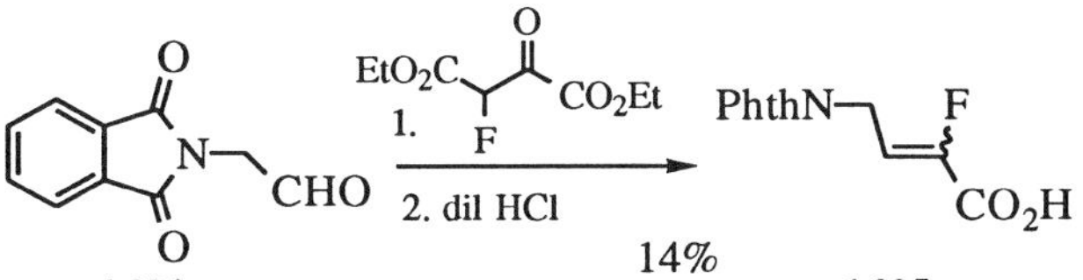

1.224 **1.225**

1.3.D. Phenylalanine and Leucine

Just as in the example just cited, phenylalanine can be converted to an aldehyde (phenylalinal). N-Boc-Phenylalinal (***1.226***), for example, was condensed with an ylid to form an alkene, just as with any other aldehyde. In this example, Horner-Wadsworth-Emmons olefination[112] of ***1.226*** gave the alkene, and removal of the N-Boc group with trifluoroacetic acid gave methyl 4-amino-5-phenylpent-2-enoate, ***1.227***.[133]

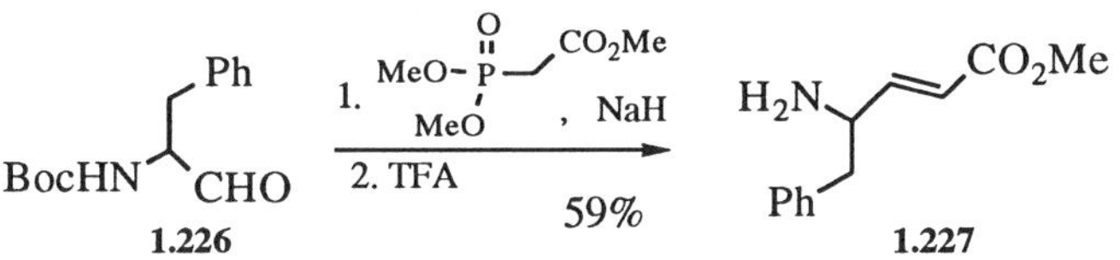

1.226 **1.227**

Amino aldehydes can also be condensed with triphenylphosphonium ylids in a Wittig olefination reaction.[134] Generation of the propargyl alcohol derivative ***1.228*** allowed conversion of the hydroxyl group to a bromide and then to the triphenylphosphonium salt. Treatment with *n*-butyllithium gave the ylid, which reacted with

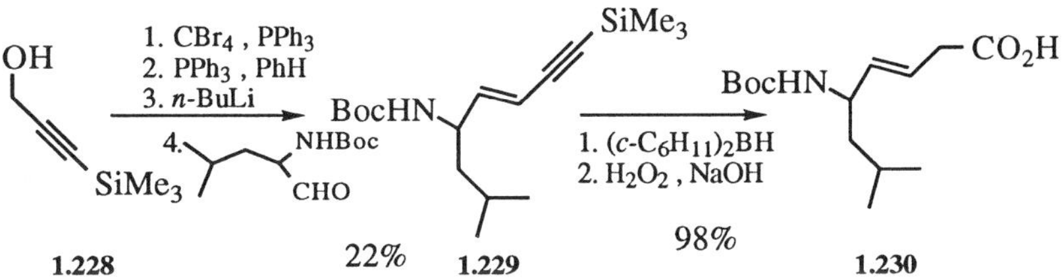

1.228 **1.229** **1.230**

N-Boc leucinal (the aldehyde derivative of leucine) to give ***1.229***.[135] Hydroboration of the alkyne moiety and oxidation led to 5-(N-Boc amino)-7-methyloct-3-enoic acid (***1.230***), which was shown to be an active analog of a renin inhibitory compound (see statine in chapter six, section 6.3).

1.3.E. Miscellaneous Amino Acids

There are, of course, many amino acids that can be used to prepare new amino acids. One is β-arginine (***1.231***,),which reacted with nitrous acid (HONO, generated *in situ* by reaction with sodium nitrite and acid) to convert one amino group to an alcohol moiety. The final product was 2-hydroxy-4-aminobutanoic acid (***1.232***).[136]

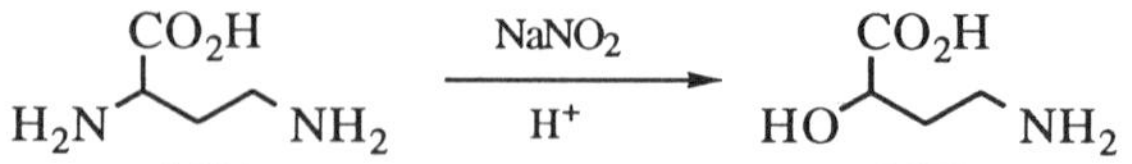

1.231 **1.232**

In a different approach, N-tosyl-β-arginine (***1.233***) was converted to ***1.234***.[137] Protection of nitrogen and esterification with diazomethane was followed by an Arndt-Eistert[138] reaction to give the 3-amino derivative. This was hydrolyzed to give 5-amino-3-(N-trityl-N-methylamino)pentanoic acid (***1.235***).

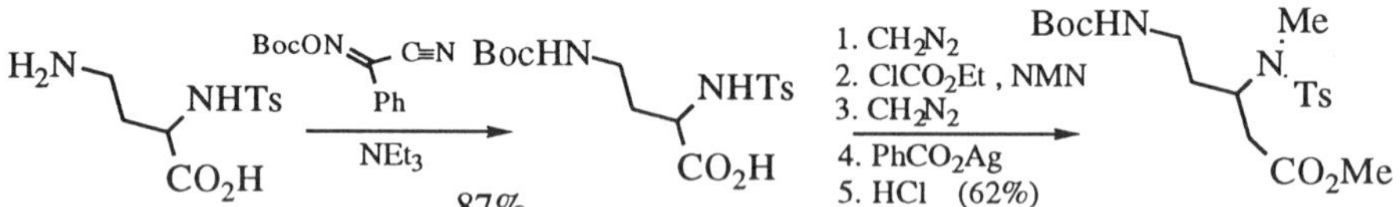

1.233 **1.234** **1.235**

An approach that was not very successful saw a carbamate isoasparagine derivative (***1.236***) converted to a nitrile via dehydration of the amide moiety. Removal of the protecting group gave 3-amino-3-cyano-propanoic acid (***1.237***).[139] This approach was also used to prepare 4-amino-4-cyanobutanoic acid, but the yield of amino acid was again very poor. Introduction of a cyano group in this particular way does not appear to be a good synthetic procedure.

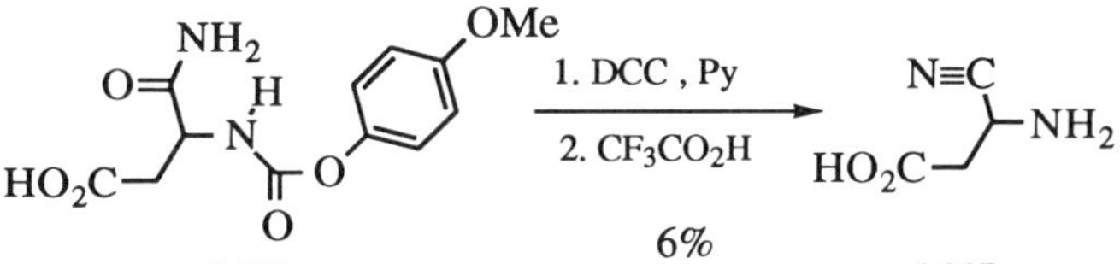

1.236 **1.237**

Other amino acids can be important synthetic precursors. Methyl nipecotate (***1.238***) was protected (giving ***1.239***) and then converted to 5-amino-2-mercapto-methylpentanoic acid (***1.241***) in about eight steps, via ***1.240***.[140]

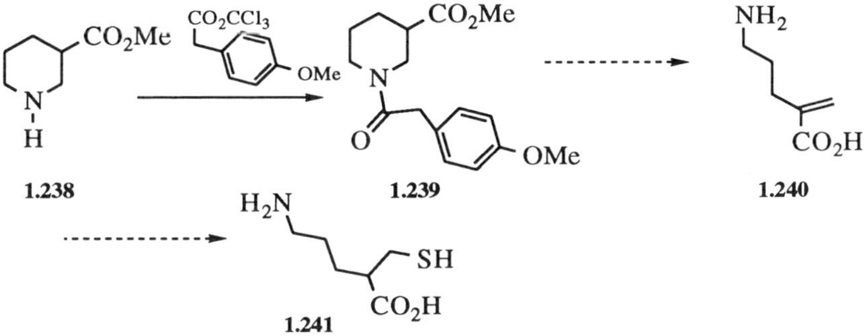

1.238 **1.239** **1.240** **1.241**

A saturated amino acid can be functionalized to serve as a good synthetic precursor to an unsaturated amino acid, via a *syn*-elimination reaction. Reaction of ***1.242*** with lithium hexamethyldisilazide gave the ester enolate and reaction with

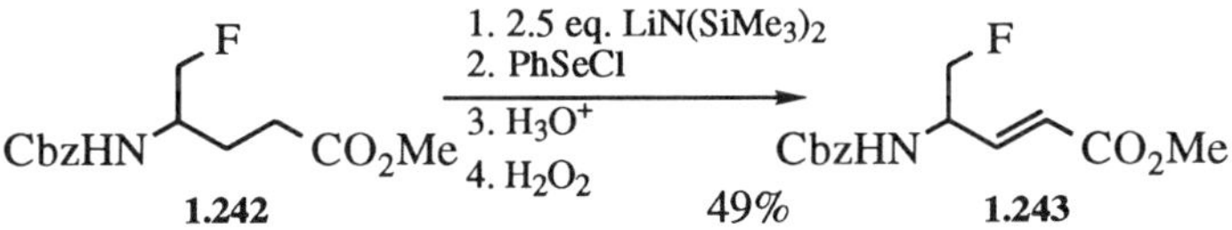

1.242 **1.243**

benzeneselenyl chloride gave the phenylselenide. Oxidation with hydrogen peroxide gave the selenoxide, which spontaneously eliminated PhSeOH to give methyl 4-(N-Cbz amino)-5-fluoropent-2-enoate, ***1.243***.[113]

Such conjugated amino acids can be functionalized in other ways. 4-Aminobut-2-enoic acid (***1.27***), for example, reacted with thiophenol to give 4-amino-3-phenylthiobutanoic acid, ***1.244***. This phenyl sulfide was then oxidized with hydrogen peroxide to give 4-amino-3-phenylsulfonylbutanoic acid, ***1.245***.[141]

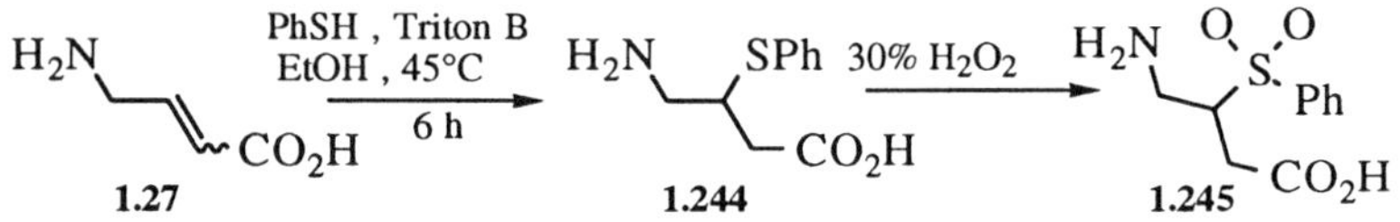

Although it is most common for alkynes to be converted to alkenes in the preparation of alkenyl amino acids, alkenes can also be converted to alkynes.

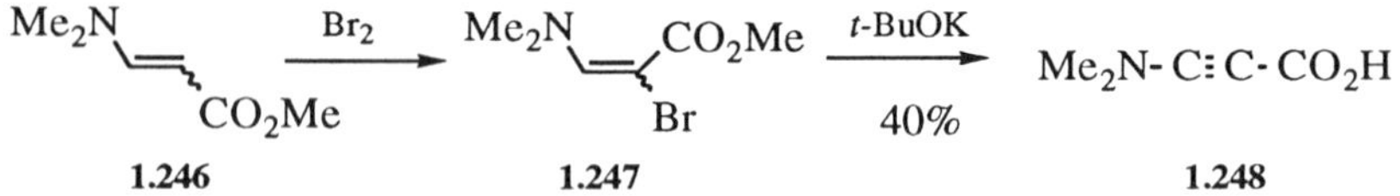

Bromination of ***1.246***, for example, gave ***1.247***, and this gave ***1.248*** [methyl 4-(N,N-dimethylamino)prop-2-ynoate] upon treatment with potassium *t*-butoxide.[142] Such moderate yields of alkynes from vinyl halides under basic conditions are common.

1.4. Specialized Methods

There are, of course, other methods for the preparation of amino acids. Often, these methods are specific to one system and although they can sometimes be used for other syntheses, there is no obvious or general linkage to one of the categories listed above. Several such strategies are collected into this section.

1.4.A. Rearrangement Strategies

There are several synthetic approaches that rely on a molecular rearrangement to incorporate the appropriate functionality. A classical reaction is the Curtius rearrangement,[143] in which an azido ketone rearranges to an isocyanate. Subsequent hydrolysis converts the isocyanate to an amine. This approach can be applied to the synthesis of amino acids. When 2,2,3,3-tetramethylbutanedioic acid (***1.249***) was converted to azido-ketone ***1.250***, heating led to formation of isocyanate ***1.251***[144] via

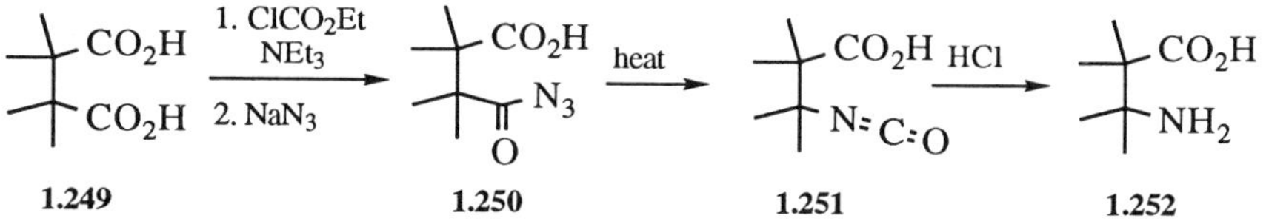

a Curtius rearrangement. Subsequent reaction with hydrochloric acid gave 2,2,3-trimethyl-3-aminopropanoic acid, ***1.252***. In a similar manner, the mono methyl ester of ***1.253*** was converted to 3-amino-4,4-difluorobutanoic acid (***1.254***)[145] via a Curtius rearrangement.

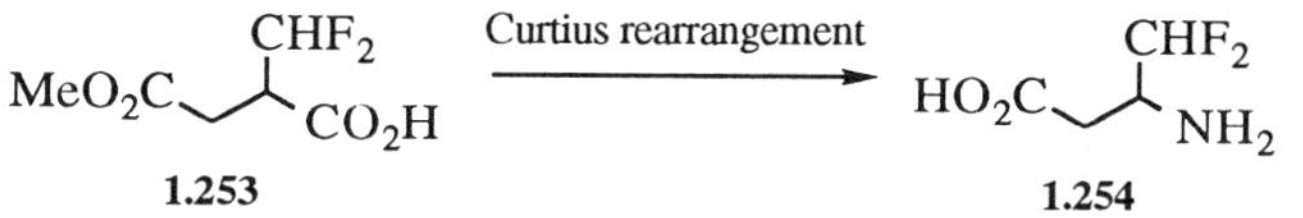

Modifications of this basic approach include functionalization and rearrangement of trichloroacetamide derivatives. Dienoic amino acids are possible, as when ***1.255*** was converted to ***1.256***. Thermal rearrangement led to ***1.257***[146] and deprotection followed by oxidation gave 3-(N-Boc amino)hexa-2E,4-dienoic acid, ***1.258***.

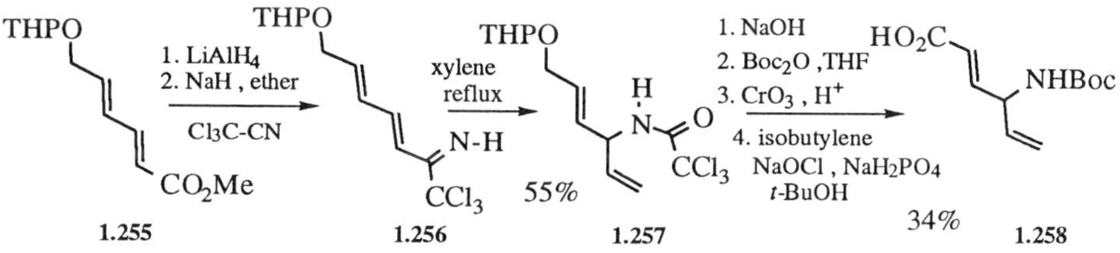

1.4.B. Oxidative Strategies

There are several other interesting methods that convert alkenes to amino acid derivatives. All include somewhat novel oxidative methods. Ruthenium complex ***1.259***, for example, was converted to methyl 4-(N,N-dimethylamino)-2-methylbut-anoate (***1.260***).[147] In addition, methyl 2-ethyl-4-(N,N-dimethylamino)butanoate was prepared by this method, in 40% yield. When the ruthenium complex was derived from a secondary amine, the amino-ester product spontaneously closed to give a 2-pyrrolidinone derivative.[147]

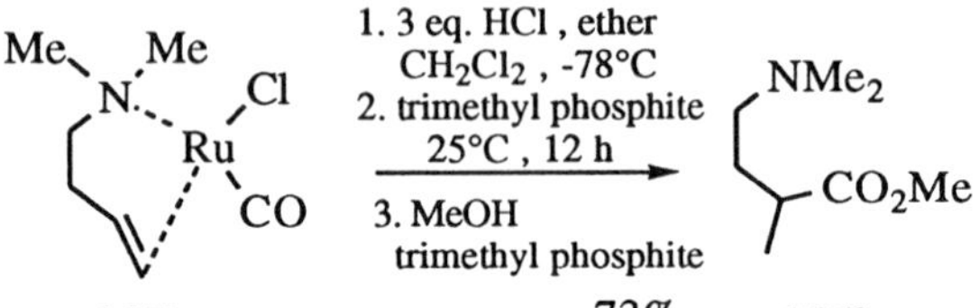

1.259 **1.260**

Non-conjugated alkenes can be used to prepare amino acids via oxidative carboxylation. Formic acid adds to alkenes in the presence of oxygen and hydrogen to give poor yields of the corresponding acid. If an amine group is present, amino acids can be prepared. Allyl amine, for example, gave a 38% yield of 2-methyl-3-aminopropanoic acid upon reaction with formic acid and oxygen[148] and ***1.261*** gave 2-methyl-4-aminobutanoic acid (***1.262***) under the same conditions. This procedure is valuable for the direct conversion of an alkenyl amine to an amino acid. The yields are generally poor, however, as seen. In another example using this method, 4-amino-2-phenylbutanoic acid was prepared in 16% yield from 3-phenyl-1-amino-2-propene.[148]

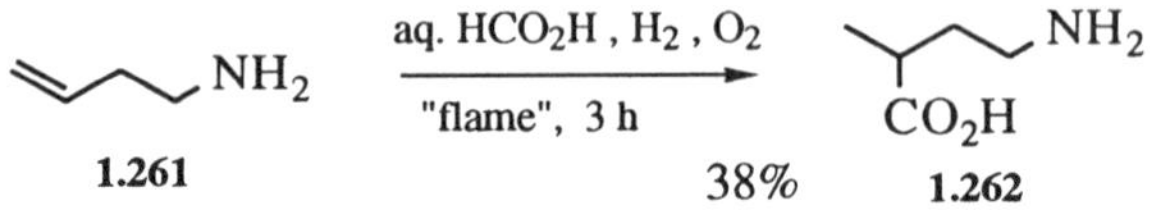

1.261 **1.262**

Alkenes can be oxidatively cleaved by a variety of methods. Addition of allylmagnesium bromide to α-fluoro-acetonitrile gave an imine and its reduction (*in situ*) with methanolic sodium borohydride gave ***1.263***.[149] Protection of the amino group allowed oxidative cleavage of the alkene moiety with $KMnO_4$ to give the carboxylic acid, ***1.264***. Acid hydrolysis gave 3-amino-4-fluoro-butanoic acid (***1.265***).

Silverman modified this approach by condensing lithio *t*-butyl acetate with α-fluoroacetate to give ***1.266***. Subsequent reduction of the enamine moiety gave ***1.265***.[150]

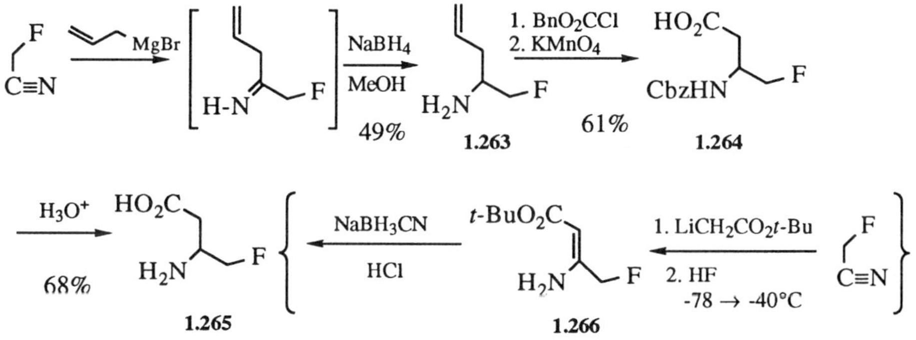

1.4.C. Reductive Methods

Just as there are oxidative methods that produce the acid moiety in an amino acid, there are reductive methods that produce the amino moiety. Reduction of amine surrogates to amines was described in section 1.1.B. This section will describe specialized methods that produce an amine in the synthesis of amino acids.

When ***1.88*** was treated with *n*-butylamine and a rhodium catalyst, 11-(N-butylamino)undecanoic acid (***1.267***) was obtained.[151] This rhodium catalysis method allows an amine to add to an alkene in an anti-Markovnikov manner, as shown. The formation of ***1.260*** from the previous section also involves a rhodium species.

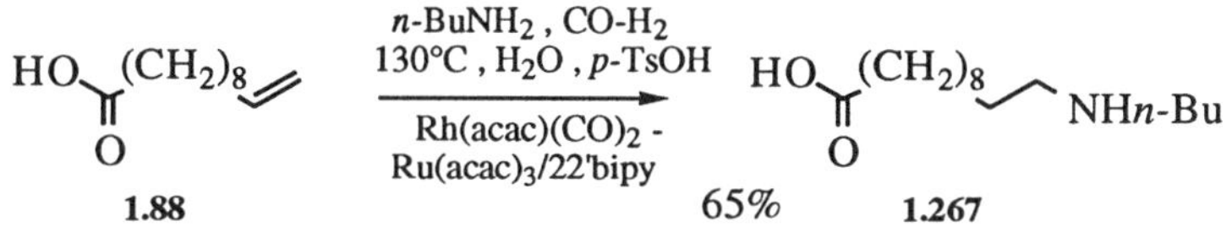

There is a specific but interesting synthesis of amino acids that used functionalized thiophene derivatives. The thiophene moiety acted as a template for incorporating both the acid and amine functional groups. The thiophene was removed via reduction to "liberate" the acyclic amino acid. An example of this thiophene-template approach is treatment of ***1.288*** with Raney nickel in aqueous ammonia

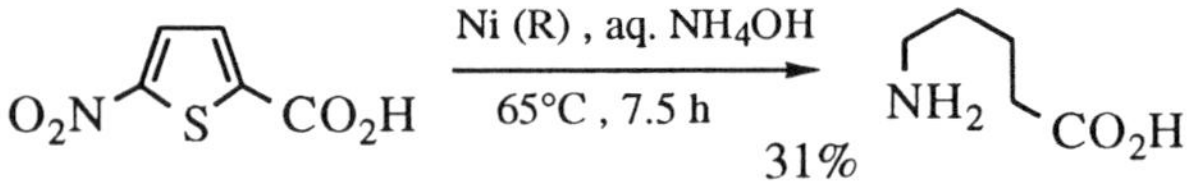

1.76 **1.3**

to give 5-aminopentanoic acid (***1.3***).[152] Similarly prepared was 4-aminoheptanoic acid (also called γ-aminoenanthic acid). Oxime derivatives can also be attached to the thiophene template, further extending the chain in the final amino acid. An example is the conversion of ***1.269*** to 12-aminododecanoic acid (***1.85***).[153] Both of these thiophene-based syntheses suffer from poor yields. Since a functionalized thiophene starting material is required, the scope of the procedure is somewhat limited. Another example of this reaction is presented in chapter two, section 2.5.E.

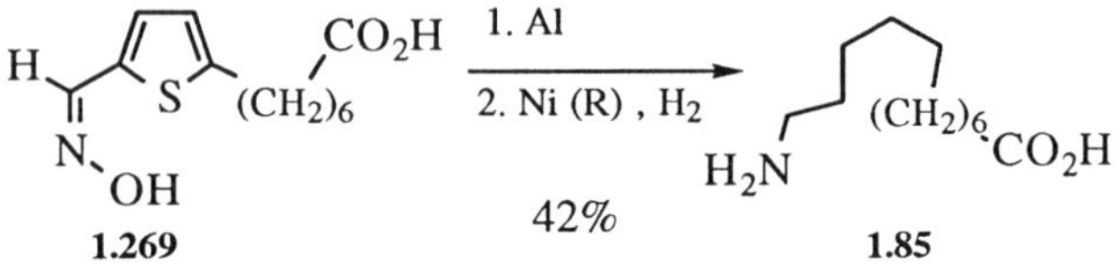

1.269 **1.85**

1.4.D. Coupling Strategies

This section contains a collection of unusual amino acid syntheses that generally involve coupling of two fragments, but they use different strategies. In one example, nitromethane reacted with sodium ethoxide and then with triethyloxonium tetrafluoroborate (Meerwein's reagent)[154] to give a nitroso intermediate that cyclized with methyl acrylate to give a mixture of isoxazolidine ***1.270*** and oxime ***1.271***.[155] Heating this mixture to 100°C converted ***1.270*** to ***1.271***, quantitatively. Catalytic hydrogenation and acid hydrolysis gave 4-amino-2-hydroxybutanoic acid (***1.232***). In general, nitroso compounds are interesting reaction partners in 1,3-dipolar addition reactions such as this and the products are useful 1,2-oxazolidine derivatives.[156]

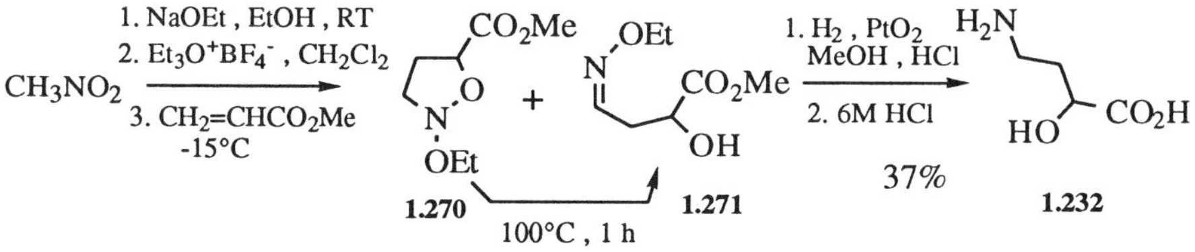

1.270 1.271 1.232

Another interesting coupling reaction reacted acrolein with pyran *1.272* to give 2-hydroxyl-6-aminohexanoic acid (α-hydroxy-ε-amino caproic acid, *1.273*), after reductive amination of the initial product.[157]

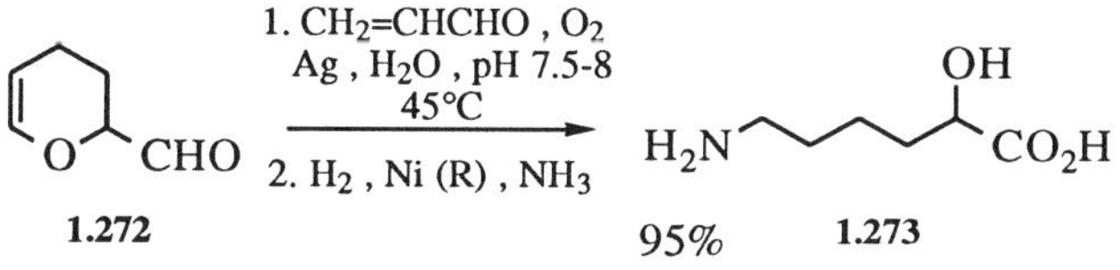

1.272 1.273

A specialized synthesis used ethyl N-thioacyl urethane *1.275* in a Wittig olefination[134] reaction to give *1.276*.[158] The requisite starting material (*1.275*) was

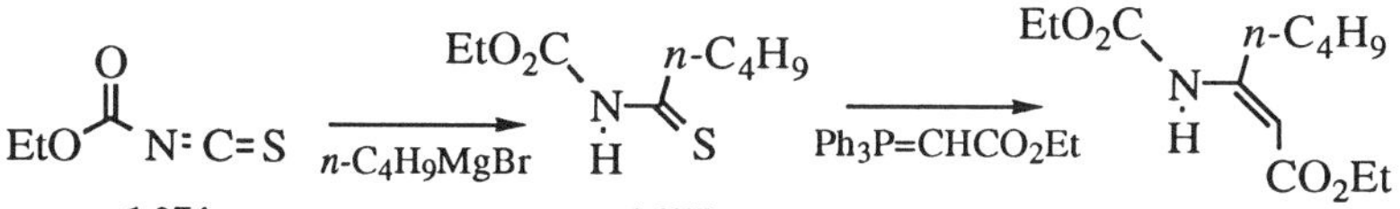

1.274 1.275 1.276

prepared by reaction of *n*-butylmagnesium bromide and isothiocyanate *1.274*.[159] This sequence was also used with methylmagnesium bromide to give the 3-aminobut-2-enoate derivative; with ethylmagnesium bromide to give the 3-aminopent-2-enoate derivatives; and, with *n*-propylmagnesium bromide to give 3-aminohex-2-enoate.

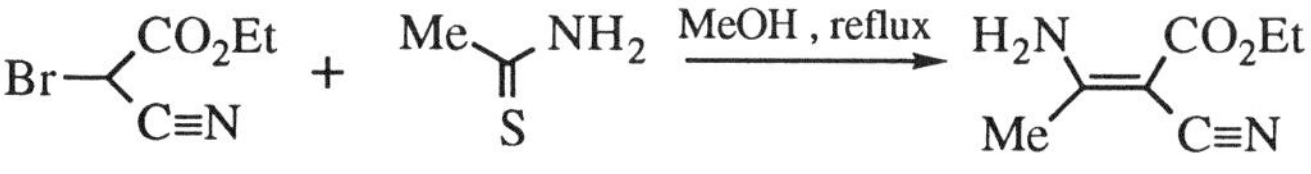

1.277 1.278 1.279

A slightly related sequence involved the reaction of ethyl α-bromo-α-cyanoacetate (*1.277*) with thioamide ***1.278*** to give ethyl 3-amino-2-cyanobut-2Z-enoate, *1.279*.[160]

Dienoic acid derivatives such as ***1.281*** can be prepared by modification of other amino acids. The condensation of ethyl 3-(N-benzylamino)but-2-enoate (***1.280***) and ethyl propiolate gave ethyl 5-(N-benzylamino)-4-carboethoxyhexa-2Z,4E-dienoate, ***1.281***.[161] Changing the 3-methyl group in the alkenyl amino acid to phenyl or hydrogen allowed the synthesis of other dienoic acid derivatives. In this study, the N-alkyl group was also modified from benzyl to hydrogen or methyl.[161]

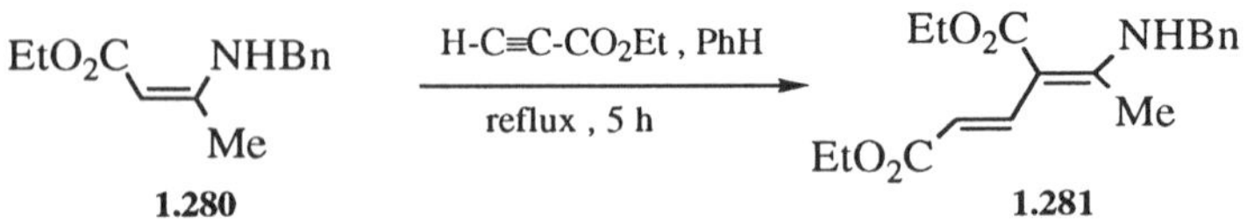

1.280 **1.281**

A last specialized method involved the reaction of methyl 3-aminopent-2-enoate (***1.282***)[162] and DMF-$POCl_3$ under Vilsmeier conditions.[163] In this particular example, Vilsmeier salt ***1.283*** was formed and subsequent reaction with sodium hydrogen sulfide gave methyl 3-aminopent-2Z-enoate 2-thioaldehyde, ***1.284***.[163,164]

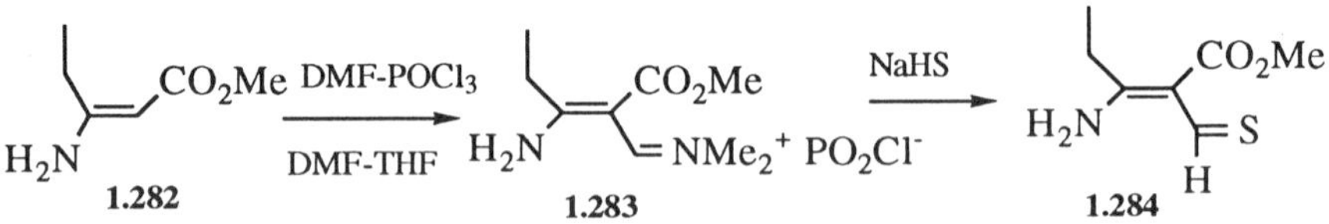

1.282 **1.283** **1.284**

1.4.E. Specialized Addition Reaction Strategies

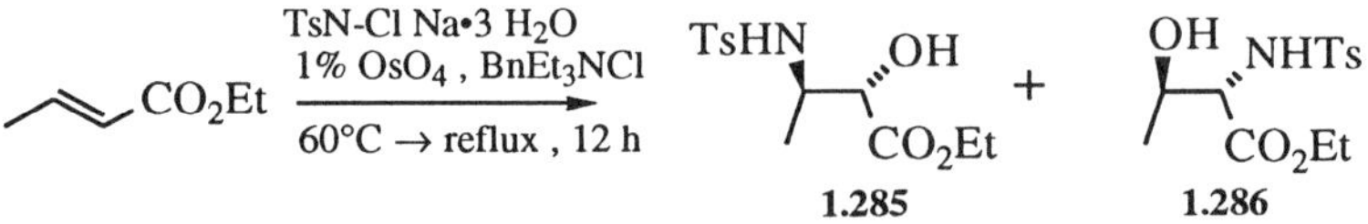

1.285 **1.286**

All of the examples found in this section involve a reaction of one reagent with an alkene or an alkyne. They are generally classified as addition reaction, although they are unusual. In what is generally a "normal" addition reaction, ethyl crotonate reacted with chloramine-T in the presence of osmium tetroxide to give a mixture of diasteromeric ethyl 3-(N-tosyl)-2-hydroxybutanoate, ***1.285*** (36% yield) and ***1.286*** (22% yield).[165] This reaction is included here because the amino group is incorporated at both the "conjugated" position and also the "unconjugated" position in roughly equal amounts.

An interesting synthesis involved the addition of N-hydroxymethylphthalimide (***1.287***) to trichloroethylene, giving 2-chloro-3-(N-phthalimido)propanoic acid, ***1.288.***[166] Addition of ***1.287*** to the alkene led to a *gem*-dichloro intermediate that was hydrolyzed to produce the acid moiety.

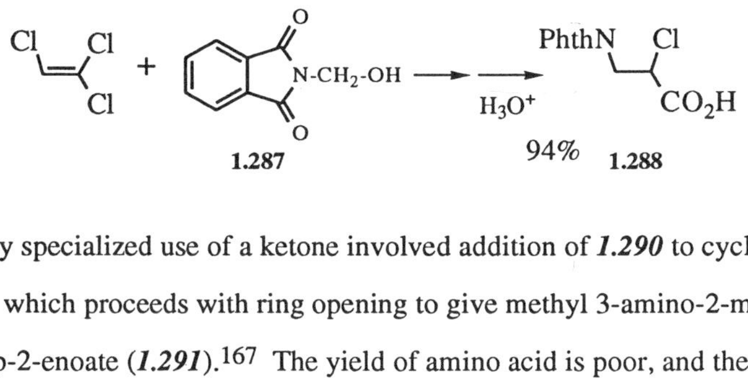

1.287 **1.288**

A highly specialized use of a ketone involved addition of ***1.290*** to cyclopropenone ***1.289***, which proceeds with ring opening to give methyl 3-amino-2-methyl-3-phenylprop-2-enoate (***1.291***).[167] The yield of amino acid is poor, and the scope of the process is limited to forming only a few amino acids due to the difficulties in preparing cyclopropenone derivatives.

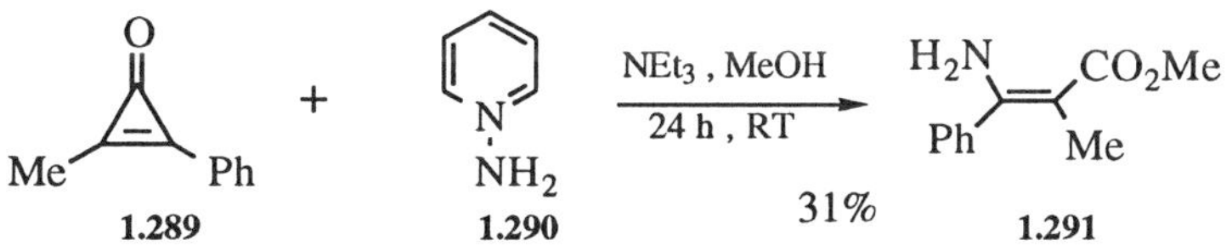

1.289 **1.290** **1.291**

Hydrocyanic acid (HCN) added to alkyne ***1.292*** in the presence of a nickel (0) catalyst to give a 4:1 mixture of carboxylic acids ***1.293:1.294***, after separation.[94] Hydrolysis of the phthalimidoyl group in ***1.293*** gave 2-aminomethylbut-2-enoic acid (***1.295***). The last reaction in this section is another example of how alkynes can be used as synthetic precursors. The [2+2]-cycloaddition reaction of phenylacetylene

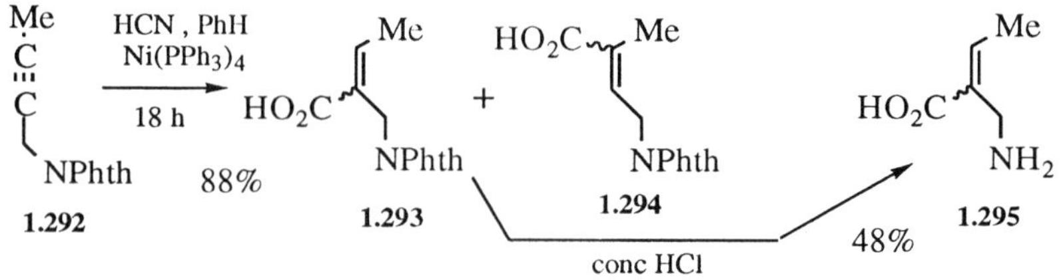

with benzyl isocyanate gave 3-amino-3-phenylprop-2-enoic acid (***1.296***), after hydrolysis.[168]

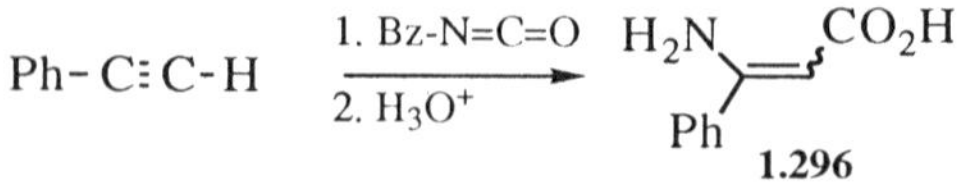

REFERENCES

1 Temple Jr., C.; Elliott, R.D.; Montgomery, J.A. *J. Med. Chem.*, ***1988***, *31*, 697.

2 Baker, J.A.; Harper, J.F. *J. Chem. Soc., C*, ***1967***, 2148.

3 Allan, R.D. *Aust. J. Chem.*, ***1979***, *32*, 2507.

4 Bailey, W.J.; Fujiwara, E. *J. Am. Chem. Soc.*, ***1955***, *77*, 165.

5 Beart, P.M.; Johnston, G.A.R. *Aust. J. Chem.*, ***1972***, 25, 1359.

6 Allan, R.D.; Johnston, G.A.R.; Twitchin, B. *Aust. J. Chem.*, ***1980***, *33*, 1115.

7 (a) Giese, B. *Angew. Chem. Int. Ed., Engl.*, ***1983***, *22*, 753; (b) *Ibid*, ***1985***, *24*, 553; (c) Giese, B. *Radicals in Organic Synthesis: Formation of Carbon-Carbon Bonds*, Pergamon Press, Oxford, ***1986***.

8 (a) Freidlina, R. Kh.; Vasil'eva, T.T.; Velichko, F.K.; Terent'ev, A.B. *Fr. Demande* 2,336,381 [*Chem. Abstr.*, ***1978***, *88*: 120614s]; (b) Idem *Ger. Offen.* 2,558,743 [*Chem. Abstr.*, ***1978***, *88*: 6332y]; (c) Idem *Brit.* 1,500,633 [*Chem. Abstr.*, ***1978***, *89*: 129057r.

9 Walborsky, H.M.; Baum, M.; Loncrini, D.F. *J. Am. Chem. Soc.*, ***1955***, *77*, 3637.

10 Katayama, K.; Ohashi, T. *Jpn. Kokai Tokkyo Koho* JP 02,262,545 [*Chem. Abstr.*, ***1991***,*114*: P123081g].

11 Naito, T.; Nakagawa, S. *Jpn. Kokai* 74 24,914 [*Chem. Abstr.*, ***1974***, *81*:

P4264j].
12 (a) Denki Kagaku Kogyo K.K. *Jpn. Kokai Tokkyo Koho* 81 68,649 [*Chem. Abstr.*, ***1981***, *95*: P169788k]; (b) Idem, *Jpn. Kokai Tokkyo Koho* JP 82 28,035 [*Chem. Abstr.*, ***1982***, *97*: P23301r].
13 Leibman, K.C.; Fellner, S.K. *J. Org. Chem.*, ***1962*** *27*, 438.
14 Ozeki, T.; Kasano, K.; Nakamura, T.; Hashimoto, T.; Kusaka, M. *Kogyo Kagaku Zasshi*, ***1971***, *74*, 2316 [*Chem. Abstr.*, ***1972***, *76*: 24647m].
15 Pinza, M.; Pifferi, G. *J. Pharm. Sci.*, ***1978***, *67*, 120.
16 (a) Allan, R.D.; Twitchin, B. *Aust. J. Chem.*, ***1978***, *31*, 2283; (b) Onda, M.; Konda, Y.; Ömura, S.; Hata, T. *Chem. Pharm. Bull.*, ***1971***,*19*, 2013.
17 Olomucki, M. *Bull. Soc. Chim. Fr.*, ***1963***, 2067.
18 Vessiere, R. *Ann. Fac. Sci. Univ. Clermont Chim.*, ***1960***, 1 [*Chem. Abstr.*, ***1963***, *59*: 13811g].
19 Ziv, D.; Olomucki, M. *Bull. Soc. Chim. Fr.*, ***1970***, 1025.
20 Bongini, A.; Cardillo, G.; Orena, M.; Porzi, G.; Sandri, S. *Tetrahedron*, ***1987***, *43*, 4377.
21 (a) Soffer, L.M.; Katz, M. *J. Am. Chem. Soc.*, ***1956***, *78*, 1705; (b) Gribble, G.W.; Switzer, F.L.; Soll, R.M. *J. Org. Chem.*, ***1988***, *53*, 3164; (c) Rylander, P.N. *Catalytic Hydrogenation in Organic Synthesis*, Academic Press, New York, ***1979***, pp. 140-141
22 Gibson, M.S.; Bradshaw, R.W. *Angew. Chem. Int. Ed., Engl.*, ***1968***, *7*, 919.
23 Ing, H.R.; Manske, R.H.F. *J. Chem.. Soc.*, ***1926***, 2348.
24 Osby, J.O.; Martin, M.G.; Ganem, B. *Tetrahedron Lett.*, ***1984***, *25*, 2093.
25 (a) Mitsui Toatsu Chemicals Inc. *Jpn. Kokai Tokkyo* 81 20,556 [*Chem. Abstr.*, ***1981***, *95*: P62707f]; (b) Idem, *Jpn. Kokai Tokkyo Koho* 80,147,246 [*Chem. Abstr.*, ***1981***, *94*: P157263b].
26 Uchimaru, F.; Sato, M.; Kosasayama, E.; Shimizu, M.; Takashi, H. *Jpn.* 70 16,682 [*Chem. Abstr.*, ***1970***, *73*: P77617w].
27 (a) Bowden, K.; Heilbron, I.M.; Jones, E.R.H.; Weedon, B.C.L. *J. Chem. Soc.*, ***1946***, 39; (b) Heilbron, I.M.; Jones, E.R.H.; Sondheimer, F. *Ibid*, ***1949***, 604; (c) Haynes, L.J.; Heilbron, I.M.; Jones, E.R.H.; Sondheimer, F. *Ibid*, ***1947***, 1583; (d) Heilbron, I.M.; Jones, E.R.H.; Sondheimer, J. *Ibid*, ***1947***, 1586.
28 Tsuda, T.; Horii, Y.; Nakagawa, Y.; Ishida, T.; Saegusa, T. *J. Org. Chem.*, ***1989***, *54*, 977.

29 Takano, S.; Ogasawara, K.; Sekiguchi, Y.; Kasai, N.; Sakaguchi, K. *Jpn. Kokai Tokkyo Koho* JP 63,174,957 [*Chem. Abstr.*, ***1989***, *110*: P24308x].

30 Testa, E.; Fontanella, L. *Brit.* 829,663 [*Chem. Abstr.*, ***1962***, *56*: P1430c].

31 Garritsen, J.W.; Penders, J.M. *Neth. Appl.* 72 16,661 [*Chem. Abstr.*, ***1974***, *81*: P151570j].

32 Smirnov, Yu.D.; Tomilov, A.P. *Zh. Org. Khim.*, ***1969***, *5*, 864 (Engl. 850).

33 Lukeš, R.; Kloubek, J. *Coll. Czech. Chem. Commun.*, ***1960***, *25*, 607.

34 Howe, R.K.; Gruner, T.A.; Carter, L.G.; Franz, J.E. *J. Heterocyclic Chem.*, ***1978***, *15*, 1001.

35 Ram, S.; Ehrenkaufer, R.E. *Synthesis*, ***1986***, 133.

36 Ohta, H.; Kobayashi, N.; Ozaki, K. *J. Org. Chem.*, ***1989***, *54*, 1802.

37 Ram, S.; Ehrenkaufer, R.E. *Tetrahedron Lett.*, ***1984***, *25*, 3415.

38 Knouzi, N.; Vaultier, M.; Toupet, L.; Carrie, R. *Tetrahedron Lett.*, ***1987***, *28*, 1757.

39 Bartra, M.; Vilarrasa, J. *J. Org. Chem.*, ***1991***, *56*, 5132.

40 Kitatsume, T. *Jpn. Kokai Tokkyo Koho* JP 63 60,954 [*Chem. Abstr.*, ***1988***,*109*: P190870y].

41 (a) Brown, H.C.; Krishnamurthy, S. *Aldrichimica Acta*, ***1979***, *12*, 3; (b) Schenker, R. *Angew. Chem.*, ***1961***, *73*, 81; (c) Fieser, L.F.; Fieser, M. *Reagents for Organic Synthesis*, Vol. 1, Wiley, New York, ***1967***, p. 1049.

42 Williams, T.M.; Crumbie, R.; Mosher, H.S. *J. Org. Chem.*, ***1985***, *50*, 91 and references cited therein.

43 (a) Melikoff, P. *Ber.*, ***1879***, *12*, 2227; (b) Erlenmeyer, E. *Ibid*, ***1880***, *13*, 1077; (c) Fisher, E.; Leuchs, H. *Ibid*, ***1902***, *35*, 3787; (d) Nakajima, Y.; Kinishi, R.; Oda, J.; Inouye, Y. *Bull. Chem. Soc., Jpn.*, ***1977***, *50*, 2025.

44 Taber, D.F.; Hoerrner, R.S. *J. Org. Chem.*, ***1992***, *57*, 441.

45 Allan, R.D.; Dickenson, H.W.; Johnston, G.A.R.; Kazlauskas, R.; Tran, H.W. *Aust. J. Chem.*, ***1985***, *38*, 1651.

46 Talley, J.J. *Eur. Pat. Appl.* EP 396,526 [*Chem. Abstr.*, ***1991***, *114*: 229382w].

47 (a) Effenberger, F.; Drauz, K.; Förster, S., Müller, W. *Chem. Ber.*, ***1981***, *114*, 173; (b) Effenberger, F.; Drauz, K. *Angew. Chem.*, ***1979***, *91*, 504.

48 Siclari, F.; Rossi, P.P.; De Gaetano, M. *Ger. Offen.* 2,608,203 [*Chem. Abstr.*, ***1977***, *77*:4923b].

49 Mileo, J.-C.; Sillion, B.; de Gaudemaris, G. *C.R. Acad. Sci., Ser. C*, ***1968***,

267, 93.
50 Diaper, D.G.M.; Mitchell, D.L. *Can. J. Chem.*, ***1962***, *40*, 1189.
51 Vaccher, C.; Berthelot, P.; Flouquet, N.; Debaert, M. *Synth. Commun.*, ***1989***, *19*, 2049.
52 Merger, F.; Fischer, R.; Harder, W.; Priester, C.U.; Vagt, U. *Ger. Offen.*, DE 3,843,792 [*Chem. Abstr.*, ***1990***, *113*: P230789v].
53 Ladouceur, G.; Mais, D.E.; Jakubowski, J.A.; Utterback, B.G.; Robertson, D.W. *Bioorg. and Medicinal Chemistry Lett.*, ***1991***, *1*, 173.
54 Charles, J.; Descotes, G. *Bull. Soc. Chim. Fr.*, ***1966***, 102.
55 Brannock, K.C.; Bell, A.; Burpitt, R.D.; Kelly, C.A. *J. Org. Chem.*, ***1964***, *29*, 801.
56 Weintraub, L.; Wilson, A.; Goldhamer, D.L.; Hollis, D.P. *J. Am.Chem. Soc.*, ***1964***, *86*, 4880.
57 Arata, Y ; Wada, H. *Kanazawa Daigaku Yakugakuba Kenyu Nempo*, ***1963***, *13*, 72 [*Chem. Abstr.*, ***1964***, *60*: 7907b].
58 Arata, Y.; Ohashi, T.; Okumura, Z.; Wada, Y.; Ishiwata, M. *Yakugku Zasshi*, ***1962***, *82*, 782 [*Chem. Abstr.*, ***1963***, *58*: 2432h].
59 Weiler, L. *J. Am. Chem. Soc.*, ***1970***, *92*, 6702.
60 (a) Shibuya, M.; Kuretani, M.; Kubota, S. *Tetrahedron Lett.*, ***1981***, *22*, 4453; (b) Shibuya, M.; Kubota, S. *Heterocycles*, ***1980***, *14*, 601.
61 Tomoeda, M.; Tani, Y.; Okada, H. *Yakugaku Zasshi*, ***1966***, *86*, 1213 [*Chem. Abstr.*, ***1967***, *67*: 2960v].
62 Cane, D.E.; Shiao, M.-S. *J. Am. Chem. Soc.*, ***1978***, *100*, 3203.
63 Seke, V.; Fleš D. *Croat. Chem. Acta*, ***1966***, *38*, 9 [*Chem. Abstr.*, ***1966***, *65*: 13536e
64 (a) Arndt, F.; Eistert, B. *Ber.*, ***1935***, *68*, 200; (b) Bachmann, W.E.; Struve, W.S. *Org. React.*, ***1942***, *1*, 38.
65 Fleš D.; Seke, V.; Dadić, M. *J. Polym. Sci., Part C*, ***1969***, *22*, 971.
66 (a) Stork, G.; Terrell, R.; Szmuszkovicz, J. *J. Am. Chem. Soc.*, ***1954***, *76*, 2029; (b) Mundy, B.F.; Ellerd, M.G. *Name Reactions and Reagents in Organic Synthesis*, Wiley, New York, ***1988***, pp. 206-207.
67 Conrad, M.; Epstein, W. *Deutsh. Geschel. Ber.*, ***1887***, *20*, 3052.
68 Buckler, R.T.; Hartzler, H.E. *J. Med. Chem.*, ***1975***, *18*, 509.
69 Combes, A.; Combes, C. *Bull. Soc. Chim., Fr.*, ***1892***, *7*, 778.

70 (a) Glickman, S.A.; Cope, A.C. *J. Am. Chem. Soc.*, ***1945***, *67*, 1017; (b) Michaelis, A. *Ann.*, ***1909***, *366*, 324 (see p. 337); (c) Brühl, J.W. *Z. Physik. Chem.*, ***1895***, *16*, 216.

71 (a) Gasteiger, J.; Strauß, U. *Chem. Ber.*, ***1981***, *114*, 2336; (b) also see Züblin, J. *Ber.*, ***1878***, *11*, 1417; (c) Bülow, C.; Neber, P. *Ibid*, ***1912***, *45*, 3732; (d) Parmerter, S.M. *Org. React.*, ***1959***, *10*, 1; (e) Prager, B. *Ber.*, ***1901***, 34, 3600.

72 Slopianka, M.; Gossauer, A. *Liebigs Ann. Chem.*, ***1981***, 2258.

73 Furukawa, M.; Okawara, T.; Noguchi, Y.; Terawaki, Y. *Chem. Pharm. Bull.*, ***1979***, *27*, 2223.

74 Potěšil, T.; Potěšilová H. *J. Chromatogr.*, ***1982***, *249*, 131.

75 Aberhart, D.J.; Lin, H.-J. *J. Org. Chem.*, ***1981***, *46*, 3749.

76 Veronese, A.C.; Gandolfi, V.; Basato, M.; Corain, B. *J. Chem. Res., Synop.*, ***1988***, 246.

77 Fukuda, K.; Sato, J.; Ito, K.; Suzuki, K.; Nawamaki, T.; Watanabe, S. *Jpn. Kokai Tokkyo Koho* JP 03,215,476 [*Chem. Abstr.*, ***1992***, *116*: P59391v].

78 Boosen, K.-J. *Helv. Chim. Acta*, ***1977***, *60*, 1256.

79 Boosen, K.-J. *Chimia*, ***1973***, *27*, 541.

80 Sano, T.; Horiguchi, Y.; Toda, J.; Imafuku, K.; Tsuda, Y. *Chem. Pharm. Bull.*, ***1984***, *32*, 497.

81 Van Sant, K.; South, M.S. *Tetrahedron Lett.*, ***1987***, *28*, 6019.

82 Gelin, S.; Rouet, J. *Bull. Soc. Chim. Fr.*, ***1971***, 2179.

83 Birkofer, L.; Ritter, A.; Richter, P. *Chem. Ber.*, ***1963***, *96*, 2750.

84 Kloek, J.A.; Leschinsky, K.L. *J. Org. Chem.*, ***1978***, *43*, 1460.

85 Célérier, J.P.; Deloisy, E.; Kapron, P.; Lhommet, G.; Maitte, P. *Synthesis*, ***1981***, 130.

86 (a) Meldrum, A.N. *J. Chem. Soc.*, ***1908***, *93*, 598; (b) Hanford, W.E.; Sauer, J.P. *Org. React.*, ***1946***, *3*, 108 (see p. 124).

87 (a) Pinner, A.; Klein, F. *Ber.*, ***1889***, *10*, 1889; (b) *Ibid*, ***1878***, *11*, 4, 1475; (c) Pinner, A., *Ibid*, ***1883***, *16*, 352,1643; (d) Roger, R.; Neilson, D. *Chem. Rev.*, ***1961***, *61*, 179; (e) Gray, A.P.; Dipinto, V.M.; Solomon, I.J. *J. Org. Chem.*, ***1977***, *41*, 2428.

88 Prelog, V. *Coll. Czech. Chem. Commun.*, ***1930***, *2*, 712.

89 Suyama, T.; Kanao, S. *Nippon Kagaku Zasshi*, ***1962***, *83*, 1058 [*Chem. Abstr.*, ***1963***, *59*: 2638d].

90 Lartillot, S.; Baron, C. *Bull. Soc. Chim., Fr.*, ***1966***, 3798.

91 Pfaltz, A.; Anwar, S. *Tetrahedron Lett.*, ***1984***, *25*, 2977.

92 Mekhtiev, S.I.; Safarov, Yu.D.; Polchaev, R.A. *USSR* 445,945 [*Chem. Abstr.*, ***1975***, *83*: 10851f].

93 Delpierre, G.R.; Eastwood, F.W.; Gream, G.E.; Kingston, D.G.I.; Sarin, P.S.; Todd, L.; Williams, D.H. *J. Chem. Soc., C*, ***1966***, 1653.

94 Jackson, W.R.; Perlmutter, P.; Smallridge, A.J. *Tetrahedron Lett.*, ***1988***, *29*, 1983.

95 Peet, N.P.; Burkhart, J.P.; Angelastro, M.R.; Giroux, E.L.; Mehdi, S.; Bey, P.; Kolb, M.; Neises, B.; Schirlin, D. *J. Med. Chem.*, ***1992***, *33*, 394.

96 Neidlein, R.; Kikelj, D. *Chem. Ber.*, ***1988***, *121*, 1817.

97 Caubère, P. *Bull. Soc. Chim., Fr.*, ***1964***, 148.

98 (a) Sneeden, R.P.A. in *The Chemistry of Carboxylic Acids and Esters*; Patai, S., Ed.; Interscience: New York, ***1969***; pp. 137-175; (b) Kharasch, M.S.; Reinmuth, O. *Grignard Reactions of Nonmetallic Substances*, Prentice-Hall: New York, ***1954***; pp. 913-948; (c) Eberson, L. *Acta Chem. Scand.*, ***1962***, *16*, 781; (d) Oppolzer, W. Kündig, E.P.; Bishop, P.M.; Perret, C. *Tetrahedron Lett.*, ***1982***, *23*, 3901.

99 Marszak, I.; Olomucki, M. *Bull. Soc. Chim., Fr.*, ***1959***, 182.

100 (a) Olomucki, M.; Marszak, I. *Compt. Rend.*, ***1956***, *242*, 1338; (b) Olomucki, M.; Ziv, D. *Bull. Soc. Chim, Fr.*, ***1968***, 4923.

101 MacInnes, I.; Walton, J.C. *J. Chem. Soc., Perkin Trans. 2*, ***1987***, 1077.

102 Ziv, D.; Olomucki, M.; Marszak, I. *Bull. Soc. Chim. Fr.*, ***1970***, 150.

103 (a) Olomucki, M.; Marszak, I. *C.R. Acad. Sci.*, ***1961***, *253*, 2239; (b) Lövgren, K.; Nilsson, J.L.G. *Acta Pharm. Suec.*, ***1975***, *12*, 379; (c) also see Corriu, R.J.P.; Huynh, V.; Moreau, J.J.E. *Tetrahedron Lett.*, ***1984***, *25*, 1887; and also (d) Bestmann, H.J.; Wolfel, G. *Chem. Ber.*, ***1984***, *117*, 1250.

104 (a) Olomucki, M.; Marszak, I. *Bull. Soc. Chim., Fr.*, ***1964***, 1767; (b) *Ibid*, ***1959***, 315.

105 Ohoka, M.; Asada, K.; Yanagida, S.; Okahara, M. *J. Org. Chem.*, ***1973***, *38*, 2287.

106 Ohoka, M.; Yanagida, S.; Komori, S. *J. Org. Chem.*, ***1972***, *37*, 3030.

107 Jakobiec, T. *Acta Polon. Pharm.*, ***1966***, *23*, 111 [*Chem. Abstr.*, *65*: 10492b ***1966***].

108 Edwards, D.; Hamer, D.; Stewart, W.H. *J. Pharm. Pharmacol.*, ***1964***, *16*, 618.

109 Shono, T.; Kise, N.; Sanda, F.; Ohi, S.; Tsubata, K. *Tetrahedron Lett.*, ***1988***, *29*, 231.

110 (a) Meyers, A.I.; Nabeya, A.; Adickes, H.W.; Politzer, I.R. *J. Am. Chem. Soc.*, ***1969***, *91*, 763; (b) Meyers, A.I.; Nabeya, A.; Adickes, H.W.; Politzer, I.R.; Malone, G.R.; Kovelesky, A.C.; Nolen, R.L.; Portnoy, R.C. *J. Org. Chem.*, ***1973***, *38*, 36.

111 Thompson, W.J.; Tucker, T.J.; Schwering, J.E.; Barnes, J.L. *Tetrahedron Lett.*, ***1990***, *31*, 6819.

112 (a) Horner, L.; Hoffmann, H.; Wippel, J.H.; Klahre, G. *Ber.*, ***1959***, *92*, 2499; (b) Wadsworth,W.S. Jr.; Emmons, W.D. *J. Am. Chem. Soc.*, ***1961***, *83*, 1733; (c) Boutagy, J.; Thomas, R. *Chem. Rev.*, ***1974***, *74*, 87.

113 Silverman, R.B.; Invergo, B.J.; Mathew, J. *J. Med. Chem.*, ***1986***, *29*, 1840.

114 Liu, S.; Hanzlik, R.P. *J. Med. Chem.*, ***1992***, *35*, 1067.

115 Allmendinger, T.; Furet, P.; Hungerbühler, E. *Tetrahedron Lett.*, ***1990***, *31*, 7297, 7301.

116 Wiberg, K.B. In *Oxidation in Organic Chemistry, Part A*; Wiberg, K.B., Ed.; Academic: New York, ***1965***; pp. 174.

117 Allan, R.D.; Johnston, G.A.R.; Kazlauskas, R. *Aust. J. Chem.*, ***1985***, *38*, 1647.

118 Ceika, A.; Kupetis, G.; Mozolis, V. *Liet. TSR Mokslu Akad. Darb., Ser. B*, ***1969***, 101 [*Chem. Abstr.*, ***1970***, *72*: 89715u].

119 Ayres, D.C. *J. Chem. Soc., Perkin Trans. I*, ***1978***, 585.

120 Grzonka, Z.; Lammek, B. *Synthesis*, ***1974***, 661.

121 Honoré, T.; Hjeds, H.; Krogsgaard-Larsen, P.; Christiansen, T.R. *Eur. J. Med.Chem. - Chim. Ther.*, ***1978***, *13*, 429.

122 Emmons, W.D.; Levy, J.F. *Fr.* 1,507,036 [*Chem.Abstr.*, ***1969***, *70*: P19556c].

123 Sakai, S.; Miyaji,Y.; Furutani, H.; Kobata, M.; Hachisuka, T.; Nakayama, A.; Takada, S.; Hayashi, T. *Jpn.* 12,264 ('62) [*Chem. Abstr.*, ***1963***, *59*: P9805e].

124 Suzuki, Y.; Ogura, Y.; Takahashi, H. *Jpn. Kokai* 73 52,721 [*Chem. Abstr.*, ***1974***, *80*: P47465x].

125 (a) Bergmann, E.D.; Cohen, A. *Isr. J. Chem.*, ***1967*** *5*, 15; (b) Nefkens, G.H.L.; Tesser, G.I.; Nivard, R.J.F. *Recuil. Trav. Chim.*, ***1960***, *79*, 688.

126 Krogsgaard-Larsen, P.; Larsen, A.L.N.; Thyssen, K. *Acta Chem. Scand.*, ***1978***, *B32*, 469.

127 Falch, E.; Hedegaard, A.; Nielsen, L.; Jensen, B.R.; Hjeds, H.; Krogsgaard-Larsen,P. *J. Neurochem.*, ***1986***,*47*, 898.

128 see Enna, S.J. *The GABA Receptors*, The Humana Press, Clifton, New Jersey, ***1983***, pp. 5, 8-9, 14-16, 39, 63, 70, 75, 132-133, 152-153, 195, 198-200, 305.

129 Silverman, R.B.; Invergo, B.J.; Levy, M.A.; Andrew, C.R. *J. Biol. Chem.*, ***1987***, *262*, 3192.

130 Evans, D.A.; Sidebottom, P.J. *J. Chem . Soc., Chem. Commun.*, ***1978***, 753.

131 Siedel, W.; Nahm, H. *Ger. Patent* 928,711 [*Chem. Abstr.*, ***1958***, *52*: P5471g].

132 Bergmann, E.D.; Cohen. A. *Tetrahedron Lett.*, ***1965***, 2085.

133 Thompson, S.A.; Andrews, P.R.; Hanzlik, R.P *J. Med. Chem.*, ***1986***, *29*, 104.

134 (a) Wittig, G.; Rieber, M. *Ann.*, ***1949***, *562*, 187; (b) Wittig, G.; Geissler, G. *Ibid*, ***1953***, *580*, 44; (c) Wittig, G.; Schöllkopf, U. *Chem. Ber.*, ***1954***, *87*, 1318; (d) Gensler, W.J. *Chem. Rev.*, ***1957***, *57*, 191 (see p. 218).

135 Johnson, R.L. *J. Med. Chem.*, ***1984***, *27*, 1351.

136 Horiuchi, Y.; Akita, E.; Ito, T. *Agr. Biol. Chem.*, ***1976***, *40*, 1649.

137 Nomoto, S.; Shiba, T. *Chem. Lett.*, ***1978***, 589.

138 (a) Arndt, F.; Eistert, B. *Ber.*, ***1935***, *68*, 200; (b) Bachmann, W.E.; Struve, W.S. *Org. React.*, ***1942***, *1*, 38.

139 Ressler, C.; Nagarajan, G.R.; Kirisawa, M.; Kashelikar, D.V. *J. Org. Chem.*, ***1971***, *36*, 3960.

140 Ondetti, M.A.; Condon, M.E. *Ger. Offen.*, 2,801,911 [*Chem. Abstr.*, ***1978***, *89*: P180372s].

141 Hayashi, T.; Kondo, M.; Tanaka, M.; Ishiyama, N. *Jpn.* 72 24,007 [*Chem. Abstr.*, ***1972***, *77*: P151679u].

142 Ostroumov, I.G.; Tsil'ko, A.E.; Maretina, I.A. *Zh. Org. Khim.*, ***1988***, *24*, 2321 (Engl., 2092).

143 (a) Curtius, T. *J. Prakt. Chem.*, ***1894***, *50*, 275; (b) Smith, P.A.S. *Org. React.*, ***1946***, *3*, 337; (c) Saunders, J.H.; Slocombe, R.J. *Chem. Rev.*, ***1948***, *43*, 203.

144 Shadbolt, R.S.; Stephens, F.F. *J. Chem. Soc., C*, ***1971***, 1665.

145 Schirlin, D.; Baltzer, S.; Heydt, J.G.; Jung, M.J. *J. Enzyme Inhib.*, ***1987***,*1*, 243 [*Chem. Asbtr.*, ***1987***, *107*: 213975m].

146 Bey, P.; Gerhart, F.; Jung, M. *J. Org. Chem.*, ***1986***, *51*, 2835.

147 Krafft, M.E.; Wilson, L.J.; Onan, K.D. *Tetrahedron Lett.*, ***1988***,*29*, 6421.

148 Namoto, S.; Harada, K. *Chem. Lett.*, ***1985***, 145.

149 Bey, P.; Jung, M.; Gerhart, F. *Eur. Pat. Appl.* 24,965 [*Chem. Abstr.*, ***1981***, *95*: P62710b].
150 Mathew, J.; Invergo, B.J.; Silverman, R.B. *Synth. Commun.*, ***1985***, *15*, 377.
151 Drent, E.; Breed, A.J.M. *Eur. Patent* EP 457, 386 [*Chem. Abstr.*, ***1992***, *116*:P83212h].
152 Fabrichnyi, B.P.; Zurabyan, S.E.; Shalavina, I.F.; Gol'dfarb, Ya.L. *Izv. Akad. Nauk. SSSR, Ser Khim.*, ***1967***, 2102
153 Fabrichnyi, B.P.; Shalavina, I.F.; Gol'dfarb, Ya.L. *Zh. Obshch.Khim.*, ***1964***, *34*, 3878 (Engl. 3938).
154 Meerwein, H. *Org. Synth. Collect. Vol. 5*, ***1973***, 1080.
155 (a) Sato, H.; Kusumi, T.; Imaye, K.; Kakisawa, H. *Bull. Chem. Soc., Jpn.*, ***1976*** *49*, 2815 (b) Idem, *Chem. Lett.*, ***1975***, 965.
156 Fleming, I. *Frontier Orbitals and Organic Chemical Reactions,* Wiley, London, ***1976***, pp. 93, 109-110, 148-161.
157 Ottenheym, J.H.; Verstrijden, U.; Pesch, W. *US Patent* 3,120,562 [*Chem. Abstr.*, ***1964***, *61*: P1761a].
158 Slopianka, M.; Gossauer, A. *Synth. Commun.*, ***1981***, *11*, 95.
159 Gossauer, A.; Roessler, F.; Zilch, H. *Liebigs Ann. Chem.*, ***1979***, 1309.
160 Singh, H.; Cheema, A.S. *Chem. Ind. (London)*, ***1976***, 413.
161 Anghelide, N.; Draghici, C.; Raileanu, D. *Tetrahedron*, ***1974***, *30*, 623.
162 Muraoka, M.; Yamamoto, T.; Takeshima, T. *Chem. Lett.*, ***1982***, 101.
163 (a) Vilsmeier, A.; Haack, A. *Ber.*, ***1927***, *60*, 119; (b) Mundy, B.P.; Ellerd, M.G. *Name Reactions and Reagents in Organic Synthesis*, Wiley, New York, ***1988***, pp. 214-215; (c) Witiak, D.T.; Williams, D.R.; Kakodkar, S.V.; Hite, G.; Shen, M.-S. *J. Org. Chem.*, ***1974***, *39*, 1242; (d) de Maheas, M.R. *Bull. Soc. Chim. Fr.*, ***1962***, 1989.
164 Muraoka, M.; Yamamoto, T.; Enomoto, K.; Takeshima, T. *J. Chem. Soc., Perkin Trans. I*, ***1989***, 1241.
165 Herranz, E.; Sharpless, K.B. *J. Org. Chem.*, ***1978***, *43*, 2544.
166 Bott, K. *Chem. Ber.*, ***1973***, *106*, 2513.
167 Kascheres, A.; Marchi, D. Jr.; Rodriques, J.A.R. *J. Org. Chem.*, ***1978***, *43*, 2892.
168 Arbuzov, B.A.; Zobova, N.N. *Dokl. Akad. Nauk. SSR*, ***1967***, *172*, 845 (Engl., 103).

2

Cyclic Precursors

The preparation of acyclic molecules from cyclic precursors is a well-known strategy. The ability to control relative stereochemistry and regiochemistry in the reactions of cyclic molecules is important since cleavage of that ring transfers the resident stereochemistry and/or regiochemistry to the acyclic product. Several methods will be described in this chapter that lead to the synthesis of amino acids, particularly substituted amino acids, from various cyclic precursors.

2.1. From Ketones

Cyclic ketones are readily available and can be functionalized to produce amino and carboxyl groups. The ring can be opened, after functionalization or prior to it, to generate an acyclic amino acid. This section will discuss synthetic strategies based on cyclic ketone precursors.

2.1.A. Rearrangement Strategies

Cyclic ketones can be readily converted to lactams, which can then be hydrolyzed to the corresponding amino acid. Although this could formally be included in the lactam section (2.3), it is presented here to illustrate that cyclic ketones are excellent starting materials for the synthesis of amino acids. There are two major reactions of cyclic ketones that lead to a lactam, each involving a rearrangement: the Beckmann rearrangement and the Schmidt rearrangement.

2.1.A.i. The Beckmann Rearrangement

The Beckmann rearrangement[1] is a classical method for converting a cyclic ketone to a lactam. It involves treatment of the oxime of a cyclic ketone with a strong acid such as sulfuric acid or polyphosphoric acid (PPA), but reagents such as phosphorus pentachloride (PCl_5) can also be used. In a simple example, the oxime of cyclohexanone was treated with sulfuric acid to give caprolactam. Subsequent hydrolysis gave 6-aminohexanoic acid (see section 2.3 for further discussion of lactam intermediates). A modification was used in the Beckmann rearrangement of oxime ***2.1***, which trapped the Beckmann product with potassium benzene sulfinate to give a O-benzenesulfonate lactim ether (***2.2***). Subsequent reaction of this lactim with the morpholino enamine of cyclohexanone was followed by ring opening to give 12-amino-7-oxododecanoic acid (***2.3***).[2] Oximes of several different cyclic ketones were similarly treated to produce a variety of oxo-amino acids, including: 7-oxo-13-aminotridecanoic acid, 7-oxo-14-aminotetradecanoic acid, 7-oxo-15-aminopentadecanoic acid, and 7-oxo-16-aminohexadecanoic acid.[2]

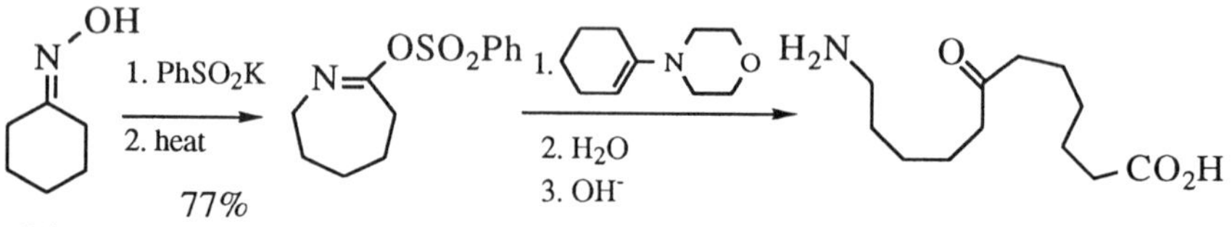

2.1 2.2 2.3

2.1.A.ii. The Schmidt Rearrangement

Cyclic ketones can be converted directly to a lactam by using a mixture of sodium azide and an acid, which generates hydrazoic acid (HN_3) *in situ*. This reaction is known as the Schmidt reaction[3] and gives good yields of lactams, which are then hydrolyzed to the corresponding amino acid. When 2-isobutylcyclopentanone (*2.4*) reacted with polyphosphoric acid and sodium azide, 5-isobutyl-2-piperidone (*2.5*) was obtained.[4] Acid hydrolysis gave a near quantitative yield of 5-amino-7-methyloctanoic acid (*2.6*). In this case, there was excellent selectivity for a single regioisomer, for *2.5* rather than for 3-isobutyl-2-piperidone. In general, the Schmidt rearrangement (and also the Beckmann rearrangement) leads to attachment of nitrogen to the more highly substituted carbon of the cyclic ketone. Despite this

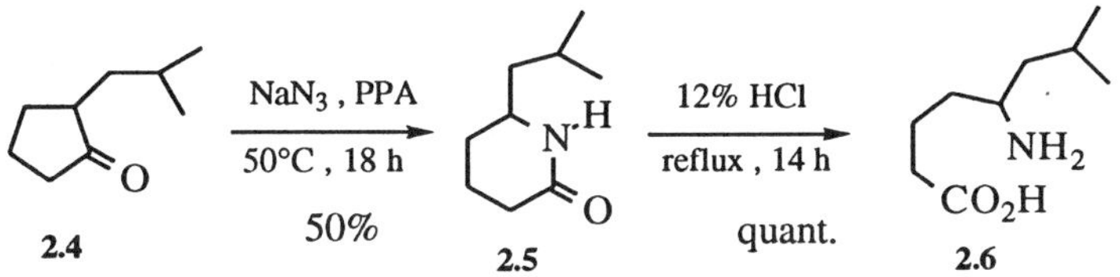

preference for one isomer, regioisomeric mixtures of lactams are usually produced by Schmidt rearrangement (and also Beckmann rearrangement) since two carbons are adjacent to the carbonyl group in the starting material. An typical example of a

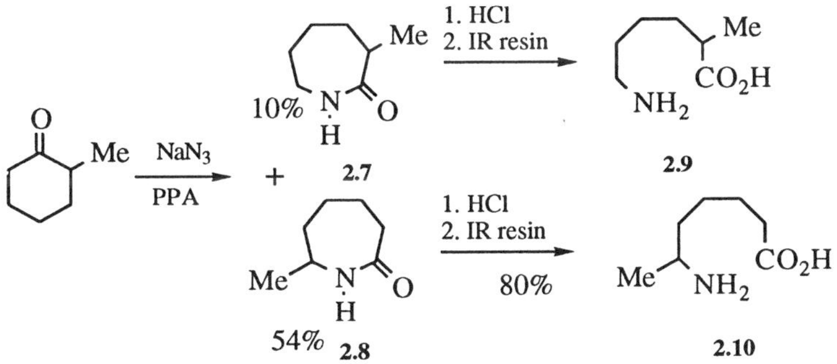

rearrangement that generates two regioisomers is the reaction of 2-methylcyclohexanone and sodium azide, in the presence of polyphosphoric acid.[5] The two lactam products were 7-methylhexahydroazepin-2-one (7-methylcaprolactam, *2.8*) and *2.7*.[6]

Acid hydrolysis of ***2.8*** gave 6-aminoheptanoic acid(***2.10***). Chromatographic separation of ***2.8*** and ***2.7*** was necessary in order to obtain 2-methyl-7-aminoheptanoic acid (***2.9***) via hydrolysis of ***2.7***.[6] If the target is ***2.9***, this is a good synthetic method. If it is ***2.227***, however, the regioselectivity of the Schmidt rearrangement works against the synthesis. The clear lesson here is that the regioselectivity of this reaction dictates its synthetic use.

2.1.B. Cleavage Methods

Another important method for the preparation of amino acids involves direct cleavage of a cyclic ketone. In section 2.1.A, a lactam product (a cyclic amide) was produced from the ketone and then cleaved by hydrolysis, but in this section the ketone moiety is cleaved directly. A cyclic ketone such as cyclopentanone can be oxidatively cleaved with sulfuric acid. When this cleavage was followed by treatment with nitrosyl sulfate, an oximino acid (***2.11***) was formed.[7] Catalytic hydrogenation of the oximino group gave 4-aminobutanoic acid (***2.12***). Similar treatment of cycloheptanone gave 6-aminohexanoic acid (***2.15***) and cyclononanone led to 8-aminooctanoic acid.[7]

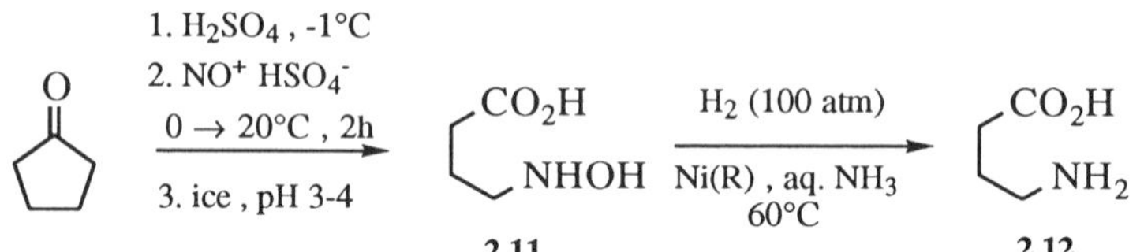

2.11 **2.12**

Nitro-ketones, particularly cyclic derivatives, are important sources of amino acids since the nitro group can be reduced to an amino group (see chapter one, section 1.1.B). In one example, cyclohexene was treated with nitrogen dioxide (N_2O_4) and the product was oxidized with chromium trioxide to give 2-nitrocyclohexanone (***2.13***).[8] When ***2.13*** was treated with aqueous bicarbonate, cleavage of the ring gave 6-nitrohexanoic acid (*2.14*). Catalytic hydrogenation of the nitro group generated 6-aminohexanoic acid (*2.15*).[8,9]

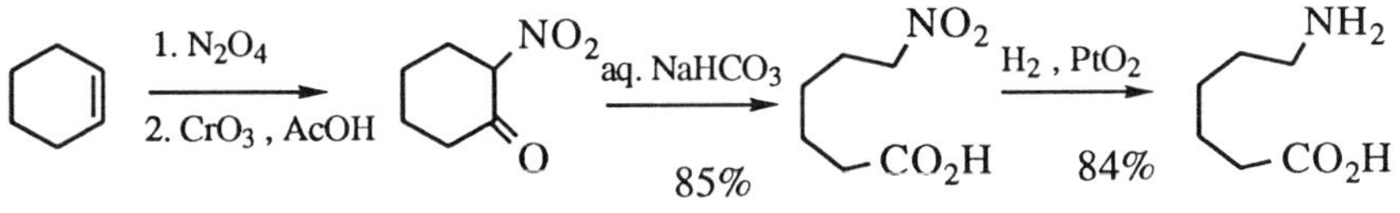

2.13 **2.14** **2.15**

Ammonium formate, in the presence of methanolic Pd on carbon, is also effective for the reduction of a nitro group to an amine.[10] Both methyl 4-aminobutanoate and methyl 3-aminopropanoate were produced by reduction of the appropriate nitro-ester with this reagent.[10]

An alternative preparation of 2-nitro-ketones was reported by Lang, who reacted cyclodecene with nitric oxide and air to give oxime *2.16*.[11] Hydrolysis with aqueous HCl converted the oxime moiety in *2.16* to a ketone moiety in *2.17*. Cleavage occurred upon treatment with aqueous NaOH and gave 12-nitrododecanoic acid, *2.18*.[11] Catalytic hydrogenation of the nitro group in *2.18* gave 12-aminododecanoic acid, *2.19*.

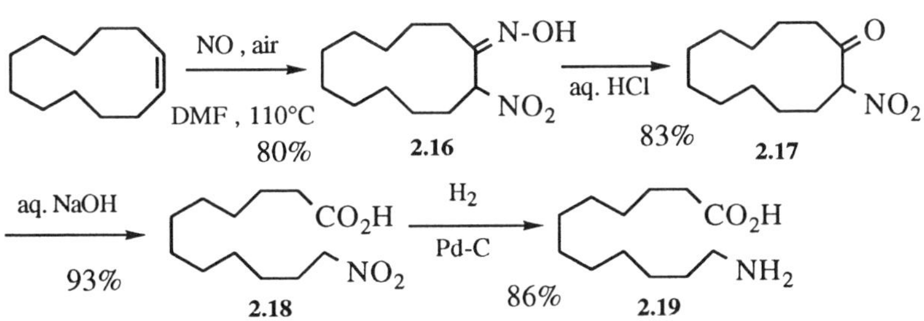

2.16 **2.17** **2.18** **2.19**

2.2. From Lactones and Anhydrides

Lactones and anhydrides each react with ammonia or amines to give nitrogen derivatives. In some cases, lactones are converted to lactams and anhydrides to imides. In these cases, hydrolysis will open the ring and generate an amino acid (see sections 2.3 and 2.4). In other cases, the lactone or anhydride is converted directly to an amino acid. Reaction of five- and six-membered ring lactones with amines tend to give the lactam, whereas the same reaction with larger ring lactones leads to the acyclic amino acid. Anhydrides show similar behavior. This section will focus attention on the synthesis of amino acids that use a lactone or an anhydride as a precursor or a key intermediate.

2.2.A. From Lactones

Just as amines react with esters to form amides, lactones react with amines to give lactams or are opened to give amino acid derivatives directly. When β-propiolactone (***2.20***) reacted with ammonia, for example, 3-aminopropanamide (***2.21a***) was obtained.[12] Similar reaction with methylamine gave no ***2.21b*** but rather a 97% yield of 3-aminopropanoic acid, ***2.22***. Dimethylamine, however, converted ***2.20*** to ***2.21c*** in 79% yield, along with 13% of ***2.22***.[12]

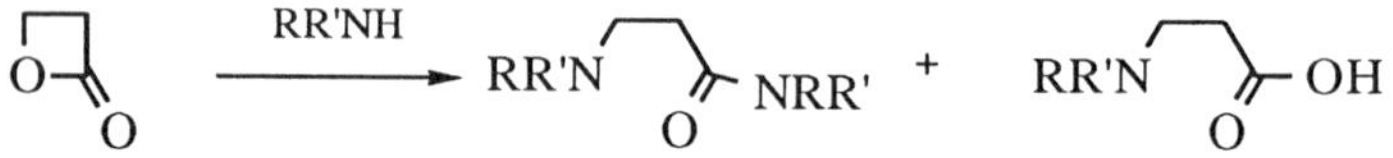

2.20 **2.21** **a** R = R' = H **2.22**
b R = Me ; R' = H
c R = R' = Me

Reaction of γ-butyrolactones with ammonia is a convenient method for preparing five-membered ring lactams (2-pyrrolidinone derivatives). When 3-benzyl-γ-butyrolactone (***2.23***) was heated with ammonia, 4-benzyl-2-pyrrolidinone (***2.24***) was formed.[13] Hydrolysis led to 3-benzyl-4-aminobutanoic acid (***2.25***), an antispasmodic sedative.[13] Biologically active amino acids are discussed in chapter six.

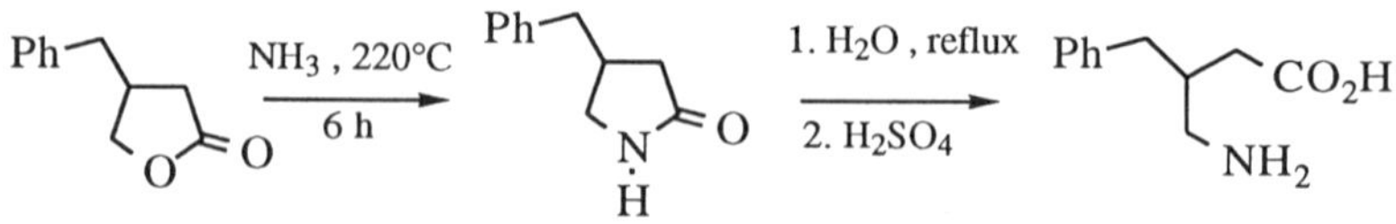

2.23 **2.24** **2.25**

Other lactones can also be used. α-Bromo-γ-butyrolactone (***2.26***) was converted to hydroxylactone ***2.27***.[14] The lactone ring was opened by potassium phthalimide to give ***2.28*** and removal of the phthalimido group gave 2-hydroxy-4-aminobutanoic acid (***2.29***).[14a]

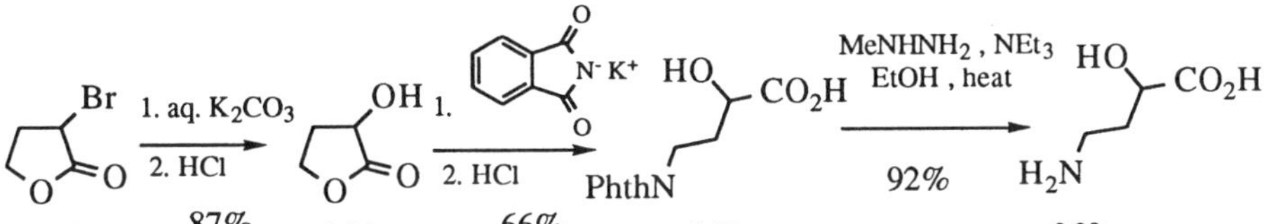

2.26 **2.27** **2.28** **2.29**

Aminolactone ***2.30*** (Trt = trityl; $-CPh_3$) offered another possibility for synthesis via reduction of the carbonyl moiety, giving a 1:5 mixture of ***2.31:2.32***. After chromatographic separation of these diastereomers, Wittig olefination[15] of ***2.31*** (via formation of an aldehyde from the hemi-acetal, *in situ*) gave ***2.33*** [methyl 4-(N-tritylamino)-6-hydroxyhex-2-enoate] in 80% overall yield.[16]

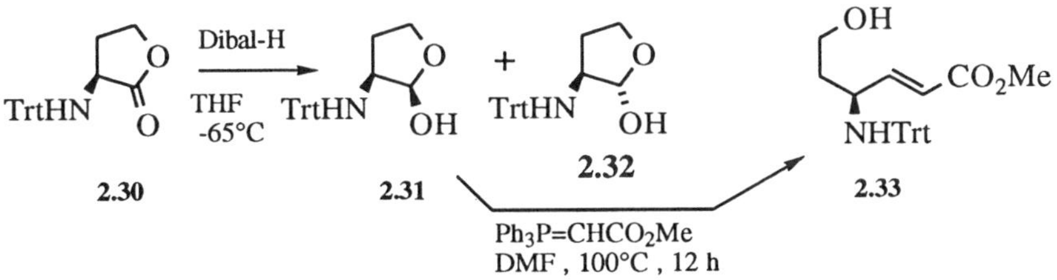

2.30 **2.31** **2.32** **2.33**

Lactone ***2.34*** (paraconic acid) is another lactone derivative that can be converted to 4-hydroxy-3-aminobutanoic acid (***2.36***) by first generating the amide (***2.35***).

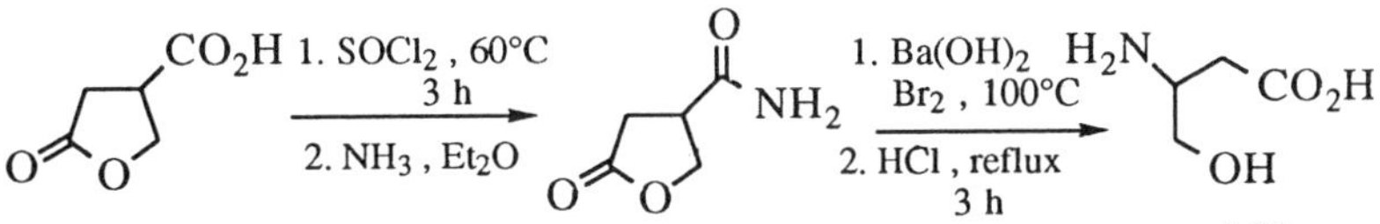

2.34 **2.35** **2.36**

Treatment of ***2.35*** with a basic solution of bromine converted the amide to the amine via a Hofmann rearrangement,[17] and acid hydrolysis gave 3-amino-4-hydroxybutanoic acid, ***2.36***.[18]

2.2.B. From Anhydrides

Anhydrides can be converted to ω-amino acids via reaction with amines and reduction of the resultant imide or acyclic amide product. When succinic anhydride was treated with two equivalents of diethylamine, amido acid ***2.37*** was produced.[19]

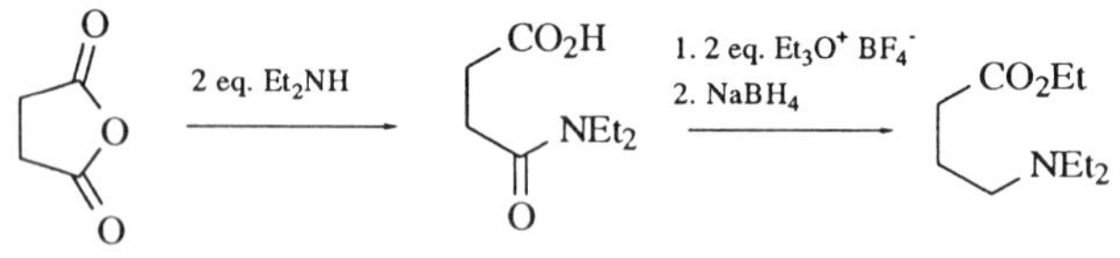

2.37 **2.38**

Reaction with triethyloxonium tetrafluoroborate[20] generated an iminium ester that was reduced with sodium borohydride to give ethyl 4-(N,N-diethylamino)butanoate (***2.38***) in 83% overall yield.[19] Using maleic anhydride leads to an alkenyl amino acid as a product. Reaction with ammonia gave 3Z-aminoprop-2-enoic acid (***2.39***, also known as maleamic acid).[21]

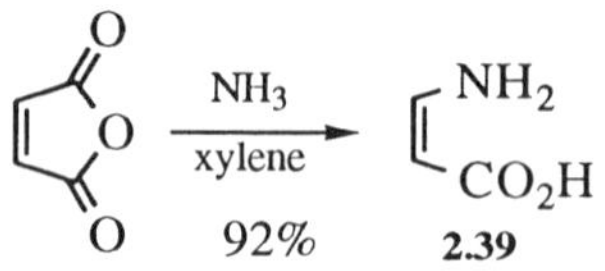

2.39

Both anhydrides and imides have been used as precursors for substituted amino acids. β-Amino acids in particular can be prepared from these reagents. Weinreb showed that ***2.40*** was converted to methyl 2R-methyl-3-aminobutanoic acid (***2.41***) via formation of an intermediate azide.[22] This sequence utilized a Curtius rearrangement[23] (see chapter one, section 1.5.A) after opening the anhydride and formation of a diazo-ketone. Although imide precursors will not be formally discussed until section 2.3.B, an example is presented here because the reaction of an anhydride and ammonia or an amine gives an imide. This constitutes another synthetic route from anhydrides to amino acids. Treatment of ***2.42*** with potassium chlorate followed by acid hydrolysis led to ***2.43*** (2S-methyl-3-aminobutanoic acid).[24] It is noted that N-tosyl aziridino esters can give β-amino acid derivatives via reaction with Grignard reagents, but α-amino acids were formed as the major product.[25]

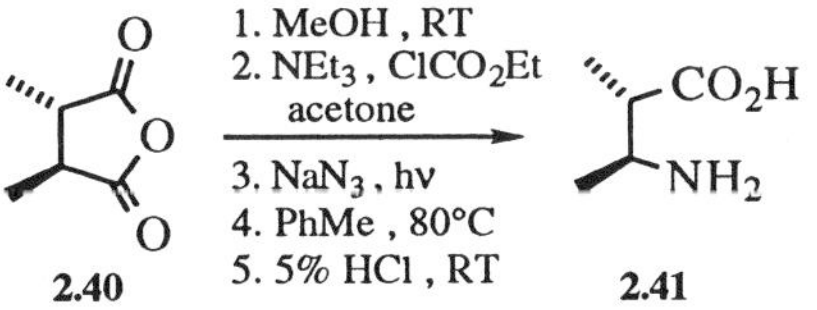

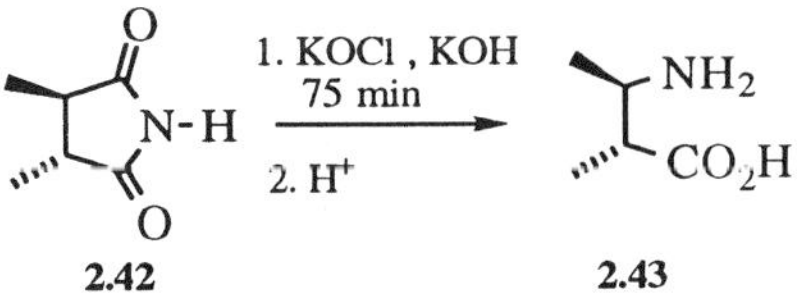

2.3. From Lactams and Imides

Just as lactones and anhydrides can be used to prepare amino acids, their nitrogen analogs (lactams and imides) can also be used. As just seen, and in several previous sections, lactams were generated in an initial reaction and then hydrolyzed to give an amino acid. This section will focus primarily on lactam starting materials and on those reactions for which lactams or imides are key intermediates.

2.3.A. From Lactams

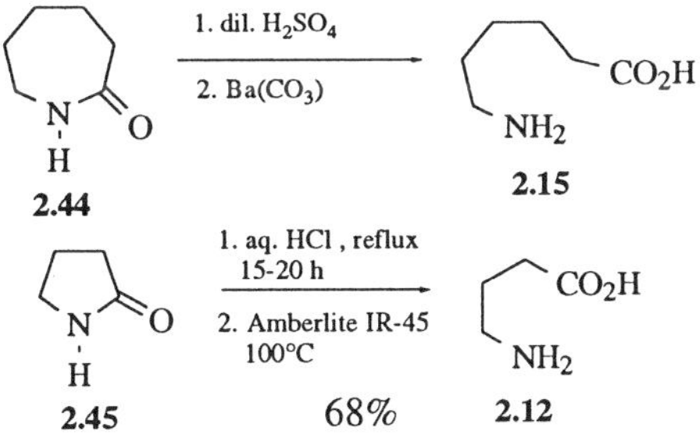

A very old and useful technique for the synthesis of amino carboxylic acids is the hydrolysis of lactams, as in the conversion of caprolactam (***2.44***) to 6-aminohexanoic acid (***2.15***),[26] which is an *Organic Syntheses* preparation.[27] The hydrolysis of 2-pyrrolidinone (***2.45***) to 4-aminobutanoic acid (***2.12***) is another simple example.[28] These examples illustrate different conditions for acid hydrolysis and that passage through an ion exchange resin is often required for the actual isolation of the amino acid product. There are many variations of this reaction, especially in the patent literature.[29] Virtually any lactam opened to the corresponding amino acid

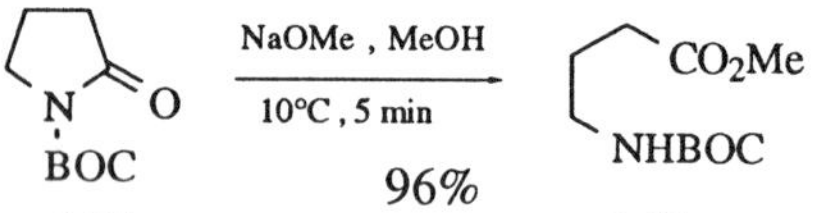

2.46 **2.47**

under these conditions.[28] Base induced ring opening is also possible, as seen in the milder procedure for N-Boc lactams such as ***2.46***, reported by Grieco, which gave methyl 4-(N-Boc amino)butanoate, ***2.47***.[30] Hydrolysis of larger ring lactams leads

Table 2.1. Preparation and hydrolysis of lactam precursors to amino acids.

Reaction	Transformation	Reference
1.	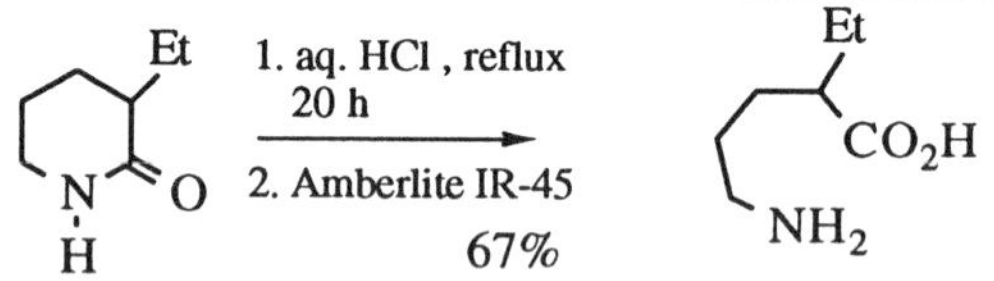 **2.48** **2.49**	31
2.	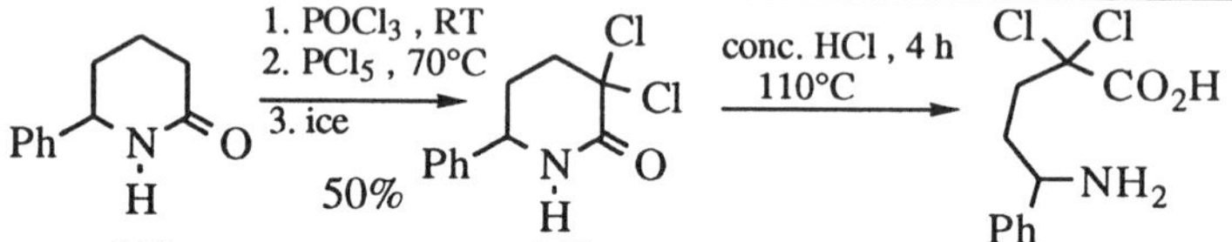 **2.50** **2.51** **2.52**	32
3.	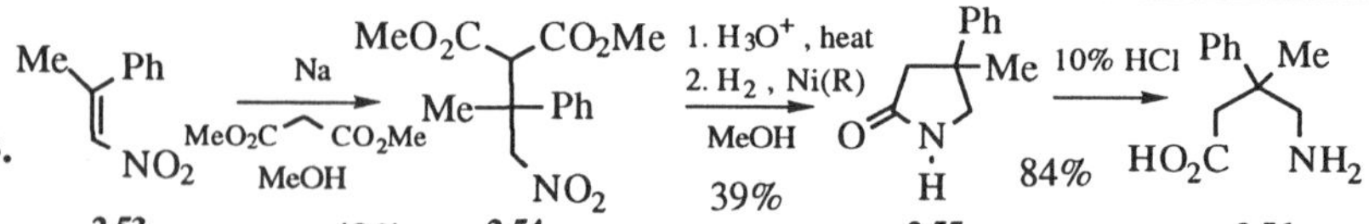 **2.53** **2.54** **2.55** **2.56**	33

Table 2.1. (Continued).

Reaction	Transformation	Reference
4.	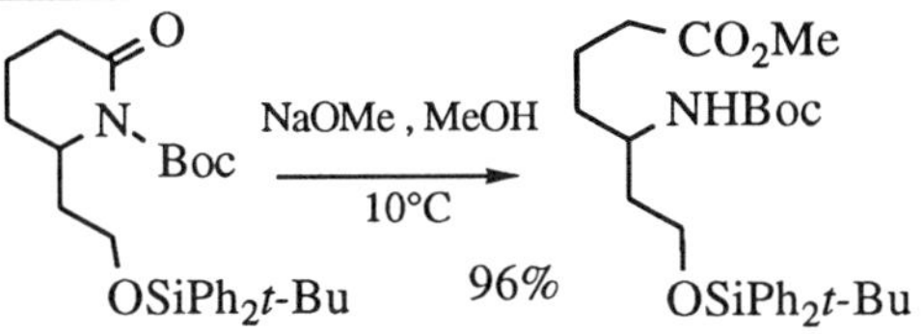 **2.57** **2.58**	30
5.	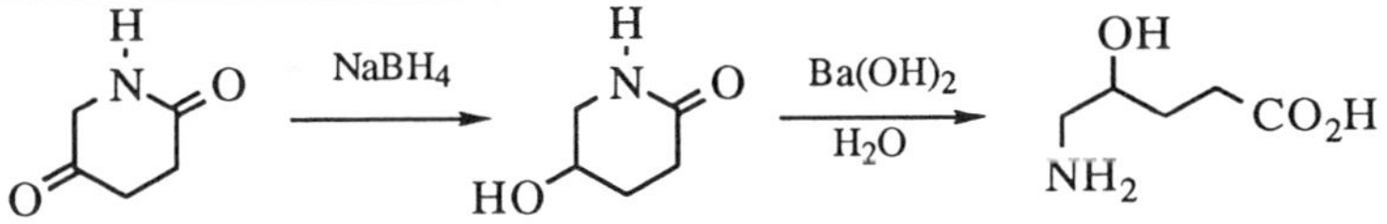 **2.59** **2.60** **2.61**	34
6.	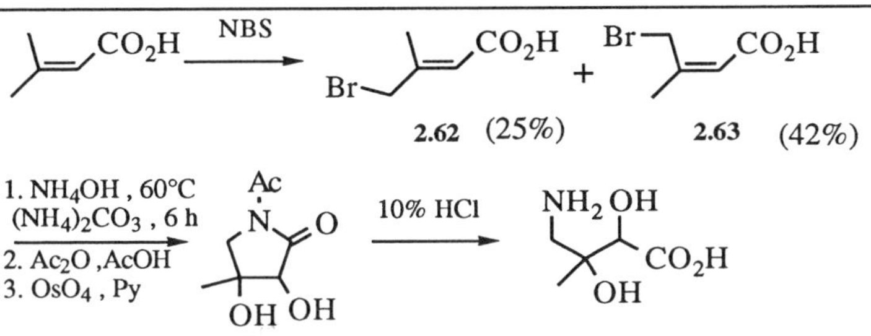 **2.64** **2.65**	36
7.	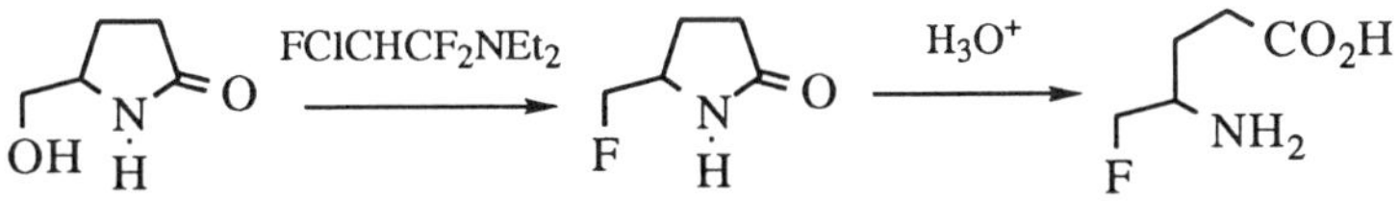 **2.66** **2.67** **2.68**	37
8.	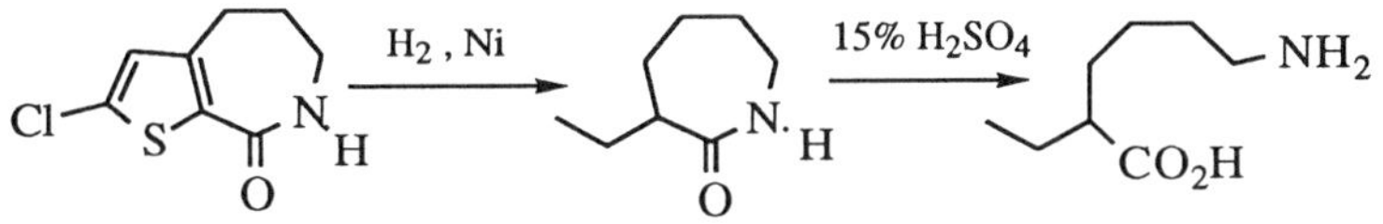 **2.69** **2.70** **2.71**	38

to longer chain amino acids . Hydrolysis of 2-pyrrolidinone derivatives gives 4-amino-butanoic acid derivatives and 2-piperidones lead to 5-aminopentanoic acid derivatives. Several additional examples are collected into Table 2.1 to illustrate the scope of the process as well as a few of the many structural features that can be incorporated into the amino acid.

The hydrolysis of lactams is rather straightforward and most of the variety in this table results from different structural features and different synthetic manipulations of the lactams. Reaction 1 shows a simple hydrolysis of 3-methyl-2-pyrrolidinone (***2.48***) to give 2-ethyl-4-aminobutanoic acid (***2.49***).[31] Several substituted derivatives were prepared, including 2-methyl-4-aminobutanoic acid, 3-ethyl-4-aminobutanoic acid, 3-propyl-4-aminobutanoic acid, 3-(1-methylethyl)-4-aminobutanoic acid, 2,3-diethyl-4-aminobutanoic acid, 2-ethyl-5-aminopentanoic acid, 3-methyl-5-aminopentanoic acid, 3-ethyl-5-aminopentanoic acid, 3-propyl-5-aminopentanoic acid, and 3-(1-methylethyl)-5-aminopentanoic acid.[31]

α-Chlorination of lactams is an important route to halogenated amino acids, and this is illustrated in reaction 2. Treatment of 6-phenyl-2-piperidone (***2.50***) with phosphorus oxychloride gave the 3,3-dichloro lactam, ***2.51***.[32] Acid hydrolysis led to 5-amino-2,2-dichloro-5-phenylpentanoic acid, ***2.52***, quantitatively.

Nitro-alkenes are amine surrogates and function as receptor molecules in Michael-type additions. An example is reaction 3 in which ***2.53*** reacted with sodium malonate to give ***2.54***.[33] Aqueous hydrolysis of the ester, which was accompanied by decarboxylation of the resulting diacid, was followed by catalytic hydrogenation (over Raney nickel) of the nitro group to give an amino-ester. This product spontaneously cyclized to give lactam ***2.55***. This was hydrolyzed with acid to give 3-methyl-3-phenyl-4-aminobutanoic acid (***2.56***).[33a]

There are many ways to prepare functionalized lactams. Reaction 4 shows an example in which ***2.57*** was treated with sodium methoxide to give 5-(N-Boc amino)-7-(*t*-butyldiphenylsilyloxy)heptanoic acid, ***2.58***.[30] Aqueous acid hydrolysis and deprotection of ***2.58*** gave the corresponding amino acid, 5-amino-7-hydroxy-heptanoic acid. Another hydroxy amino acid is shown in reaction 5, but the precursor was a ketone moiety. Sodium borohydride reduction of the ketone moiety in lactam ***2.59*** gave ***2.60***. Basic hydrolysis of the lactam gave 4-hydroxy-5-amino-pentanoic acid, ***2.61***.[34,35]

A different approach to hydroxy amino acids used the allylic bromination of 3-methyl-2-butenoic acid to give a mixture of ***2.62*** and ***2.63***, as shown in reaction 6.[36] Reaction of this mixture with ammonium hydroxide led to formation of lactam ***2.64***, which was hydrolyzed to 4-amino-2,3-dihydroxy-3-methylbutanoic acid, ***2.65***. Hydroxylactams can be converted to functionalized amino acids other than hydroxy amino acids. An example is shown in reaction 7, where 5-fluoromethyl-2-pyrrolidinone (***2.67***) was prepared from 5-hydroxymethyl-2-pyrrolidinone (***2.66***) by reaction with chlorotrifluorodiethylaminoethane.[37] Hydrolysis gave 4-amino-5-fluoropentanoic acid, ***2.68***.

There are several different ways in which a lactam precursor can be generated. An interesting but relatively specific synthesis is shown in reaction 8 and employs thiophene derivatives. Catalytic hydrogenation of bicyclic thiophene ***2.69*** gave 2-ethylhexahydroazepine-2-one (***2.70***) and acid hydrolysis gave 2-ethyl-7-amino-hexanoic acid, ***2.71***.[38]

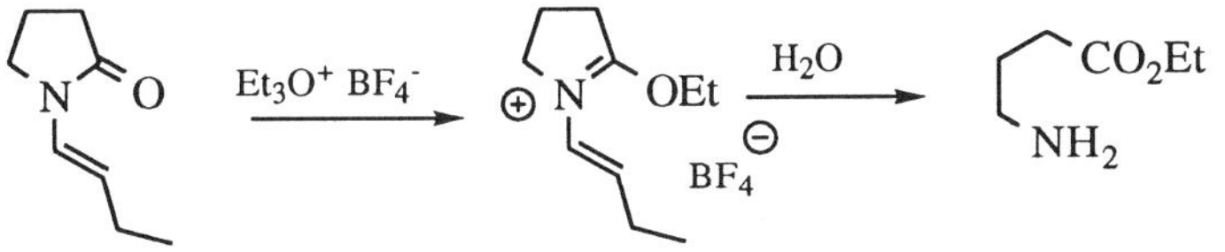

2.72 **2.73**

An interesting method that does not strictly follow the reaction types and hydrolysis procedures used in Table 2.1 is the conversion of lactam ***2.72*** to the iminium salt (***2.73***) by reaction with triethyloxonium tetrafluoroborate (Meerwein's reagent).[20]

Dissolution in neutral water gave ethyl 4-aminobutanoate.[39] The iminium salt is much more reactive to ring opening than the lactam, giving the amino acid under very mild conditions. A number of lactams unsubstituted at nitrogen were converted to the iminium salt and were opened in neutral water to give the corresponding

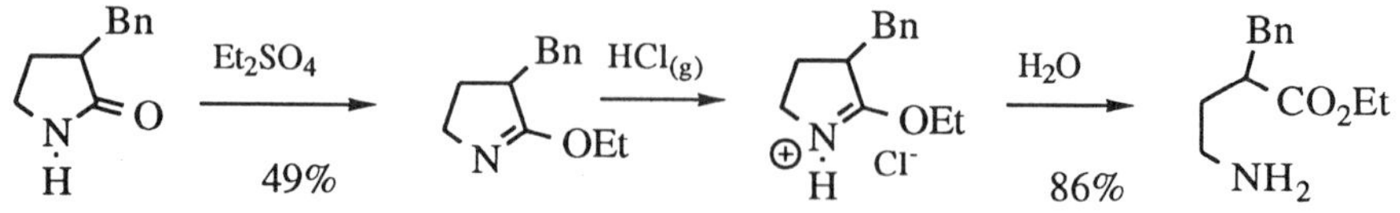

2.74 **2.75** **2.76** **2.77**

amino acid.[40] Reaction of 2-benzyl-2-pyrrolidinone (***2.74***) with diethyl sulfate, for example, gave lactim ether ***2.75*** (3-benzyl-2-ethoxy-1-pyrroline). Subsequent dissolution in water for less than one hour gave amino-ester ***2.77*** (ethyl 2-benzyl-4-amino-butanoate).[40] Although ***2.75*** can be isolated, that is unnecessary and the crude product from the first reaction can simply be dissolved in water after treatment with HCl_{gas} (which generates pyrrolidiniminium salt ***2.76*** *in situ*). Larger ring lactams were converted to lactim ethers using triethyloxonium tetrafluoroborate (*vide supra*, giving the iminium salt directly). This method was used to prepared ethyl amino-butanoic, aminopentanoic, aminohexanoic, aminooctanoic, aminodecanoic, and aminododecanoic esters, including ethyl 4-methyl-4-aminobutanoate, ethyl 2-methyl-4-aminobutanoate as well as ethyl 5-aminopentanoate, ethyl 6-amino-hexanoate, ethyl 7-aminoheptanoate, and ethyl 12-aminododecanoate.[40]

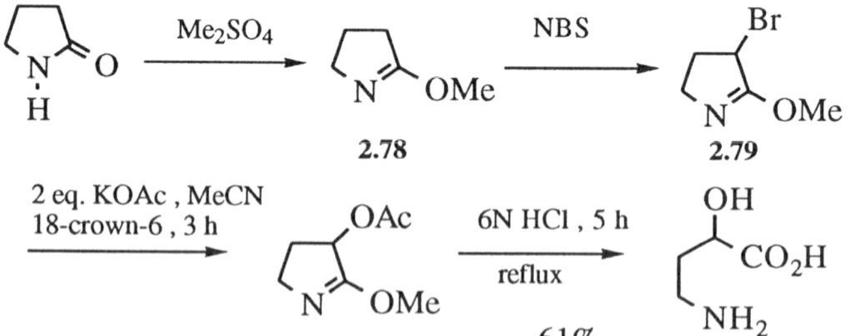

2.78 **2.79** **2.80** **2.29**

In a similar related sequence, 2-pyrrolidinone was converted to the lactim ether (2-methoxypyrroline, ***2.78***) by treatment with dimethyl sulfate.[41] This methoxy-imine was brominated[42] with N-bromosuccinimide to give ***2.79***.[43] Reaction with potassium acetate gave ***2.80*** and acid hydrolysis led to 2-hydroxy-4-aminobutanoic acid (***2.29***).[43]

2.3.B. From Imides

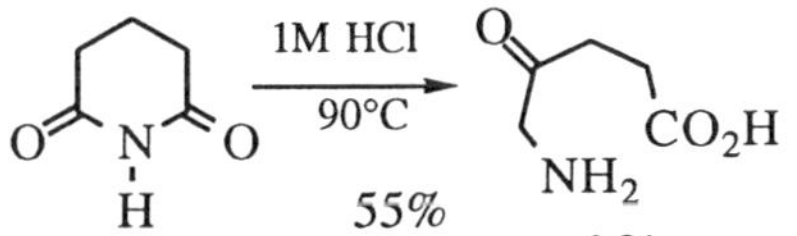

2.81

Just as a lactam can be opened to give an amino acid, an imide gives a similar reaction. A simple example is the hydrolysis of glutarimide, in this case prepared from furfural in four steps, to 5-amino-4-oxo-pentanoic acid (5-aminolevulinic acid, ***2.81***).[44] This latter compound showed relatively high herbicidal activity.[44]

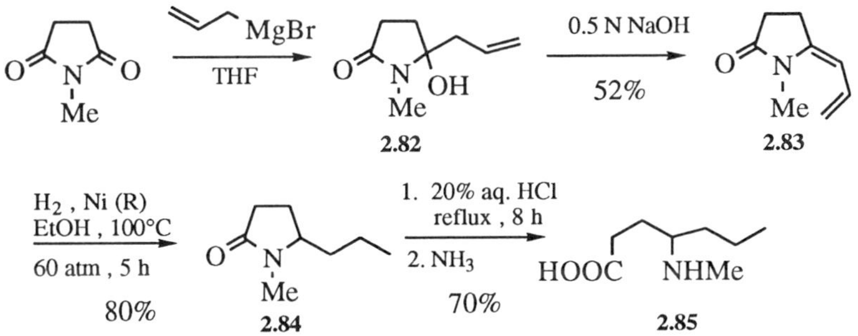

In a different approach, the reaction of allylmagnesium bromide and N-methyl-succinimide led to lactam ***2.82***. Subsequent treatment with aqueous hydroxide gave lactam diene ***2.83***.[45] In this work, Flitsch hydrogenated the diene to give N-methyl-5-propyl-2-pyrrolidinone (***2.84***) and subsequent acid hydrolysis gave 4-(N-methyl-amino)heptanoic acid, ***2.85***.[45]

2.4. From β-Lactams

As just discussed in section 2.3, hydrolysis of lactams is a convenient source of amino acids. β-Lactams (2-azetidinone derivatives) are a very interesting class of compounds and hydrolysis leads to 3-aminopropanoic acid derivatives (β-alanine derivatives). There are specialized methods that lead to β-lactams and for this reason they have been segregated into a separate section.

Just as with larger ring lactams, basic hydrolysis of a β-lactam such as 4-ethyl-2-azetidinone (***2.86***) gave 3-aminopentanoic acid (***2.87***).[46] In general, β-lactams are excellent sources of β-alanine and also substituted β-alanine derivatives.

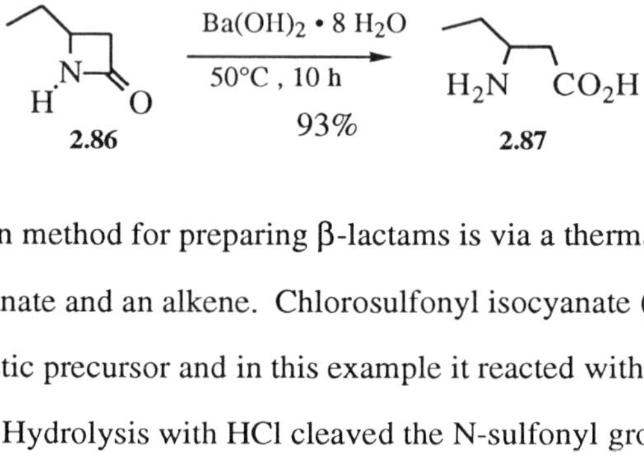

The most common method for preparing β-lactams is via a thermal [2+2]-cycloaddition of an isocyanate and an alkene. Chlorosulfonyl isocyanate (CSI, ***2.88***) is a very common synthetic precursor and in this example it reacted with 2-methyl-2-butene to give ***2.89***. Hydrolysis with HCl cleaved the N-sulfonyl group and opened

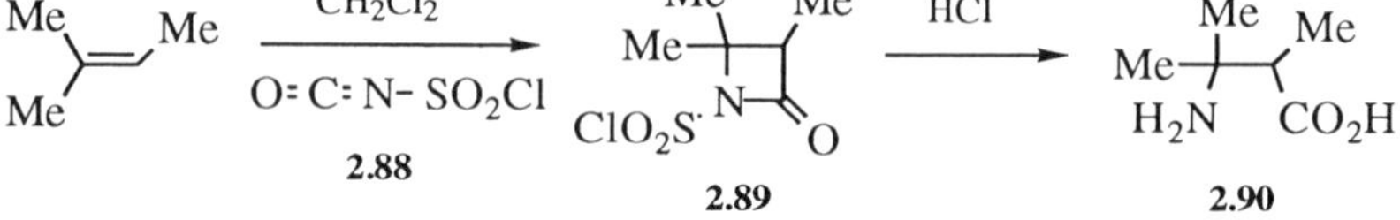

the ring to give 2,3-dimethyl-3-aminobutanoic acid, ***2.90***.[47] A similar reaction with 2,3-dimethyl-2-butene gave 2,2,3-trimethyl-3-aminobutanoic acid; 2-methyl-2-pentene gave 2,2-dimethyl-3-aminopentanoic acid; and 2,4,4-trimethyl-1-pentene gave 3,5,5-trimethyl-3-aminohexanoic acid.[47] A variation of this reaction removed the N-chlorosulfonyl group from the β-lactam by treatment with thiophenol and pyridine.[48] Reaction of *cis*-3-hexene and CSI, for example, gave ***2.91***.[48] Reaction with thiophenol•pyridine followed by reaction with concentrated HCl gave 2-ethyl-3-amino-pentanoic acid (***2.91***).[48]

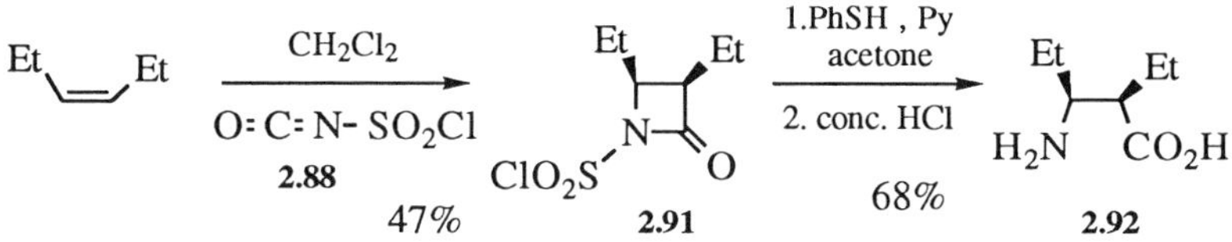

Table 2.2. Preparation and hydrolysis of β-lactams to prepare amino acids.

Reaction	Transformation	Reference
1.	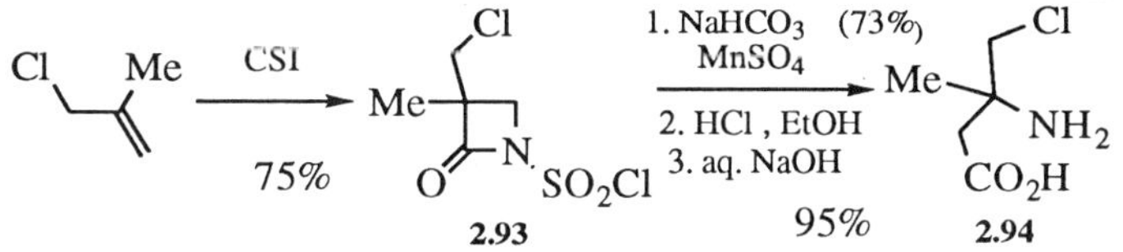	49
2.	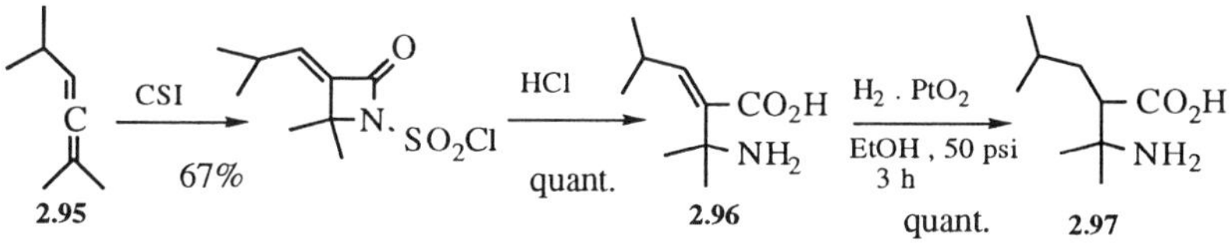	50
3.	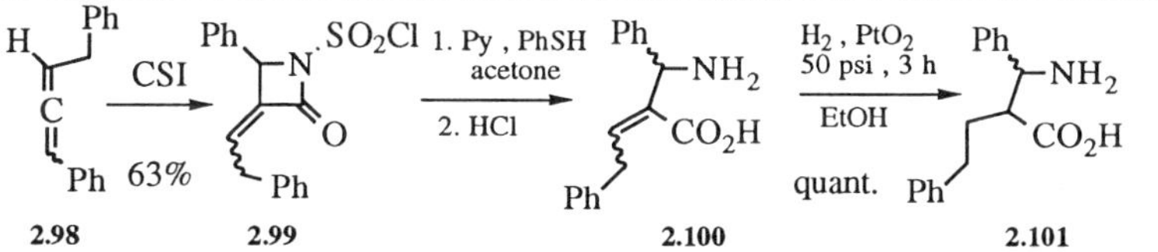	50
4.	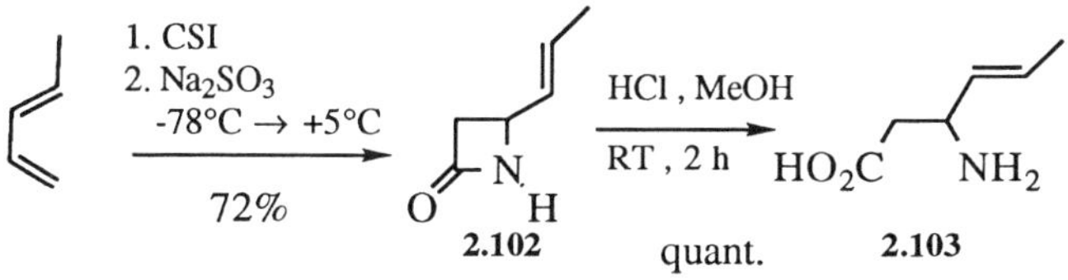	51
5.	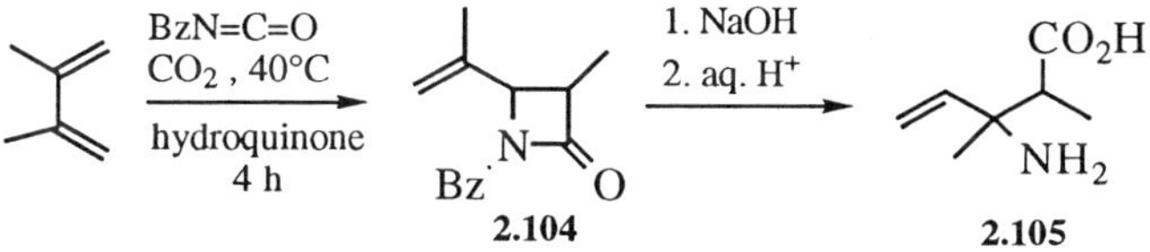	52

There are many examples of this type of cycloaddition-hydrolysis route to amino acids. Both substituted and functionalized amino acids have been prepared. Several examples of similar techniques are collected in Table 2.2.

Reactions 1-4 use N-chlorosulfonyl isocyanate (***2.88***; CSI) in reactions with various alkene moieties. When CSI reacted with 3-chloro-2-methylpropene in reaction 1, β-lactam ***2.93*** was produced.[49] Deprotection of the nitrogen was followed by hydrolysis to give 3-amino-4-chloro-3-methylbutanoic acid (***2.94***).

Reaction of CSI with an allene leads to an alkylidene β-lactam, which can be hydrolyzed and hydrogenated to give the corresponding amino acid. Allene ***2.95***, in reaction 2, reacted with CSI to give the β-lactam.[50] Treatment of this product with HCl gave 2-(1-amino-2-methylethyl)-4-methylpent-2-enoic acid (***2.96***) and catalytic hydrogenation of the alkenyl unit gave a quantitative yield of 3-amino-3-methyl-2-(2-methylpropyl)-butanoic acid, ***2.97***.[50] A variety of 2-alkyl-3-methyl-3-amino-butanoic acid derivatives were prepared by this method. Other allenes can be used in reactions with CSI, as shown in reaction 3 for ***2.98***, which gave alkylidene lactam ***2.99***. Removal of the N-sulfonyl group with pyridine and thiophenol was followed by reaction with HCl to give a good yield of ***2.100***. Catalytic hydrogenation of the double bond led to 3-amino-2-(2-phenethyl)-3-phenylpropanoic acid, ***2.101***.[50] Using an identical approach, reaction of *p*-chlorostyrene and CSI gave the β-lactam, and removal of the sulfonyl group and hydrolysis with HCl gave 3-(*p*-chlorophenyl)-3-aminopropanoic acid.[48]

Just as alkenes can be condensed with isocyanates, one alkene unit of a diene can also react, eventually leading to an alkenyl amino acid. The condensation of CSI and 1,3-pentadiene in reaction 4 is one example and it gave ***2.102***.[51] Removal of the sulfonyl group and hydrolysis gave methyl 3-aminohex-4-enoate (***2.68***),[51c] a synthetic intermediate for the preparation of daunosamine.[51a,c] Similar reaction with 1,3-

butadiene and N-benzyl isocyanate led to 3-aminopent-4-enoic acid (23% overall yield).[52] Reaction 5 also illustrates a reaction with a diene in which 2,3-dimethyl-1,3-butadiene reacted with benzyl isocyanate to give ***2.104.*** Subsequent ring opening led to 3-amino-2,3-dimethylpent-4-enoic acid, ***2.105.***[52]

Although the following reactions are not included in Table 2.2, it is noted that N-chlorosulfonyl isocyanate reacts with cyclopropane derivatives to give β-lactams. 1,1,2-Trimethylcyclopropane (***2.106***), for example, reacted with CSI to give ***2.89***.[53] Treatment with thiophenol•pyridine to remove the chlorosulfonyl group and reaction with HCl gave 2,2,3-trimethyl-3-aminobutanoic acid (***2.90***).[53] One caution is important when using this procedure. Cationic ring opening accompanied the reaction of 1,1,2,2-tetramethylcyclopropane where the identical reaction sequence led to 3-*t*-butyl-3-aminobutanoic amide.[53] Presumably, ring opening occurred to give 2,3,3-trimethyl-1-butene, which was the actual reaction partner with CSI.[53]

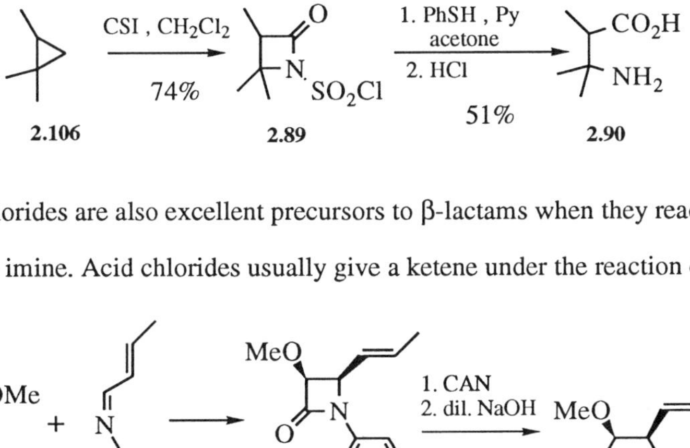

2.106 **2.89** **2.90**

Acid chlorides are also excellent precursors to β-lactams when they react with an appropriate imine. Acid chlorides usually give a ketene under the reaction conditions,

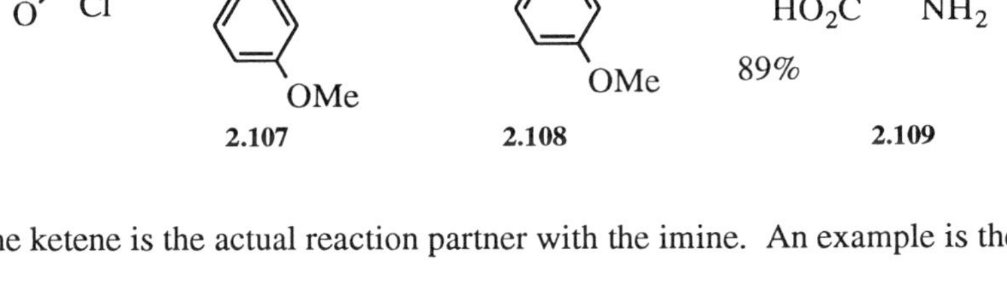

2.107 **2.108** **2.109**

and the ketene is the actual reaction partner with the imine. An example is the reaction of 2-methoxy acetyl chloride and azadiene ***2.107*** to give β-lactam ***2.108***.[54] Removal of the aryl protecting group followed by basic hydrolysis led to 3-amino-2-methoxy-hex-4-enoic acid, ***2.109.***

A ketene intermediate is not required in order to generate the β-lactam. In the case of ***2.111***, ketene formation is not possible. Reaction of the acid chloride with the imine, however, generates an intermediate where the nitrogen can displace the bromine moiety to give the lactam. The reaction of ***2.111*** and ***2.115*** proceeded, in the presence of pyridine, to give ***2.112***.[55] Reaction with HCl gave 2,2-diethyl-3-aminopropanoic acid, ***2.113***. Similarly, 2,2-dipropyl and 2,2-dibutyl-3-aminopropanoic acids were prepared from the appropriate acid chloride and ***2.110***.[55]

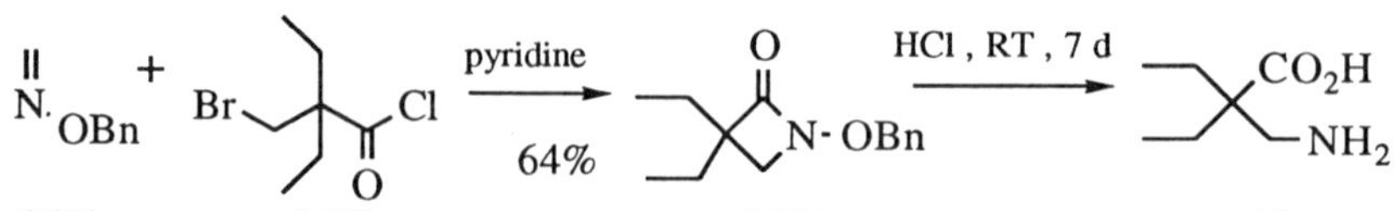

2.110 **2.111** **2.112** **2.113**

Another variation in the reaction of imines involves the titanium catalyzed condensation of a Schiff base with a ketene silyl acetal to give a β-lactam. The $TiCl_4$ catalyzed reaction of ***2.114*** (derived from propanal and phenethylamine) and ***2.115*** led to ***2.116***.[56] The reaction was mediated by a nitrogen-$TiCl_3$ intermediate. Hydrolysis of the β-lactam and reduction of the N-benzylic group with hydrogen and palladium gave 2,2-dimethyl-3-aminopentanoic acid, ***2.117***. In this study, chiral

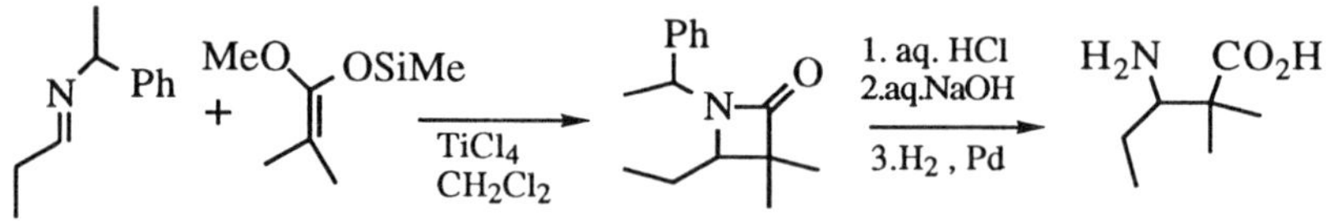

2.114 **2.115** 66% **2.116** **2.117**

non-racemic phenethylamine was used to protect the imine (also see chapter five) and amino acids were prepared. In this manner, 2,2-dimethyl-3-aminohexanoic acid,

2,2,4-trimethyl-3-aminopentanoic acid, 2,2-dimethyl-3-aminoheptanoic acid, and 2,2,5-trimethyl-3-aminohexanoic acid were prepared with asymmetric induction ranging from 44%ee to 78%ee.[56]

A different study generated the lithium imine of ***2.118*** and condensed it with ester ***2.119*** to give ***2.120***. Deprotection of the lactam nitrogen with ceric ammonium nitrate and acid hydrolysis led to formation of 3-amino-2-hydroxy-5-methylhexanoic acid, ***2.121***.[57]

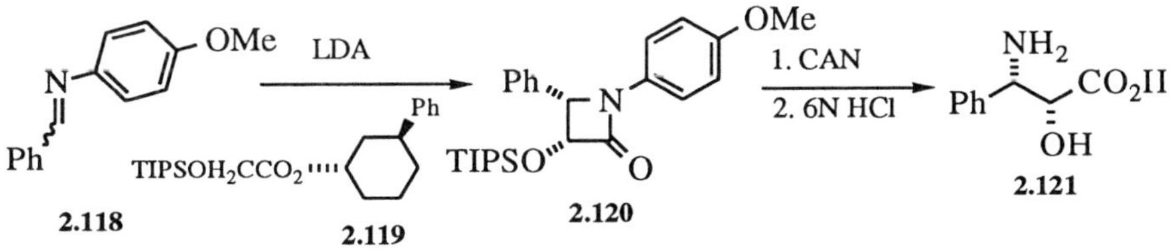

2.5. From Heterocycles

Many heterocyclic rings contain nitrogen and, in principle, reduction of such a heterocycle would produce an amino compound. A variety of functionalized heterocycles have been converted to amino acids using this fundamental idea. Several of the systems used for this purpose are discussed in this section.

2.5.A. Isoxazolidines, Isoxazoles, Oxazolidines, and Oxazine Derivatives

When a nitrone reacts with a vinyl ether, an isoxazolidine is formed via a [3+2]-cycloaddition reaction.[58] Similar reaction with an alkyne gives an isoxazole.[58]

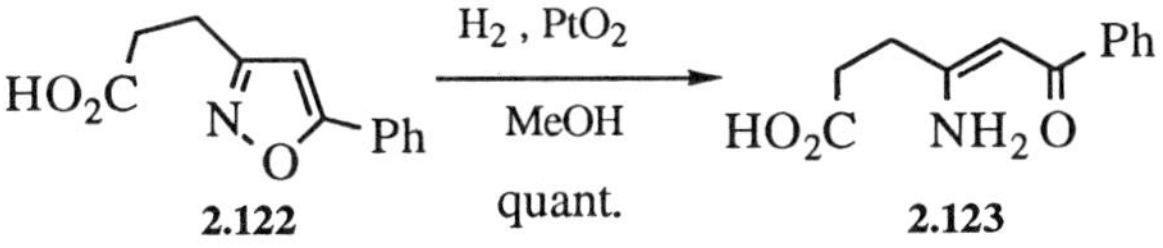

These heterocycles can then be hydrogenated to give an amino acid. Hydrogenation of isoxazole ***2.122***, for example, gave a quantitative yield of 4-amino-6-oxo-6-

phenylhex-4-enoic acid, ***2.123***.[59] Isoxazole ***2.122*** was prepared from phenyl acetylene and methyl 4-nitrobutanoate.[60] In another example, diethylamine was oxidized with hydrogen peroxide to give a nitrone, which reacted with ethyl vinyl ether to give ***2.124***.[61] Subsequent reaction with benzyl bromide gave an ammonium salt, and the ring was opened with DABCO to give ethyl 3-(N-benzyl-3-ethylamino)-butanoate, ***2.125***.

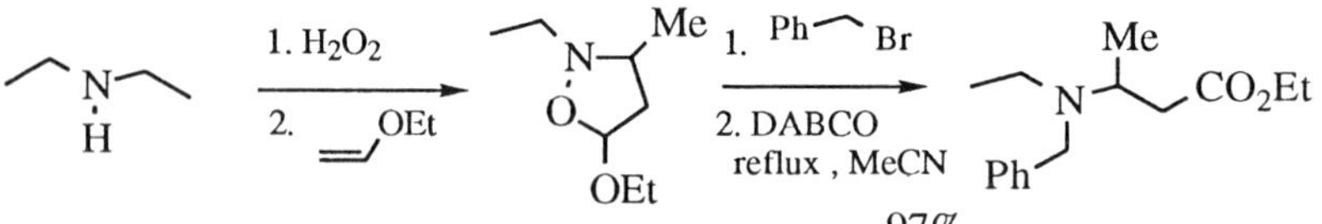

2.124 **2.125**

Other 3-aryl aminopropanoic acid derivatives have been prepared using oxazolidine derivatives derived from vinyl acetates and hydroxylamines.[62] Nitrone ***2.126*** is more sterically encumbered than the nitrones shown previously, and reaction with vinyl acetate gave a poor yield of ***2.127***.[62] Hydrolysis of ***2.337***, followed by oxidation and catalytic hydrogenation led to 4-methyl-3-aminopentanoic acid, ***2.128***.[62]

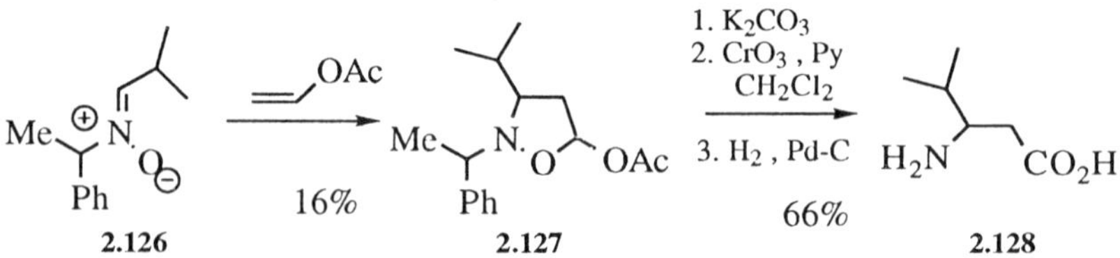

2.126 **2.127** **2.128**

An alternative strategy prepared isoxazolidinyl esters by displacement of the halide moiety in a halo-acid derivative by an amine, an amine surrogate. When

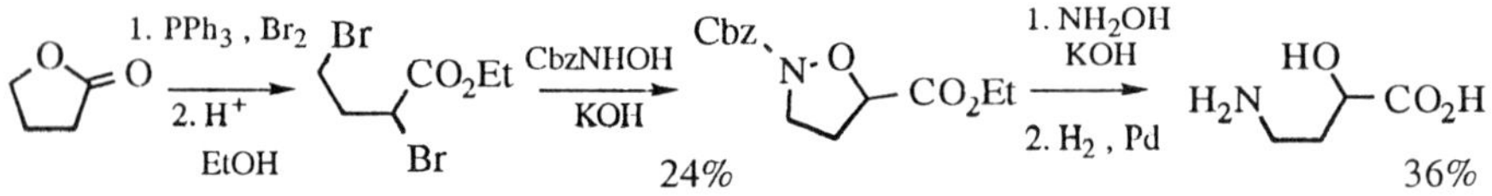

2.129 **2.130** **2.29**

γ-butyrolactone reacted with triphenylphosphine and bromine, a dibromo-acid was formed and this was esterified to give ***2.129***.[63] Reaction of ***2.129*** with Cbz-hydroxylamine gave ***2.130*** via sequential displacement of the two bromines by nitrogen and oxygen. Subsequent reaction with hydroxylamine and base, followed by catalytic hydrogenation, gave 2-hydroxy-4-aminobutanoic acid (***2.29***).[63] Dibromoesters can react with other carbamates to give hydroxyl amino acids such as 4-amino-2-hydroxypentanoic acid.[64] Isoxazolidine-5-carboxylic acid derivatives are, in general, excellent precursors to hydroxy amino acids, including 2-hydroxy-4-aminobutanoic acid.[63] Larger ring heterocycles can also be produced by this fundamental approach. α-Bromoester ***2.131***, for example, was converted to perhydrooxazine-6-carboxylic

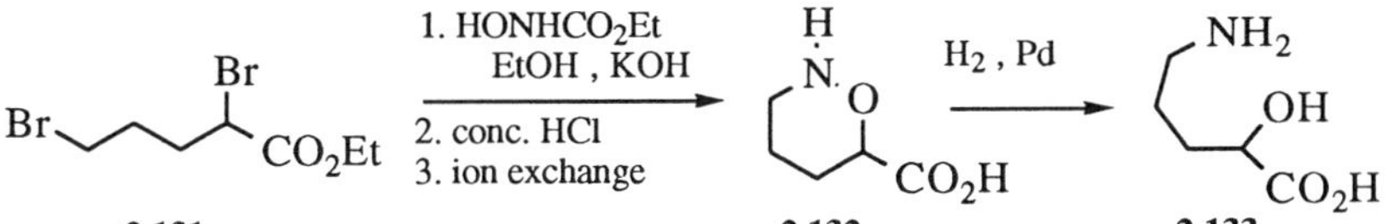

2.131 **2.132** **2.133**

acid (***2.132***).[65] Catalytic hydrogenation led to 2-hydroxy-5-aminopentanoic acid (***2.133***, 2-hydroxy-DAVA). Other six-membered ring heterocycles such as ***2.134*** can also be used to prepare amino acids. In this particular example, palladium catalyzed ring opening was accompanied by rearrangement to give ethyl 6-(N-tosylamino)-5,5-dimethylhex-3-enoate, ***2.135***.[66]

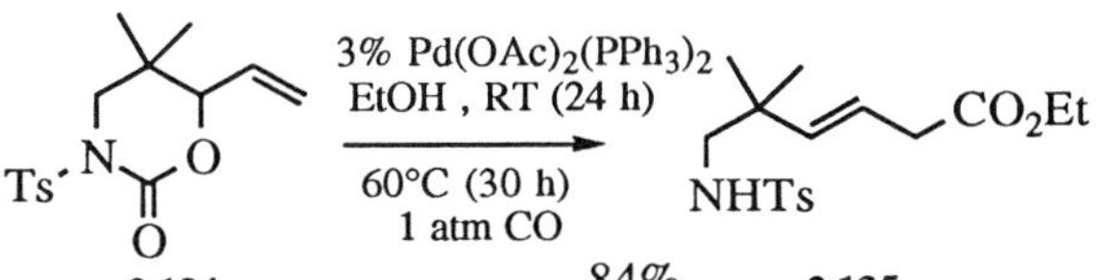

2.134 **2.135**

The strategies used in the previous examples could be loosely described as a "di-functionalization" approach. Another "di-functionalization" reaction converted 1,2-diols to oxazolidinones. When 3-amino-1,2-propanediol was treated with phosgene, oxazolidinone ***2.136*** was formed.[67] Subsequent oxidation of the primary hydroxy group with basic permanganate gave a carboxylic acid (***2.137***), and acid hydrolysis opened the ring to give isoserine (***2.138***) in good yield.

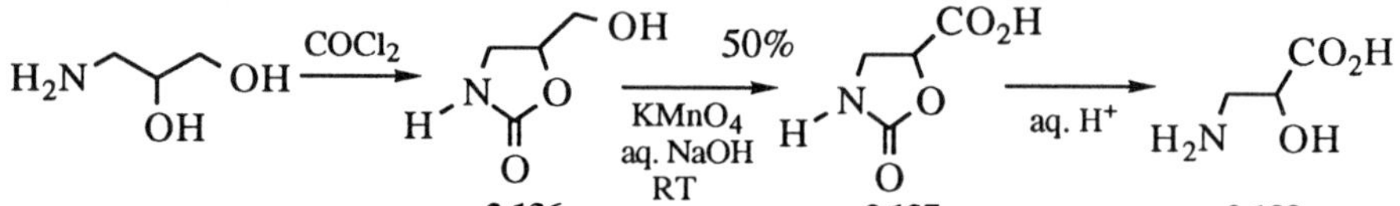

2.136 **2.137** **2.138**

Other hydroxy amino acid derivatives can be prepared by a [3+2]-cycloaddition route, and there are other methods for generating nitrones. Nitro compounds react

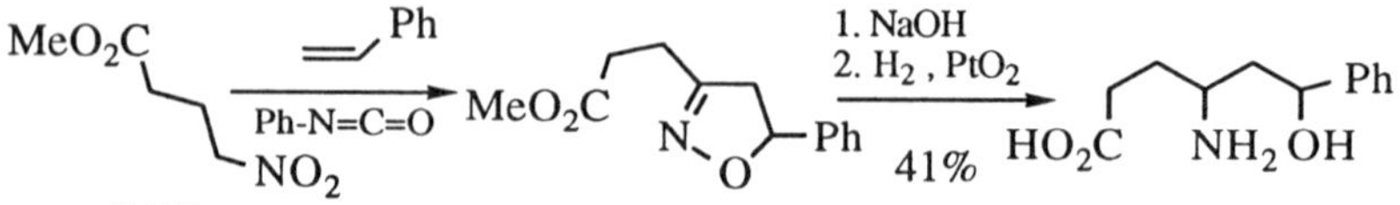

2.139 **2.140** **2.141**

with alkenes, in the presence of isocyanates, to produce dihydro-isoxazole derivatives, via a nitrile N-oxide.[68,58] In one example, styrene reacted with the nitrile N-oxide generated from ***2.139*** to give ***2.140***.[69] Treatment with base and catalytic

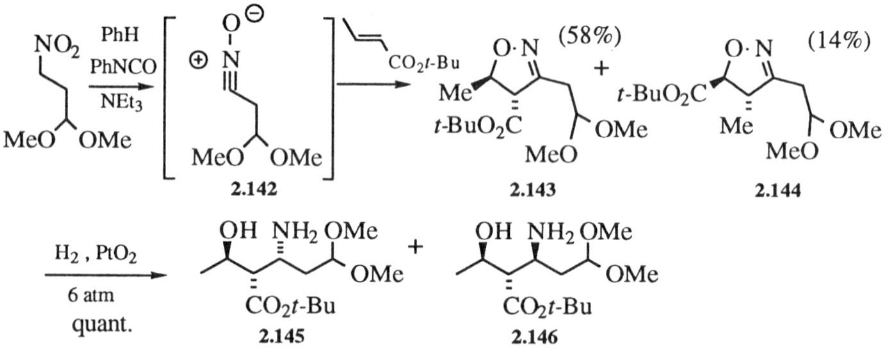

2.142 **2.143** **2.144**

2.145 **2.146**

hydrogenation led to 4-amino-6-hydroxy-6-phenylhexanoic acid (***2.141***). In another example, Kametani generated nitrile N-oxide ***2.142*** *in situ* and condensed it with *t*-butyl crotonate to give ***2.143***, along with 14% of ***2.144***.[70] Isolation of ***2.143*** was followed by catalytic hydrogenation to give a 1:1 mixture of ***2.145*** and ***2.146*** (*anti,syn:anti,anti t*-butyl 3-amino-2-(1-hydroxy-1-ethyl)-5,5-dimethoxypentanoate).

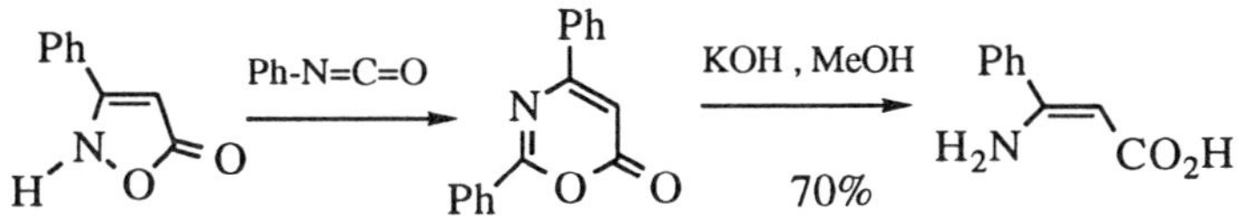

2.147 **2.148** **2.149**

Isocyanates can react with other heterocycles as well as with acyclic unsaturated compounds. An example is the reaction of isoxazolone ***2.147*** with phenyl isocyanate to give ***2.148***. Subsequent hydrolysis led to 3-amino-3-phenylprop-2-enoic acid, ***2.149***.[71] A different example involves the conversion of isoxazole acid ***2.150*** to

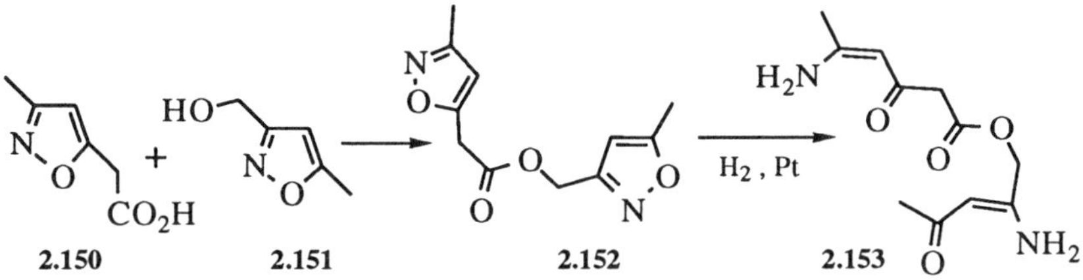

2.150 **2.151** **2.152** **2.153**

2.152 by reaction with isoxazoylcarbinol (***2.151***). Subsequent catalytic hydrogenation gave 2-amino-4-oxopentenyl 5-amino-3-oxohex-4-enoate, ***2.153***.[72] This product was used in a synthesis that targeted hydroxyphthalimides.

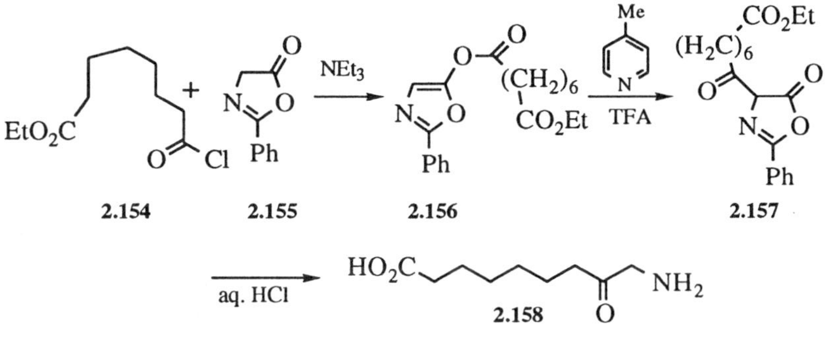

2.154 **2.155** **2.156** **2.157** **2.158**

A somewhat different approach coupled acid chloride ***2.154*** with ***2.155***[73] to give ***2.156***.[74] Treatment with γ-picoline led to the rearranged product ***2.157*** in 84% overall yield from ***2.154***. Acid hydrolysis gave 8-oxo-9-aminononanoic acid, ***2.158***.

2.5.B. Pyrazoline Derivatives

Pyrazoles such as ***2.159*** are formed by a [3+2]-cycloaddition reaction of an alkene with a diazoalkane.[75] Basic hydrolysis of such pyrazolines leads to an amino acid. Reaction of ***2.159*** with aqueous sodium hydroxide, for example, gave 2-methyl-3-aminopropanoic acid (***2.160***).[76] Similarly, 2-methyl-3-aminobutanoic acid was prepared in 52% yield; 2-ethyl-3-aminopropanoic acid in 81% yield; and, 2-methyl-3-aminopentanoic acid in 52% yield, all from the appropriate pyrazoline.[76]

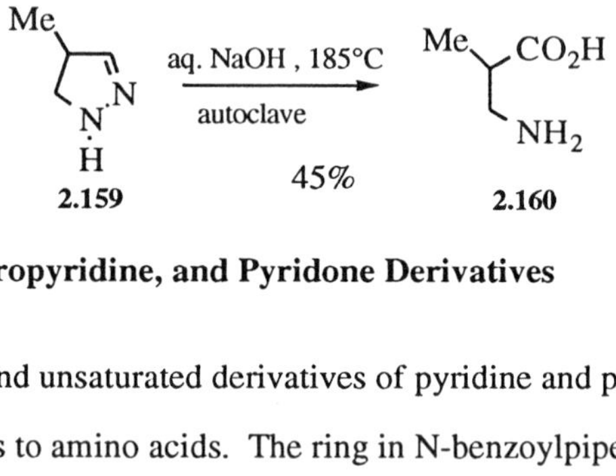

2.159 **2.160**

2.5.C. Pyridine, Hydropyridine, and Pyridone Derivatives

Several saturated and unsaturated derivatives of pyridine and pyridone can serve as synthetic precursors to amino acids. The ring in N-benzoylpiperidine (***2.161***) was oxidatively cleaved with potassium permanganate to give ***2.162***.[77] Hydrolysis of the amide moiety gave 5-aminopentanoic acid, ***2.163***.

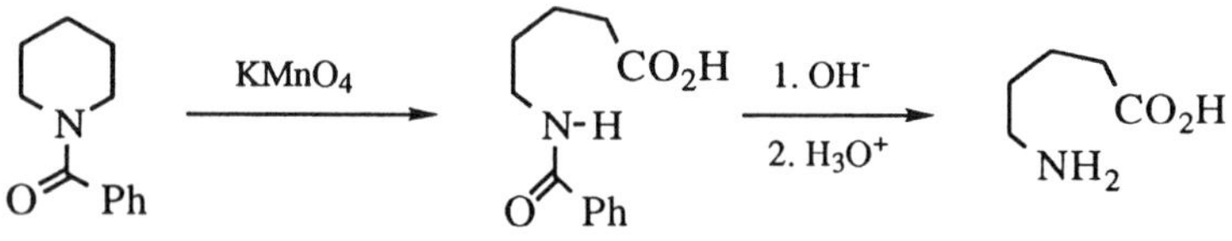

2.161 **2.162** **2.163**

Ring cleavage can occur by many routes, depending on the substituents attached to the hydropyridine ring. The reaction of 1,4,5,6-tetrahydronicotinamide (***2.164***) with hydroxylamine, for example, gave a transient heterocycle (***2.165***). This product spontaneously opened under the reaction conditions to give 5-amino-2-cyanopentanoic acid (***2.166***) as the final product.[78]

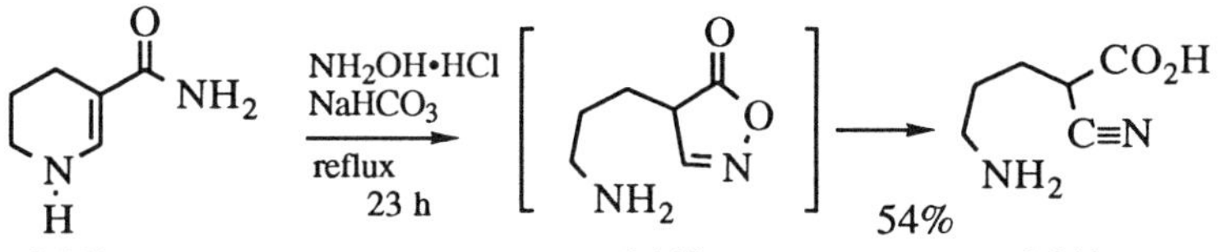

2.164 **2.165** **2.166**

2-Pyridone is another six-membered ring heterocycle, and it was converted[79] to 5-hydroxy-2-piperidone (***2.167***) by an Elbs oxidation.[80] Catalytic hydrogenation converted the pyridone ring to a lactam (***2.168***) and acid hydrolysis gave 5-amino-4-oxopentanoic acid, ***2.169***.[79,81]

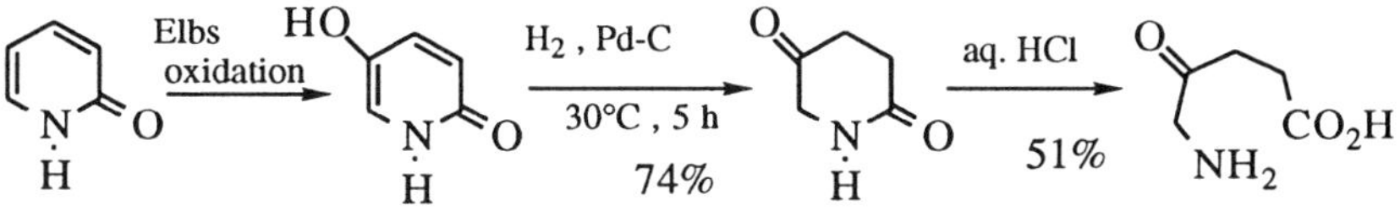

2.167 **2.168** **2.169**

2.5.D. Uracil Derivatives and Cyclic Carbamates

Another cycloaddition approach utilized the reaction of the alkene moiety of a conjugated acid such as 4-methyl-2-pentenoic acid with urea to give uracil ***2.170***.[82]

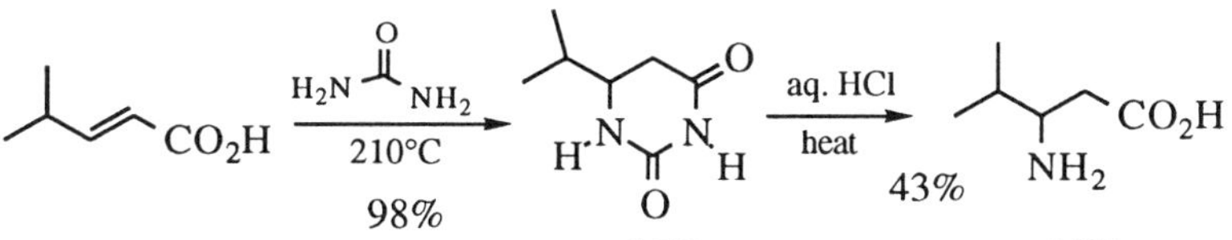

2.170 **2.171**

Acid hydrolysis of this dihydrouracil derivative led to cleavage of the ring and formation of 4-methyl-3-aminopentanoic acid, ***2.171***. Substituted dihydrouracils (such as ***2.170*** and also ***2.172***) have been converted to 2-alkyl, 2,3-dialkyl, 3-alkyl,

3,3-dialkyl, and 2,3,3-trialkyl-3-aminopropanoic acid derivatives (***2.173***).[83] A few examples of this conversion are shown in Table 2.3.

Table 2.3. Cleavage of dihydrouracil derivatives.[85]

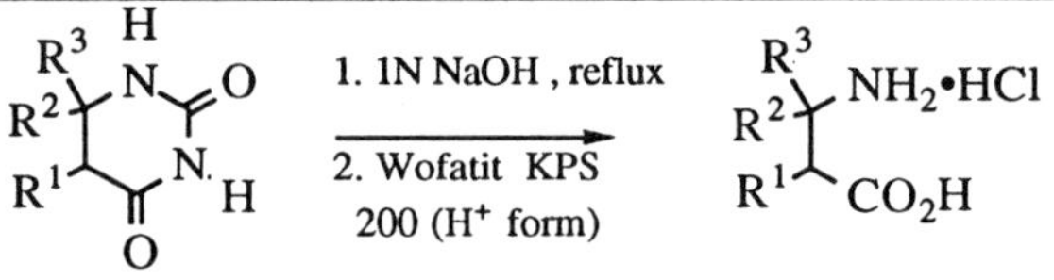

2.172 **2.173**

R^1	R^2	R^3	% yield	amino acid
H	H	H	90	3-aminopropanoic acid
Me	H	H	85	2-methyl-3-aminopropanoic acid
H	H	Me	90	3-methyl-3-aminopropanoic acid
H	Me	Me	90	3,3-dimethyl-3-aminopropanoic acid
H	H	Ph	89	3-phenyl-3-aminopropanoic acid

Cyclic five-membered ring urethane derivatives are also available, and they serve as precursors to amino acids. In one specialized synthesis, dethio-biotin(***2.174***) was treated with hot HCl to give 7,8-diaminononanoic acid (7,8-diaminopelargonic acid, ***2.175***).[84]

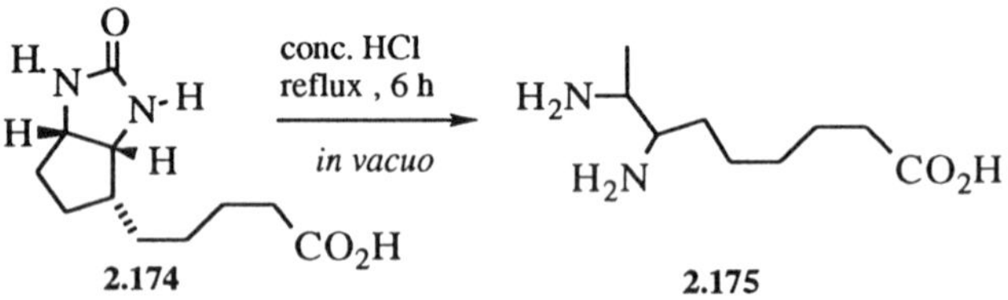

2.174 **2.175**

2.5.E. Thiophene, Thiazolidine and Oxathiolone Derivatives

The thiophene ring has been used as a template to "hold" a carbonyl moiety and an amino moiety (or its surrogate) at predetermined relative positions. Reductive cleavage of the thiophene ring "liberates" the amino acid (see the reactions of ***1.268*** and ***1.269*** in chapter one, section 1.5.C).[85] When ***2.176*** was treated with Raney nickel in aqueous ammonium hydroxide, for example, 4-aminoheptanoic acid (***2.177***) was produced (also known as 4-aminoenanthic acid).[85] In a similar manner,

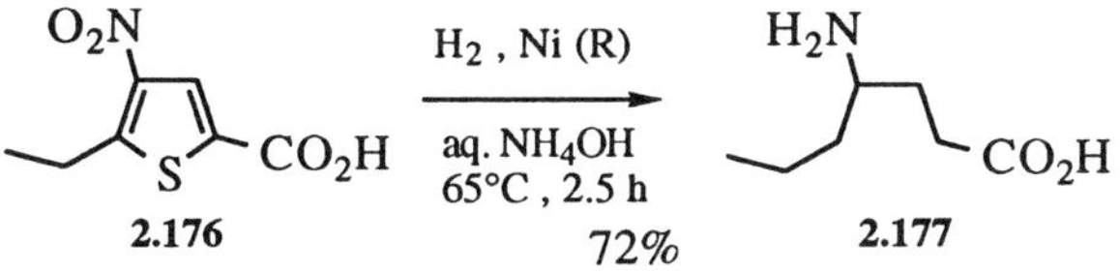

4-aminohexanoic acid was produced in 46% yield.[86] As mentioned in chapter one (section 1.5.C), oxime substituted thiophene derivatives can be used, as in the

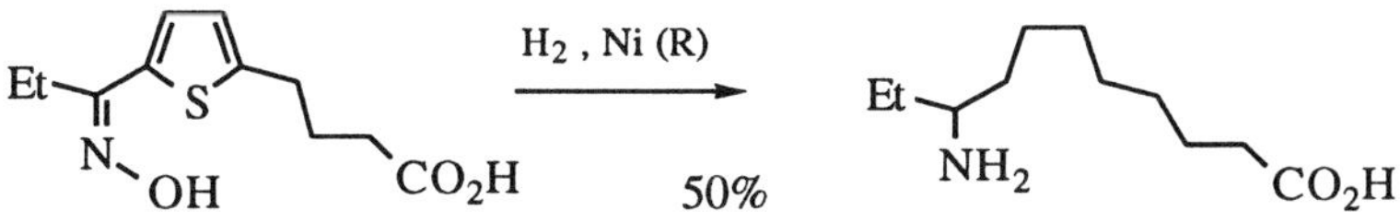

conversion of ***2.178*** to 9-aminoundecanoic acid (***2.179***).[85] It is possible to use other heterocyclic rings as templates. When dihydrothiazole ***2.180*** was condensed with malonic acid, ***2.181*** was formed.[87] Refluxing in water was sufficient to open the thiazolidine ring and give 3-amino-4-methyl-4-mercaptopentanoic acid, ***2.182***.

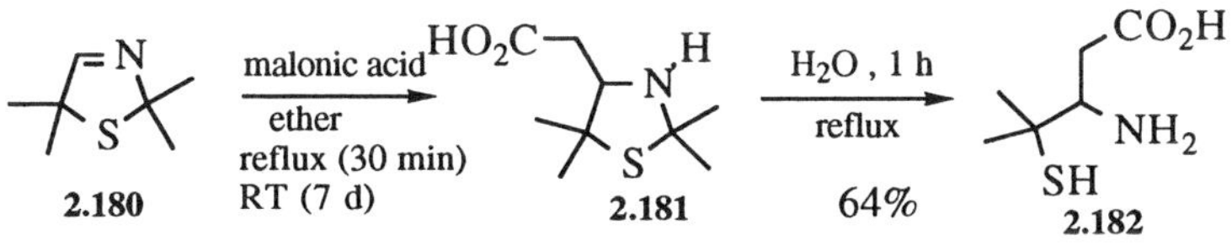

Thio-acids have been used as synthetic precursors to amino acids. Reaction of

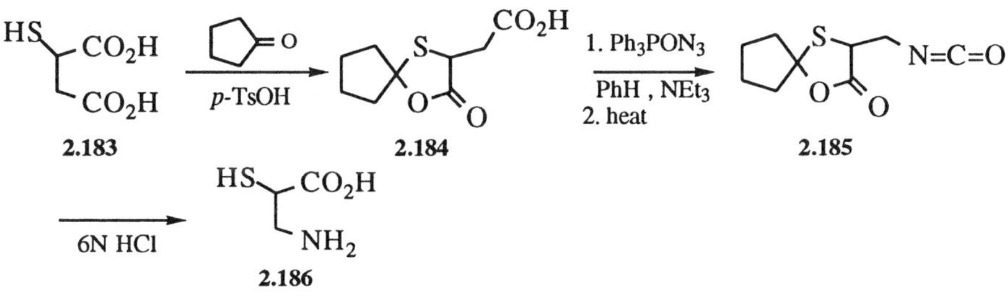

2.183 with cyclopentanone protected the thiol and one carbonyl via formation of oxathiolone derivative, ***2.184***.[88] Subsequent reaction of the "free" carbonyl with diphenylphosphoryl azide gave isocyanate ***2.183***, which led to 3-amino-2-mercapto-propanoic acid, ***2.186*** after rearrangement and hydrolysis.

2.6. Diels-Alder Strategies

A very different approach from those cited above also uses cyclic precursors for the preparation of acyclic amino acids. In work by Krief, a Diels-Alder reaction[89] between methyl acrylate and ***2.187*** gave pyran ***2.188***.[90] Catalytic hydrogenation

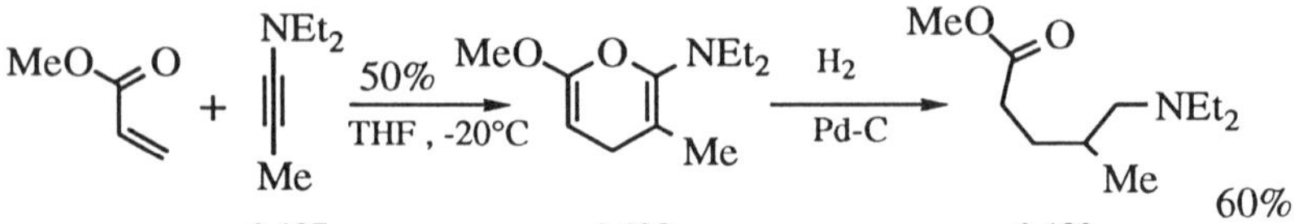

2.187 **2.188** **2.189**

opened the ring and gave methyl 5-(N,N-diethylamino)-4-methylpentanoate, ***2.189***.

The six-membered ring heterocyclic analog of the isoxazolidines (a dihydro-1,2-oxazine such as ***2.192***) is available by a Diels-Alder reaction[89] with nitroso derivatives. Reaction of methyl sorbate (***2.190***) and 1-chloro-1-nitrosocyclohexane (***2.191***) gave ***2.192***.[91] Catalytic hydrogenation followed by hydrolysis reduced the alkene, opened the ring and converted the ester to an acid; 5-amino-2-hydroxyhexanoic acid, ***2.193***.

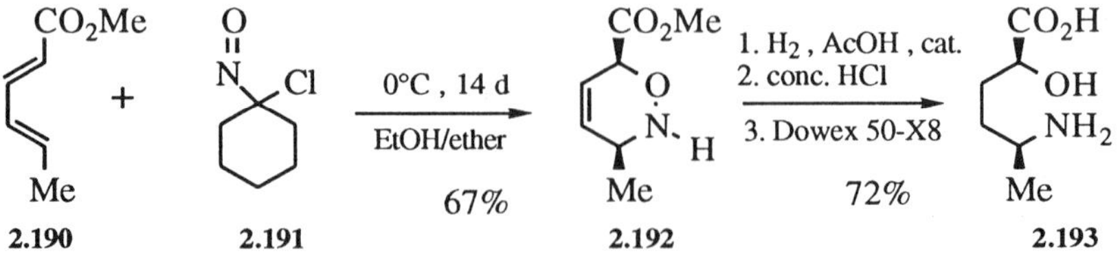

2.190 **2.191** **2.192** **2.193**

REFERENCES

1 (a) Beckmann, E. *Ber.*, ***1886***, *19*, 988; (b) Donaruma, L.G.; Heldt, W.Z. *Org. React.*, ***1960***, *11*, 1; (c) Mundy, B.P.; Ellerd, M.G. *Name Reactions and Reagents in Organic Synthesis,* Wiley, New York, ***1988***, pp. 24-25.

2 Hünig, S.; Gräßmann, W.; Meuer, V.; Lücke, E.; Brenninger, W. *Chem. Ber.*, ***1967***,*100*, 3039.

3 (a) Schmidt, R.F. *Ber.*, ***1924***, *57*, 704; (b) Wolff, H. *Org. React.*, ***1946***, *3*, 307; (c) Koldobskii, G.I. Tereshchenko, G.F.; Gerasimova, E.S.; Bagal, L.I. *Russ. Chem. Rev.*, ***1971***, p. 835; (d) Beckwith, A.L.J. in *Chemistry of Amides*, Zabicky, J. (Ed.), Interscience, London, ***1970***, pp. 137-145; (e) Mundy, B.P.; Ellerd, M.G. *Name Reactions and Reagents in Organic Synthesis*, Wiley, New York, ***1988***, p. 1909.

4 Chimiak, A. *Bull. Acad. Pol. Sci. Ser. Sci. Chim.*, ***1969***, *17*, 197 [*Chem. Abstr.*, ***1969***, *71:* 60617g].

5 Conley, R.T. *J. Org. Chem.*, ***1958***, *23*, 1330.

6 Overberger, C.G.; Parker, G.M. *J. Polym. Sci., A-1*, ***1968***, *6*, 513.

7 Lafont, P.; Thiers, M. *Fr. 1,349,281* [*Chem. Abstr.*, ***1964***, *60*:P13145b].

8 Matlack, A.S.; Breslow, D.S. *J. Org. Chem.*, ***1967***, *32*, 1995.

9 (a) Tanaka, I.; Uehara, H.; Oosaki, F.; Yamagata, M.; Tanaka, S.; Takeshita, T. *Jpn.* 71 09,810 [*Chem. Abstr.*, *1971*, ***75***: 98933c]; (b) Tanaka, I.; Yamagata, Uehara, H. *Jpn.* 70 29,965 [*Chem. Abstr.*, ***1971***, *74*: 63924c].

10 Ram, S.; Ehrenkaufer, R.E. *Tetrahedron Lett.*, ***1984***, *25*, 3415.

11 Lang, A. *Ger. Offen.* 2,062,928 [*Chem. Abstr.*, ***1971***, *75*:110690f].

12 Gresham, T.L.; Jansen, J.E.; Shaver, F.W.; Bankert, R.A.; Fiedorek, F.T. *J. Am. Chem. Soc.*, ***1951***, *73*, 3168.

13 Nabeta, S.; Kojima, S.; Sago, K. *Jpn.* 71 03 767 [*Chem. Abstr.*, ***1971***,*74*: P140928m].

14 (a) Goel, O.P.; Krolls, U.; Lewis, E.P. *Org. Prep. Proceed. Int.*, ***1985***, *17*, 91; (b) Saito, Y.; Hashimoto, M.; Seki, H.; Kamiya, T. *Tetrahedron Lett.*, ***1970***, 4863; (c) Leuenberger, H.G.; Matzinger, P.K.; Wirz, B. *Eur. Pat. Appl.* EP 272,605 [*Chem. Abstr.*, ***1988***, *109*: P188838a].

15 (a) Wittig, G.; Rieber, M. *Ann.*, ***1949***, *562*, 187; (b) Wittig, G.; Geissler, G. *Ibid*, ***1953***, *580*, 44; (c) Wittig, G.; Schöllkopf, U. *Chem. Ber.*, ***1954***, *87*, 1318; (d) Gensler, W.J. *Chem. Rev.*, ***1957***, *57*, 191 (p. 218).

16 Papaioannou, D.; Stavropoulos, G.; Sivas, M.; Barlos, K.; Francis, G.W.; Aksnes, D.W.; Maartmann-Moe, K. *Acta Chem. Scand.*, ***1991***, *45*, 99.

17 (a) Hofmann, A.W. *Ber.*, ***1881***, *14*, 2725; (b) Wallis, E.S.; Lane, J.F. *Org. React.*, ***1949***, *3*, 267; (c) Mundy, B.P.; Ellerd, M.G. *Name Reactions and Reagents in Organic Synthesis*, Wiley, New York, ***1988***, pp. 112-113.

18 Piskov,V.B. *Zh. Obshch. Khim.*, ***1962***, *32*, 3407 (Engl., 3343).

19 Bell, K.H.; Fullick, G. *Synth. Commun.*, ***1987***, *17*, 1965.

20 (a) Meerwein, H. *Org. Synth. Coll. Vol. 5*, ***1973***, 1080; (b) Borch, R.F. *Tetrahedron Lett.*, ***1968***, 61.

21 (a) Abramyan, R.K.; Filippychev, G.F. *Uch. Zap., Yaroslav. Tekhnol. Inst.*, ***1971***, 161 [*Chem. Abstr.*, ***1973***, *78*: 42826q]; (b) Takaya, T.; Ono, T.; Okuda, Y. *Jpn. Kokai* 74 35,325 [*Chem. Abstr.*, ***1974***, *81*: P119980g].

22 Garigipati, R.S.; Morton, J.A.; Weinreb, S.M. *Tetrahedron Lett.*, ***1983***, *24*, 987.

23 (a) Curtius, T. *J. Prakt. Chem.*, ***1894***, *50*, 275; (b) Smith, P.A.S. *Org. React.*, ***1946***, *3*, 337; (c) Saunders, J.H.; Slocombe, R.J. *Chem. Rev.*, ***1948***, *43*, 203.

24 Pojarlieff, I.G.; Kurtev, B.I. *Tetrahedron Lett.*, ***1963***, 525.

25 Baldwin, J.E.; Adlington, R.M.; O'Neil, I.A.; Schofield, C.; Spivey, A.C.; Sweeney, J.B. *J. Chem. Soc., Chem. Commun.*, ***1989***, 1852.

26 (a) Sulzbacher, M. *Mfg. Chemist*, ***1962***, *33*, 463; (b) Tandara, M. *Kem. Ind.*, ***1969***, *18*, 713.

27 Eck, J.C. *Org. Synth. Coll. Vol. II*, ***1943***, 28.

28 Colonge, J.; Pouchol, J.-M. *Bull. Soc. Chim. Fr.*, ***1962***, 598.

29 (a) Tandara, M. *Ger. Offen.* 1,905,530 [*Chem. Abstr.*, ***1970***, *73*: 76667a]; (b) Bosnalijek-Tvornica, *Fr. Demande* 2,027,102 [*Chem. Abstr.*, ***1971***, *75*: 21427h]; (c) Grigorescu, E.; Selmiciu, I.; Sauciuc, T.; Rusu, G.; Dumitrache, M.; Haulica, I. *Rom.* 51,420 [*Chem. Abstr.*, ***1970***, *70*:105980t].

30 Flynn, D.L.; Zelle, R.E.; Grieco, P.A. *J. Org. Chem.*, ***1983***, *48*, 2424.

31 Cologne, J.; Pouchol, J.-M. *Bull. Soc. Chim. Fr.*, ***1962***, 598.

32 DeGraaff, G.B.R.; Melger, W. Ch.; Van de Kolk, G. *Recuil. Trav. Chim., Pays-Bas*, ***1962***, *81*, 786.

33 (a) Vasil'eva, O.S.; Zobacheva, M.M.; Smirnova, A.A.; Perekalin, V.V. *Zh. Org. Khim.*, ***1978***,*14*, 1420 (Engl., 1326); (b) Smirnova, A.A.; Perekalin, V.V.; Scherbakov, V.A. *Ibid*, ***1968***, *4*,2245 (Engl., p. 2166); (c) Belostotskaya, I.S.;

Dyumaev, K.M. *Sintez Prirodn. Soedin., Ikh. Analogov i Fragmentov, Akad.Nauk. SSSR, otd. Obshch. i. Tekhn. Khim.* ***1965***, 212 [*Chem. Abstr.*, ***1966***, *65*: 5431f].

34 Herdeis, C. *Synthesis*, ***1986***, 232.

35 Herdeis, C. *Arch. Pharm (Weinheim)*, ***1983***, *316*, 719.

36 Onda, M.; Konda, Y.; Ömura, S.; Hata, T. *Chem. Pharm. Bull.*, ***1971***, *19*, 2013.

37 Bey, P.; Jung, M. *Belg. Patent* 870,796 [*Chem. Abstr.*, ***1979***, *91*: P91952e].

38 (a) Fabrichnyi, B.P.; Shalavina, I.F.; Zurabyan, S.E.; Gol'dfarb, Ya.L.; Kostrova, S.M. *Zh. Org. Khim.*, ***1968***, *4*, 680 (Engl. p. 663); (b) Fabrichnyi, B.P.; Shalavina, I.F.; Gol'dfarb, Ya.L. *Zh. Org. Khim.*, ***1967***, *3*, 2079 (Engl. p. 2027).

39 Zezza, C.A.; Kwon, T.W.; Sheu, J.L.; Smith, M.B. *Heterocycles*, ***1992***, *34*, 1325.

40 Menezes, R.; Smith, M.B. *Synth. Commun.*, ***1988***,*18*, 1625.

41 Wick, A.E.; Bartlett, P.A.; Dolphin, D. *Helv. Chim. Acta*, ***1971***, *54*, 513.

42 Yamada, Y.; Emori, T.; Kinoshita, S.; Okada, H. *Agric. Biol. Chem.*, ***1973***, *37*, 649.

43 Yamada, Y.; Okada, H. *Agric. Biol. Chem.*, ***1976***, *40*, 1437.

44 Suzuki, K.; Takeya, H. *Jpn. Kokai Tokkyo Koho* JP 03 72,450 [*Chem .Abstr.*, ***1991***, *115*: P91665c].

45 Flitsch, W. *J.L. Ann. Chem.*, ***1965***, *684*, 141.

46 Haug, Th.; Lohse, F.; Metzger, K.; Batzer, H. *Helv. Chim. Acta*, ***1968***, *51*, 2069.

47 Graf, R. *J.L. Ann. Chem.*, ***1963***, *661*, 111.

48 Moriconi, E.J.; Kelly, J.F. *Tetrahedron Lett.*, ***1968***, 1435.

49 Clauß, K. *Tetrahedron Lett.*, ***1974***, 1271.

50 Moriconi, E.J.; Kelly, J.F. *J. Org. Chem.*, ***1968***, *33*, 3036.

51 (a) Hauser, F.M.; Rhee, R.P. *J. Org. Chem.*, ***1981***,*46*, 227; (b) Moriconi, E.J.; Meyer, W.C. *J. Org. Chem.*, ***1971***, *36*, 2841; (c) Hauser, F.M.; Rhee, R.P.; Ellenberger, S.R. *J. Org.Chem.*, ***1984***,*49*, 2236.

52 Arbuzov, B.A.; Zobova, N.N. *Dokl. Akad. Nauk. SSSR*, ***1966***, *170*, 1317 (Engl. p. 993).

53 Moriconi, E.J.; Kelly, J.F.; Salomone, R.A. *J. Org. Chem.*, ***1968***, *33*, 3448.

54 Manhas, M.S.; Hegde,V.R.; Wagle, D.R.; Bose, A.K. *J. Chem. Soc., Perkin Trans. I*, ***1985***, 2045.

55 Nicolaus, B.J.R.; Bellasio, E.; Pagani,G.; Testa, E. *Gazz. Chim. Ital.*, ***1963***, *93*, 618.

56 Ojima, I.; Inaba, S. *Tetrahedron Lett.*, ***1980***, 2077.

57 Ojima, I.; Park, Y.H.; Sun, C.M.; Brigaud, T.; Zhao, M. *Tetrahedron Lett.*, ***1992***, *33*, 5737.

58 Huisgen, R. *Angew. Chem. Int. Ed., Engl.*, ***1963***, *2*, 565, 633.

59 Barco, A.; Benetti, S.; Pollini, G.P. *J. Org. Chem.*, ***1979***, *44*, 1734.

60 Stevens, R.V.; Christensen, C.G.; Edmonson, W.L.; Kaplan, M.; Reid, E.B.; Wentland, M.P. *J. Am. Chem. Soc.*, ***1971***, *93*, 6629.

61 Murahashi, S.; Kodera, Y.; Hosomi, T. *Tetrahedron Lett.*, ***1988***, *29*, 5949.

62 Keirs, D.; Moffat, D.; Overton, K.; Tomanek, R. *J. Chem. Soc., Perkin Trans. I*, ***1991***, 1041.

63 Hjeds, H.; Honoré, T. *Acta Chem. Scand., Ser. B*, ***1978***, *B32*, 187.

64 (a) Honoré, T.; Hjeds, H.; Krogsgaard-Larsen, P.; Christiansen, T.R. *Eur. J. Med. Chem. Chim.Ther.*, ***1978***, *13*, 429; (b) Hjeds, H.; Jerslev, B.; Ross-Petersen, K.J. *Dan. Tidsskr. Farm.*, ***1972***, *46*, 97 [*Chem. Abstr.*, ***1972***, *77*: 88375z]; (c) Brehm, L.; Krogsgaard-Larsen, P. *Acta Chem. Scand., B*, ***1979***, *B33*, 52.

65 Falch, E.; Hedegaard, A.; Nielsen, L.; Jensen, B.R.; Hjeds, H.; Krogsgaard-Larsen, P. *J. Neurochem.*, ***1986***, *47*, 898.

66 Bando, T.; Tanaka, S.; Fugami, K.; Yoshida, Z.; Tamaru, Y. *Bull. Chem. Soc., Jpn.*, ***1992***, *65*, 97.

67 Tsuchihashi, G.; Iriuchijima, S.; Mori, Y. *Jpn. Kokai* 76 16,660 [*Chem. Abstr.*, ***1976***, *85*: P123901g].

68 Tufariello, J.J. *Accts. Chem. Res.*, ***1979***, *12*, 396.

69 Borea, P.A.; Bonora, A.; Baraldi, P.G.; Simoni, D. *Farmaco Ed. Sci.*, ***1983***, *38*, 411.

70 (a) Kametani, T.; Nagahara,T.; Suzuki, Y.; Yokohama, S.; Huang, S.-P.; Ihara, M. *Tetrahedron*, ***1981***, *37*, 715; (b) Kametani, T.; Huang, S.-P.; Ihara, *Heterocycles*, ***1979***, *12*, 1183; (b) Kametani, T.; Nagahara, T.; Suzuki, Y.; Yokohama, S. *Ibid*, ***1980***, *14*, 403; (d) Kametani, T.; Huang, S.-P.; Yokohama, S.; Suzuki,Y.; Ihara, M. *J. Am. Chem. Soc.*, ***1980***, *102*, 2060.

71 Risitano, F.; Grassi, G.; Foti, F.; Caruso, F.; LoVecchio, G. *J. Chem. Soc., Perkin Trans. I*, ***1979***, 1522.

72 Auricchio, S.; Ricca, A.; DePava, O.V. *Tetrahedron Lett.*, ***1980***, 857.

73 Stewart, J.M.; Woolley, D.W. *J. Am. Chem. Soc.*, ***1956***, *78*, 5336.

74 Zav'yalov, S.I.; Kravchenko, N.E. *Izv. Akad. Nauk. SSSR, Ser Khim.*, ***1984***, 422 (Engl. p. 384).

75 Schneider, M.P.; Goldbach, M. *J. Am. Chem. Soc.*, ***1980***, *102*, 6114.

76 Ioffe, B.V.; Isidorov, V.A.; Stolyarov, B.V. *Dokl. Akad. Nauk SSSR*, ***1971***, *197*, 91 (Engl., p. 177).

77 (a) Schotten, C. *Chem. Gesel. Ber. Deutch.*, ***1888***, *21*, 2235 (see p. 2240); (b) *Ibid*, ***1884***, *17*, 2544 (see p. 2546).

78 Quin, L.D.; Pinion, D.O. *J. Org. Chem.*, ***1970***, *35*, 3130.

79 Franck, B.; Stratmann, H., *Heterocycles*, ***1981***, *15*, 919; (b) Franck, B. *Angew Chem. Int. Ed., Engl.* ***1982***, *21*, 343.

80 (a) Elbs, K. *J.Prakt. Chem.*, ***1893***, *48*, 179; (b) Haines, A.H. *Methods for the Oxidation of Organic Compounds*, Academic, London, ***1985***, p. 174; (c) Sethna, S.M. *Chem. Rev.*, ***1951***, *49*, 91.

81 Herdeis, C.; Dimmerling, A. *Arch. Pharm. (Weinheim)*, ***1984***, *317*, 304.

82 Birkofer, L.; Storch, I. *Chem. Ber.*, ***1953***, *86*, 529.

83 Rachina, V.; Blagoeva, I. *Synthesis*, ***1982***, 967.

84 Suyama, T.; Tsugawa, R.; Okumura, S.; Kanao, S. *Yakugaku Zasshi*, ***1968***, *88*, 223 [*Chem. Abstr.*, ***1968***,*69*: 51501w].

85 Fabrichnyi, B.P.; Zurabyan, S.E.; Shalavina, I.F.; Gol'dfarb, Ya.L. *Izv. Akad. Nauk. SSSR, Ser Khim.*, ***1967***, 2102

86 (a) Gol'dfarb, Ya.L.; Fabrichnyi, B.P.; Shalavina, I.F. *Tetrahedron*, ***1962***, *18*, 21; (b) Fabrichnyi, B.P.; Krasnyanskaya, E.A.; Shalavina, I.F.; Gol'dfarb, Ya.L. *Zh. Obshch. Khim.*, ***1963***, *33*, 2697 (Engl. p. 2627).

87 Martens, J.; Kintscher, J.; Arnold, W. *Synthesis*, ***1991***, 497.

88 Gustavson, L.M.; Srinivasan, A. *Synth. Commun.*, ***1991***, *21*, 265.

89 Desimoni, G.; Tacconi, G.; Barco, A.; Pollini, G.P. *Natural Product Synthesis through Pericyclic Reactions*, Amer. Chem. Soc., Washington, ***1983***.

90 Ficini, J.; Krief, A. *Tetrahedron Lett.*, ***1970***, 885.

91 Belleau, B.; Au-Young, Y.-K. *J. Am. Chem. Soc.*, ***1963***, *85*, 64.

3
Conjugate Addition Reactions

Amines and many amine surrogates are good nucleophiles. It is not surprising that they add to conjugated acid derivatives to give amino acids. This strategic approach to their synthesis is sufficiently novel that it has been segregated into a separate chapter. Various methods that involve conjugate addition will be presented.

3.1. Amine Nucleophiles

Ammonia is sufficiently nucleophilic that addition to the double bond of a conjugated ester is possible, even in the presence of the acyl carbon. Similarly, amines undergo conjugate addition. In both cases, the product is a 3-aminopropanoic acid derivative (a β-amino acid derivative). Both ammonia and amine nucleophiles will be discussed in this section.

3.1.A. Ammonia

The most common method for producing substituted β-amino acids is via conjugate addition to α,β-unsaturated acids or esters. Amines and ammonia add to conjugated esters, as in the reaction of ammonia with ^{14}C labeled methyl butenoate (methyl crotonate) to give radiolabeled 3-aminobutanoic acid (*3.1*), after hydrolysis.[1]

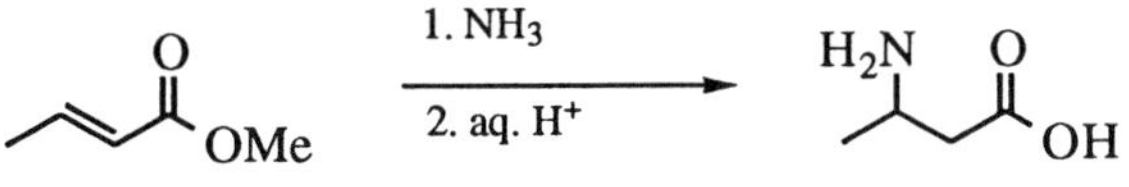

3.1

A similar route that led to a functionalized conjugated ester began with reduction of the ketone moiety in ***3.2*** to give the alcohol moiety in ***3.3***.[2] Thermal elimination gave the conjugated ester (***3.4*** along with 27% of the acid) and subsequent conjugate addition of ammonia gave 3-amino-4,4,4-trifluorobutanamide, ***3.5***.

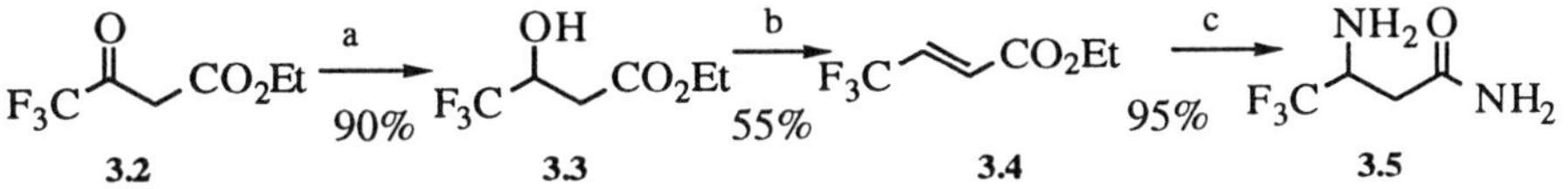

3.2 **3.3** **3.4** **3.5**

(a) H_2, Pd-C, 1500 psi, 125°C, 8 h (b) boric anhydride, 180°C → 300°C
(c) NH_3, 100°C, sealed tube, 36 h

An important synthetic route to conjugated esters involves a Wittig reaction[3] with an aldehyde or ketone. Phenylacetaldehyde was converted to ***3.6*** in one example, allowing the synthesis of ***3.7***.[4] This reaction involves conjugate addition of ammonia

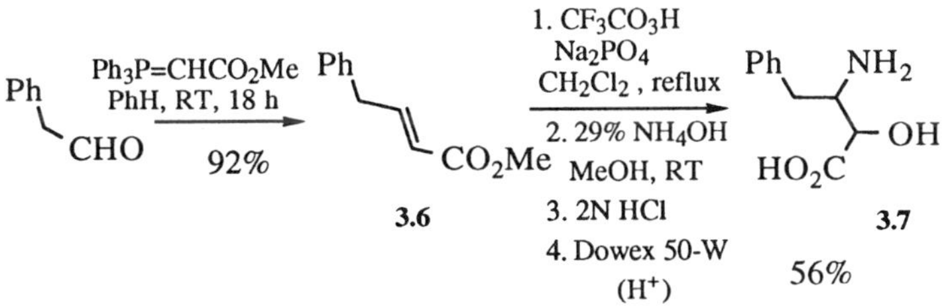

3.6 **3.7**

to the β-carbon of an α,β-epoxy ester. A similar sequence with isobutyraldehyde led to 3-amino-2-hydroxy-5-methylhexanoic acid, in 49% overall yield.[4] Another example of this is the hydrolysis of the ester moiety in ***3.8*** to give ***3.9***. Subsequent treatment with ammonia opened the epoxide ring to give 3-amino-2-cyano-2-hydroxy-3-methylbutanoic acid (***3.10***) in 71% overall yield.[5]

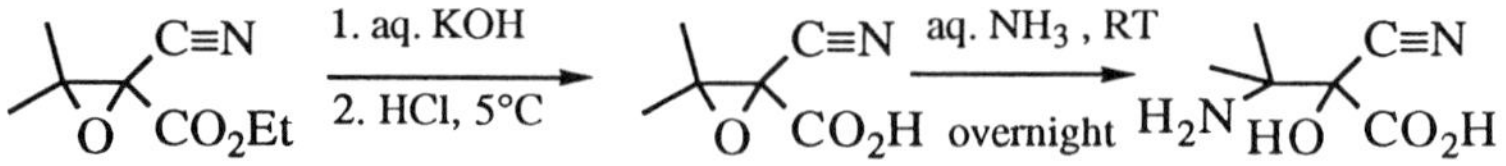

3.8 **3.9** **3.10**

Functionalized alkenes other than esters undergo Michael addition reaction with ammonia or amines. Acrylonitrile, for example, reacted with ammonia in an auto-

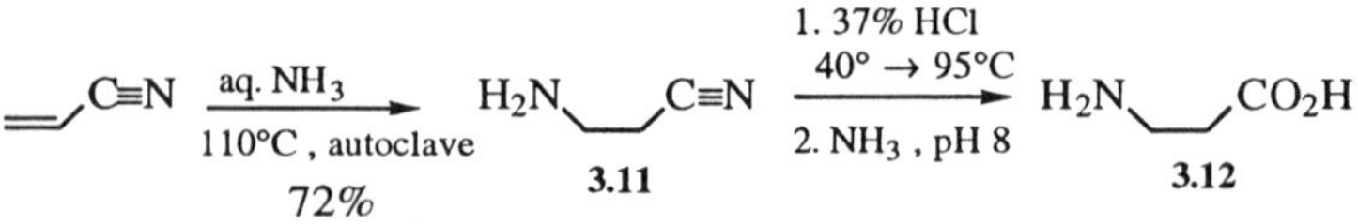

3.11 **3.12**

clave to give 3-aminopropanenitrile, ***3.11***.[6] Acid hydrolysis converted the cyano group to an acid and neutralization gave 3-aminopropanoic acid, ***3.12***. Another example converted ***3.13*** to 2-methyl-3-aminopropanoic acid (***3.14***).[7] Yet another

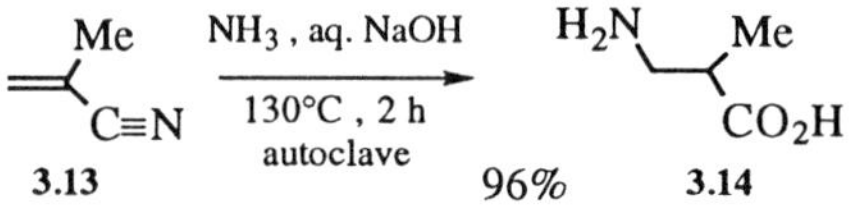

3.13 **3.14**

synthesis used the same strategy, but incorporated a chlorine into the final amino acid. The conjugate addition of ammonia to 2-chloroacrylonitrile gave ***3.15***, which was hydrolyzed to give 3-amino-2-chloropropanoic acid, ***3.16***.[8]

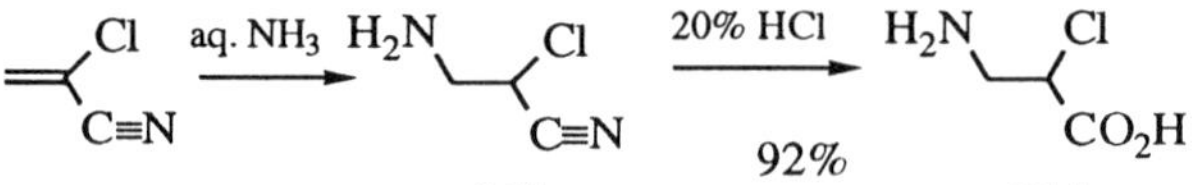

3.15 **3.16**

3.1.B. Amines

Just as ammonia undergoes a Michael addition with α,β-unsaturated carbonyl derivatives, amines also add to conjugated acids such as acrylic acid to give β-amino acids (3-aminopropanoic acids).[9] Amines such as methylamine, aniline, dimethylamine, pyrrolidine and morpholine add to acrylic acid to give the corresponding 3-N-substituted or N,N-disubstituted aminopropanoic acid derivative.[9] The ammonium carboxylate derivative of acrylic acid has been converted to β-alanine using "shockwaves" (ultrasound), but the yield was only 4%.[10] Yields in this type of reaction can be good, however, as shown by the addition of diethylamine to ethyl acrylate, which gave ethyl 3-(N,N-diethylamino)propionate (***3.17***) in 87% yield.[11]

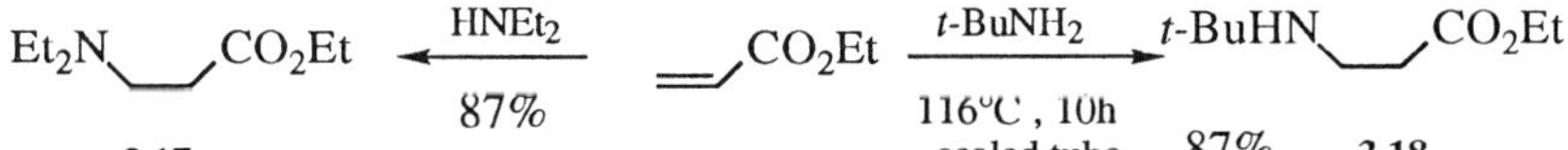

3.17 **3.18**

Similarly, *t*-butylamine was added to give an 87% yield of ***3.18***.[12] The reaction can also be done using an alcohol solvent such as ethanol rather than using the neat amine.[13] Aromatic amines such as aniline add to acrylate esters,[14] as do a wide variety of other amines.[15] It is also noted that (+)-phenethylamine added to crotonic acid to give a 27% yield of 3-aminobutanoic acid, but the extent of asymmetric induction was not reported[16] (see chapter five).

Table 3.1. Conjugate addition of amines to Michael receptors.

Reaction	Transformation	Reference
1.	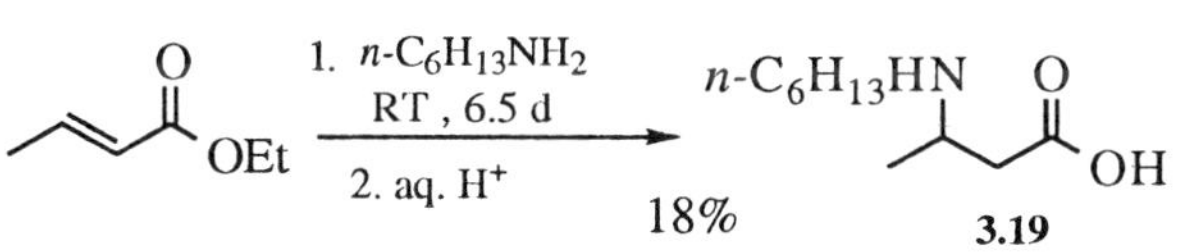 **3.19**	17
2.	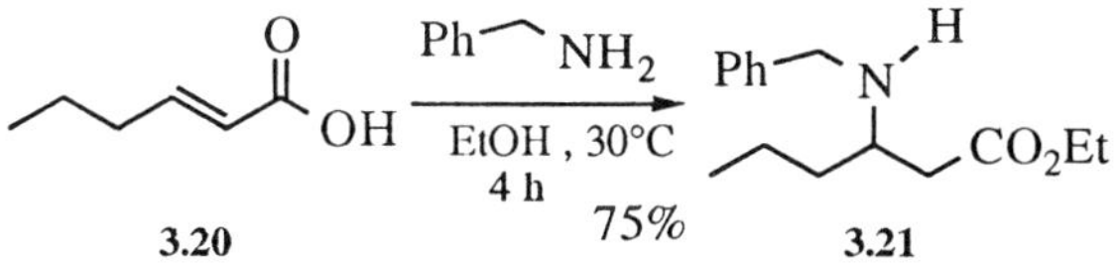 **3.20** **3.21**	18

Table 3.1. (Continued).

Reaction	Transformation	Reference
3.	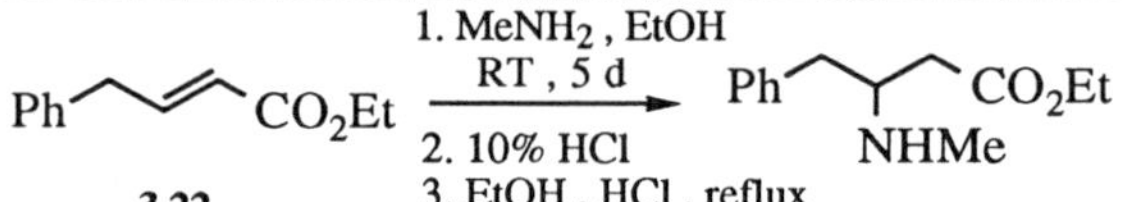 **3.22** **3.23**	20
4.	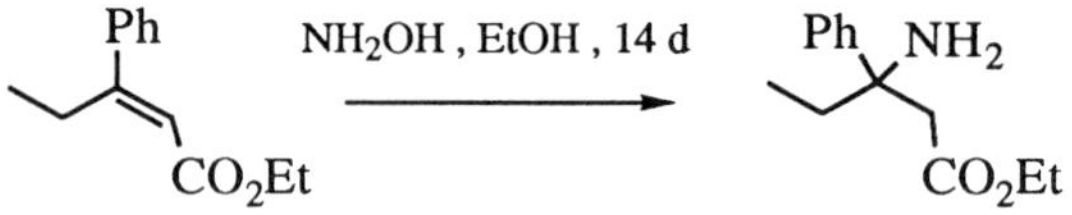 **3.24** **3.25**	21
5.	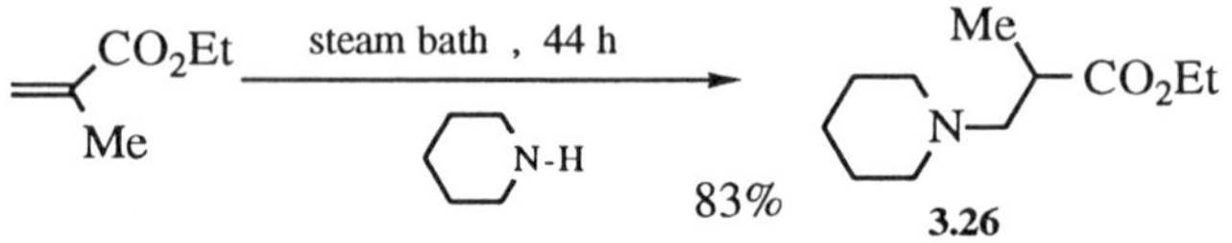 **3.26**	11
6.	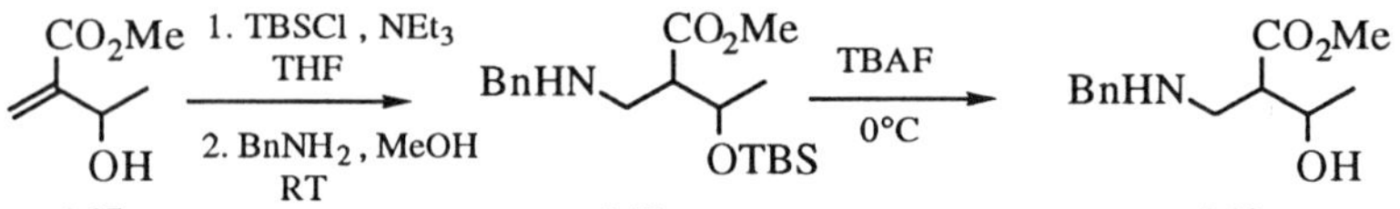 **3.27** **3.28** **3.29**	22
7.	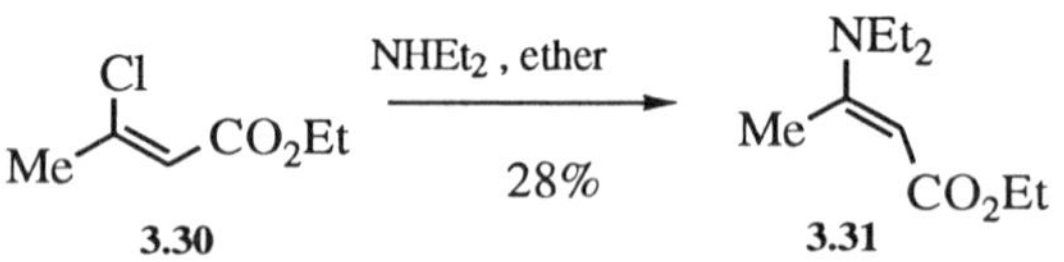 **3.30** **3.31**	23

Many of these reactions are virtually identical. Some of the variety seen in amine reactants as well as conjugated acid or ester receptors are collected in Table 3.1. This table is given to illustrate the scope of the reaction.

As seen above, acrylic acid derivatives readily add ammonia or amines to give the 3-aminopropanoic acid derivative. When the terminal carbon of the alkene is substituted, the reaction is slower and the yields of conjugate addition product are often diminished. *n*-Hexylamine, in reaction 1 from Table 3.1, reacted with ethyl but-2-enoate (ethyl crotonate) to give 3-(N-hexylamino)butanoic acid (***3.19***), but in poor yield.[17] Similarly, benzylamine added to ethyl crotonate (in neat benzylamine) to give 3-(N-benzylamino)butanoic acid in 24% yield.[17] When a solvent such as ethanol was used, rather than the neat amine, improved yields of conjugate addition product were observed. An example is the addition of benzylamine to hex-2-enoic acid (***3.20***), which produced N-benzyl-3-aminohexanoate (***3.21***) (reaction 2).[18] A variety of substituents at the terminus of the conjugated system are compatible with the amine reaction and will produce many different amino acids. One example reacted ethyl cinnamate and piperidine to give ethyl 3-phenyl-3-piperidinopropanoate[19] and, in another, methylamine was added to ethyl 4-phenylbut-2-enoate (***3.22***) to give ethyl 4-phenyl-3-methylaminobutanoate, ***3.23*** (reaction 3).[20] Reaction 4 illustrates two principles. The first is that conjugate addition can occur to somewhat hindered conjugated systems. The ethyl ester (***3.24***) has two groups at the β-carbon, but reacts to give ethyl 3-phenyl-3-aminopentanoate, ***3.25***.[21] The second point in this example by Posner is that hydroxylamine functions as an amine surrogate in the conjugate addition, generating the amine directly.

There are many examples that prepare amino acids with substituents at C_2 by using a 2-substituted conjugated acid derivative. A simple example is the addition of piperidine to ethyl methacrylate (in reaction 5), which gave ethyl 3-piperidino-2-methylpropanoate, ***3.26***.[11] Note that the yield of conjugate addition product is quite good. Even more complex substituents can be incorporated at C_2. When the alcohol moiety in ***3.27*** was protected, conjugate addition of benzylamine gave ***3.28*** (see reaction 6).[22] Removal of the silyl protecting group with tetrabutylammonium fluoride gave methyl 3-(N-benzylamino)-2-(1-hydroxyethyl)propanoate, ***3.29***. When the reaction was done using methanol as a solvent, a 4.5:1 *syn:anti* mixture of ***3.29*** was obtained in 99% yield. When the reaction was done in THF, however, a 1:2.9 ratio of *syn:anti* ***3.29*** was obtained, in 77% yield.[11]

A last example shows the reaction of a vinyl chloride such as ***3.30*** with ethylamine (see reaction 7). Addition of the amine generated the usual enolate anion intermediate, which displaced the β-chlorine to give a new conjugated ester. In this case, the product was ethyl 3-(N,N-diethylamino)but-2-enoate (***3.31***).[23] The yield was relatively poor, but this example illustrates that conjugate addition of amines can lead to either the normal saturated product or, as in this case, an unsaturated amino acid.

There are other types of conjugate addition reactions that do not readily fit into Table 3.1. Conjugate addition of amines to alkynyl esters or acid, for example, illustrate this point since they give 3-aminopropenoic acid derivatives. When ***3.32***

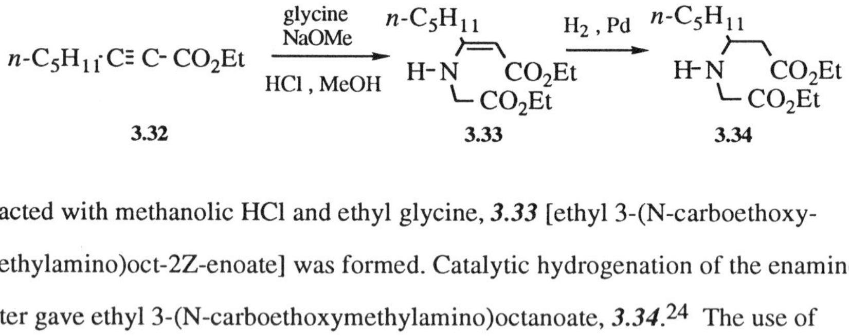

3.32 **3.33** **3.34**

reacted with methanolic HCl and ethyl glycine, ***3.33*** [ethyl 3-(N-carboethoxy-methylamino)oct-2Z-enoate] was formed. Catalytic hydrogenation of the enamino-ester gave ethyl 3-(N-carboethoxymethylamino)octanoate, ***3.34***.[24] The use of R-C≡C-CO_2Et (where R = n-C_4F_9, C_6F_{13}, C_7F_{15}, or C_8F_{17}) led to the preparation of not only the 3-aminooctanoic acid derivatives (***3.34***) but also the heptanoic, decan-oic, dodecanoic, and undecanoic acid derivatives (all as their polyfluoro analogs).[24a]

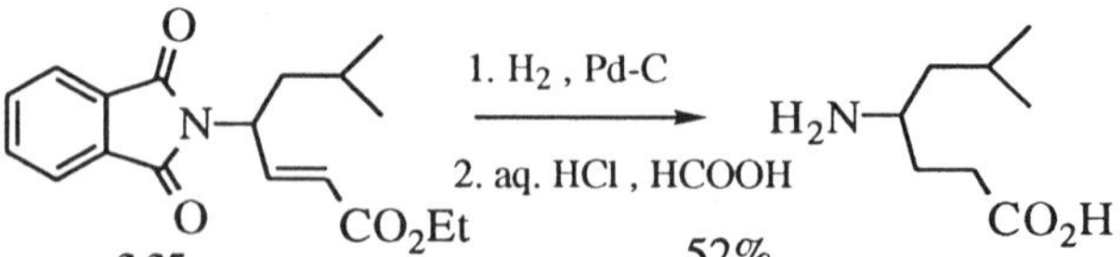

3.35 **3.36**

As seen in this example, hydrogenation of the alkenyl amino acid is an excellent route to saturated amino acids, using the conjugate addition strategy for synthesis. Hydrogenation of ***3.35*** in the presence of a palladium catalyst is another example, and the product was 6-methyl-4-aminoheptanoic acid (***3.36***, called deoxystatine; see chapter six, section 6.3) after deprotection of nitrogen and the carboxyl group.[25]

Amines add to conjugated nitriles as well as to conjugate acid derivatives, and the resulting cyano-amines can then be hydrolyzed to amino acids. In such a case, the nitrile moiety is a carboxylic acid surrogate. Benzylamine added to but-2-enenitrile (***3.37***) to give ***3.38***. Acid hydrolysis of the nitrile moiety gave N-benzyl 3-amino-butanoic acid (***3.39***) and removal of the benzyl group by palladium catalysts hydrogenation gave 3-aminobutanoic acid, ***3.1***.[26]

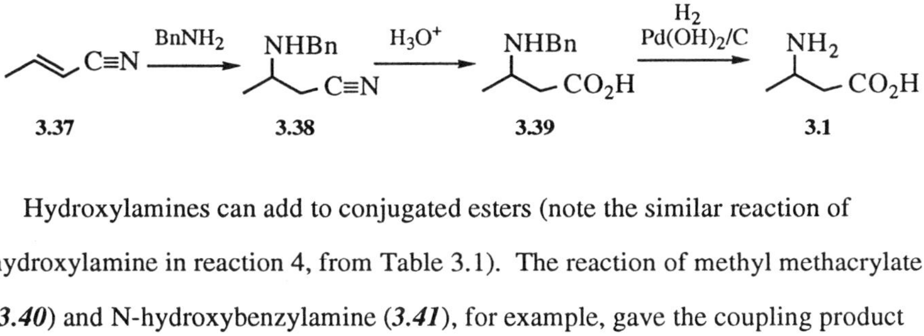

3.37 **3.38** **3.39** **3.1**

Hydroxylamines can add to conjugated esters (note the similar reaction of hydroxylamine in reaction 4, from Table 3.1). The reaction of methyl methacrylate (***3.40***) and N-hydroxybenzylamine (***3.41***), for example, gave the coupling product after treatment with lithium *bis*-trimethylsilylamide. Catalytic hydrogenation cleaved the N-benzyl group and also the hydroxylamine group to give 3-amino-2-methylpropanoic acid, ***3.14*** (62% overall yield).[27]

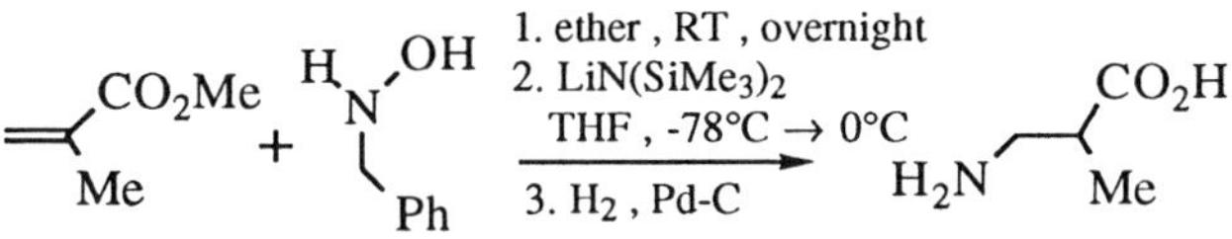

3.40 **3.41** **3.14**

Even weak nucleophiles such as amides can be added to conjugated esters if radical conditions are used, but the reaction also occurs with non-conjugated alkenes.[28] N-Ethyl acetamide added to methyl acrylate, in the presence of di-*t*-butyl peroxide, to give methyl 3-(N-acetyl-N-ethylamino)propanoate, ***3.42***. Similar reaction between methyl acrylate and N-methyl acetamide gave only 32% of the corresponding N-methyl acetamide-ester.[28] The addition to a non-conjugated alkene is illustrated by the reaction of N-butyl acetamide with methyl pent-4-enoate, under identical radical conditions, to give methyl 5-(N-acetyl-N-butylamino)pentanoate, ***3.43***.[28] The obvious limitation to this approach is that only one alkene can be present in the starting material since the intermediate amido radical will add to an alkene moiety whether it is a conjugated alkene or an unconjugated alkene.

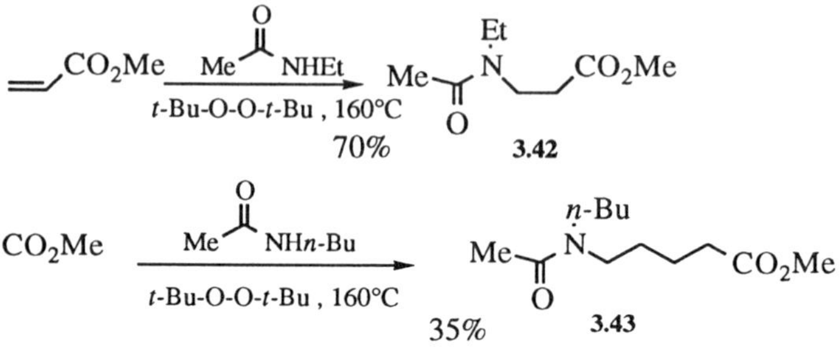

3.42 **3.43**

Another radical addition reaction pertinent to this section has been reported. In this case, however, the radial reaction was initiated photochemically. Photolysis of 2-aminopropane in the presence of ethyl crotonate led to an amino-ester (***3.44***), which spontaneously cyclized to 4,5,5-trimethyl-2-pyrrolidinone, ***3.45***.[29] Aqueous acid hydrolysis liberated 3,4-dimethyl-4-aminopentanoic acid, ***3.46***.

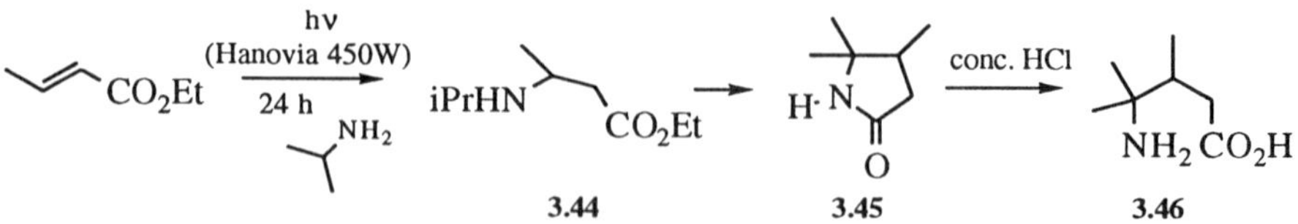

3.44 **3.45** **3.46**

3.2. Carbon Nucleophiles

Certain amine surrogates can add to conjugated systems if they can be converted to a carbanionic species. This section will focus on carbon nucleophiles that possess a nitrogen functional group. Addition of these species to conjugated acid derivatives will give an amino acid, but a second reaction is required to convert the surrogate to the amino group.

3.2.A. Addition to Conjugated Esters

This section is further subdivided by the type of Michael receptor that is used. In this part, conjugated esters will be reacted with various amine surrogates.

3.2.A.i. Nitro Enolates

Nitroalkanes react with a suitable base to form the corresponding resonance stabilized nitro-enolate anion [$^{\ominus}CHR{=}NO_2$], which is a good nucleophile and adds to conjugated systems. The anion of nitromethane, for example, added to methyl acrylate to give ***3.47***.[30] The yield of this reaction was remarkably poor, however. Catalytic hydrogenation of ***3.47*** gave the amino acid but it spontaneously cyclized to 2-pyrrolidinone, and acid hydrolysis was required to give 4-aminobutanoic acid,

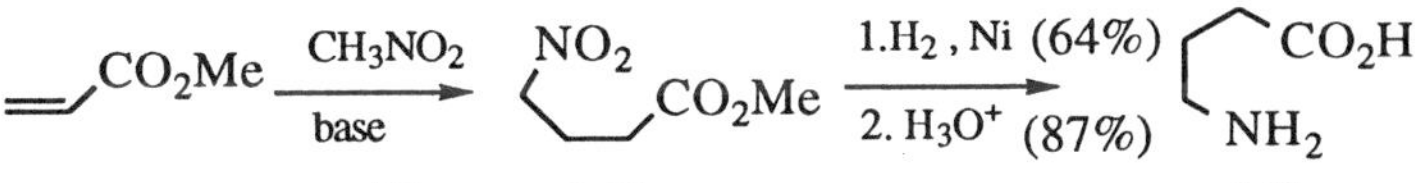

4% **3.47** **3.48**

3.48. The very poor yield of the conjugate addition in this reaction is misleading, since good yields can be obtained with proper choice of base and reaction conditions.

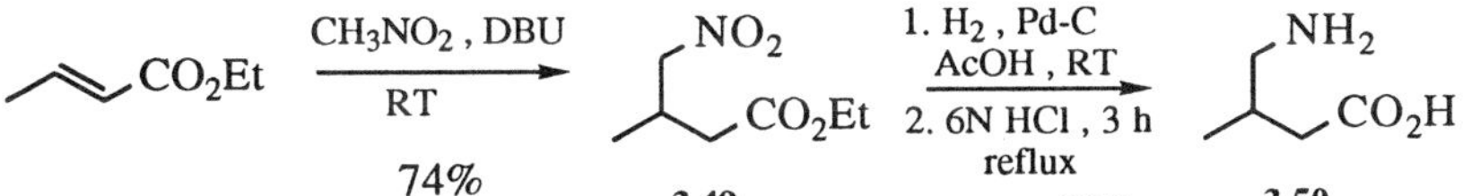

74% **3.49** 83% **3.50**

Reaction of ethyl crotonate and nitroethane in the presence of the basic 1,8-diaza-bicyclo[5.4.0]undec-7-ene (DBU) led to nitro-ester ***3.49*** in 74% yield.[31] Catalytic hydrogenation of the nitro group and acid hydrolysis produced 3-methyl-4-amino-butanoic acid, ***3.50***. Choosing different conjugated esters as starting materials led to formation of 3-ethyl- (59% overall yield); 3-propyl- (51% overall yield); 3-isopropyl- (62% overall yield); 3-butyl- (56% overall yield); 3-*sec*-butyl- (60% overall yield); 3-isobutyl- (58% overall yield); and, 3-*t*-butyl-4-aminobutanoic acid (31% overall yield).[31] The yields observed in this study clearly show that conjugate addition of nitro enolates is a synthetically useful process.

Another example shows the structural variation that is possible. When thiophene-2-carboxaldehyde (***3.51***) reacted with an appropriate ylid, the conjugated ester was formed via Wittig olefination.[3] This conjugated ester reacted with the enolate of nitromethane (Triton B is benzyltrimethylammonium hydroxide) via conjugate addition to give ***3.52***.[32] Subsequent catalytic hydrogenation of the nitro moiety gave a 1:1 mixture (19% overall yield) of ethyl 4-amino-3-(2-thiophene)-butanoate (***3.53***) and 4-(2-thiophene)-2-pyrrolidinone (***3.54***).[32] Basic hydrolysis of this mixture led to isolation of 4-amino-3-(2-thiophene)butanoic acid in 42% yield. The use of

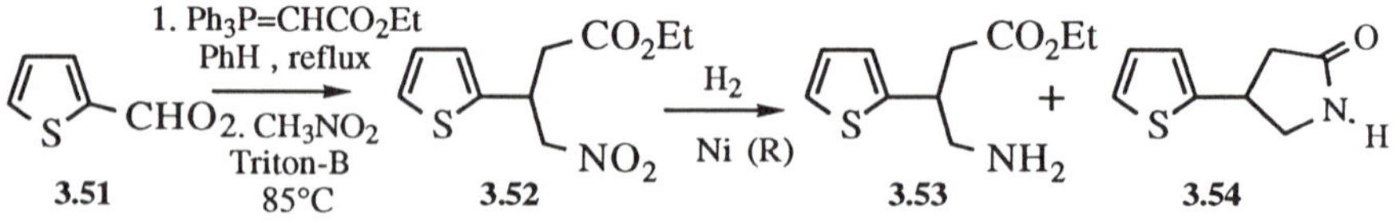

5-methyl-, 3-methyl-, or 5-bromothiophene-2-carboxaldehyde as well as thiophene-3-carboxaldehyde led to formation of several 3-thiophene-4-aminobutanoic acid derivatives.[32]

Another example also used the enolate of nitromethane, but it was generated under much milder conditions using potassium carbonate. In this case, the enolate anion added to ***3.55***, giving ***3.56***. This product was then resolved (see chapter five) using a lipase enzyme leading to ***3.57*** and ***3.58***. Each product was isolated and subjected to catalytic hydrogenation, which gave ***3.59*** and ***3.60*** (each in >98%ee). In this example, the trifluoromethyl group in ***3.55*** enhanced the ability of the conjugated ester to undergo a Michael addition (made it a better Michael acceptor), and the yields are generally good. This is a useful "trick" that can be used. If a conjugated acid derivative gives poor yields, it is possible in some cases to synthetically manipulate it to include an electron withdrawing group that will enhance the conjugate addition and that group can later be removed.

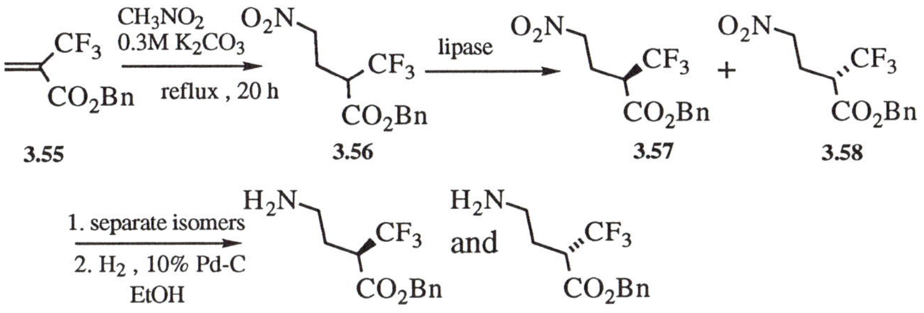

3.55 **3.56** **3.57** **3.58** **3.59** **3.60**

Nitroalkanes are also potential sources of isoxazolidines such as ***3.61***. When nitromethane was treated with triethyloxonium tetrafluoroborate (Meerwein's reagent)[33] and then with methyl crotonate, ***3.61*** was formed.[34] Heating ***3.61*** to 100°C led to a quantitative conversion to ***3.62*** (see ***1.270*** in chapter one, section 1.5.D). Catalytic hydrogenation of the oxime moiety gave methyl 4-amino-2-hydroxy-3-methylbutanote, ***3.63***.

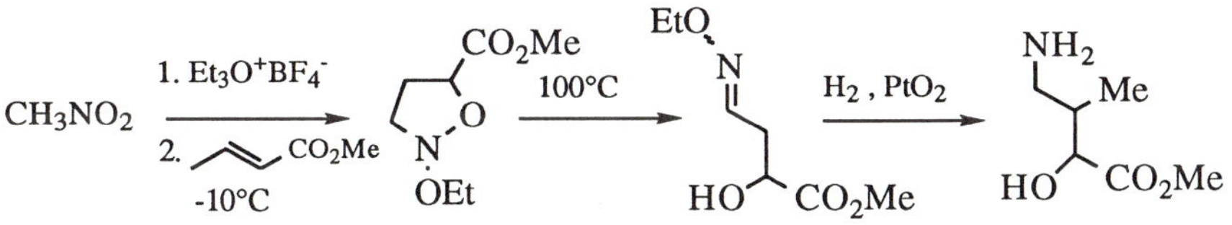

3.61 **3.62** **3.63**

3.2.A.ii. Nitrile Enolates

Nitrile anions are generated by treatment of an alkyl nitrile with a suitable base. These carbanionic reagent can undergo Michael addition with an appropriate conjugated ester to give aryl cyano-esters. Subsequent reduction of the cyano group

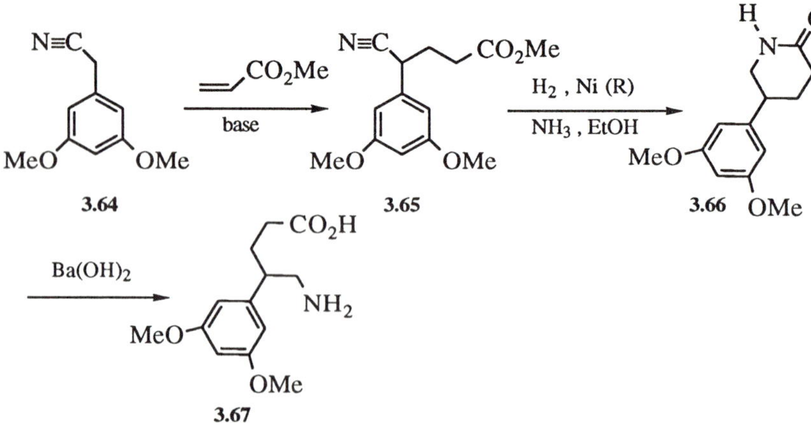

generates the corresponding amino derivatives. Lactams are sometimes formed during this process when the initial amino-ester can form a five- or six-membered ring. This process is illustrated by the reaction of ***3.64*** with acrylonitrile and base to give ***3.65***.[35] Catalytic hydrogenation initially generated an amino-ester, but it spontaneously cyclized to the lactam (***3.66***). A hydrolysis step was therefore required to give 4-(3,5-dimethylphenyl)-5-aminopentanoic acid, ***3.67***.[35]

3.2.A.iii. Radical Addition of Amines

The usual conjugate addition reactions of ammonia or amines to α,β-unsaturated esters can produce the amino acids discussed in this section. Just as amides were seen to undergo addition to conjugated esters in section 3.1.B (see ***3.42*** and ***3.43***), amine stabilized radicals add in a conjugate manner to α,β-unsaturated esters. The product of this reaction is an amino acid derivative. An example is the reaction of butylamine and butyl acrylate, in the presence of di-*t*-butyl peroxide, to give butyl 4-aminoheptanoate, ***3.68***.[36] In a similar manner, cyclohexylamine was heated with

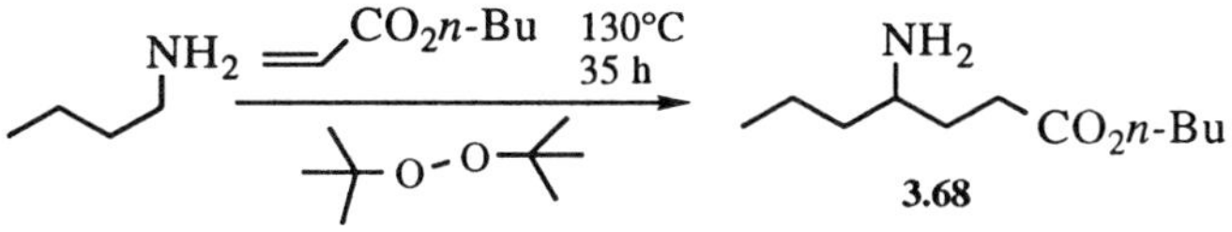

3.68

butyl acrylate in the presence of di-*t*-butyl peroxide to produce 1-amino-1-(butoxy-carbonylethyl)cyclohexane (***3.69***).[36] These radical based procedures produced 4-methyl-4-amino-pentanoic acid and amino-1-(butoxycarbonylpropyl)-cyclohexane derivatives and also 3,4-dimethyl-4-aminohexanoic acid and amino-1-(carboxy-ethyl)cyclohexane derivatives.[36]

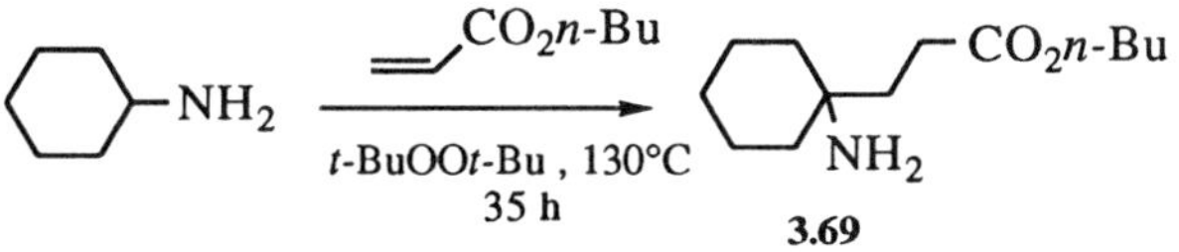

3.69

3.2.B. Addition to Conjugated Nitro Derivatives

Nitro-alkenes are good Michael acceptors and the nitro group can be reduced to an amine. For the production of amino acids, this method requires that the nucleophile contain a carbonyl moiety. When nitroethene was treated with dimethyl malonate under basic conditions, ***3.70*** was obtained.[37] As in other systems of this type, catalytic hydrogenation of ***3.70*** led to an amino-ester that cyclized to 2-pyrro-lidinone derivative (***3.71***). Acid hydrolysis (which was accompanied by decarbox-ylation) gave 4-aminobutanoic acid (***3.48***).[37,29]

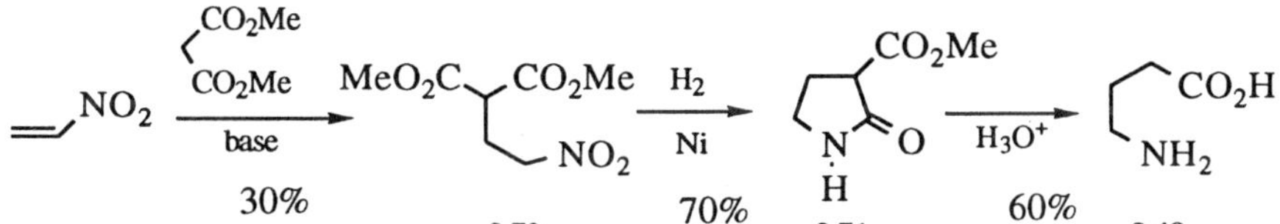

3.70 **3.71** **3.48**

REFERENCES

1 Winnacker, E.L.; Herbst, M.M.; Barker, H.A. *Biochim. Biophys. Acta,* ***1971***, *237*, 280.

2 (a) Walborsky, H.M.; Schwarz, M. *J. Am. Chem. Soc.,* ***1953***, *75*, 3241; (b) Walborsky, H.M.; Baum, M.E. *J. Org. Chem.,* ***1956***, *21*, 538.

3 (a) Wittig, G.; Rieber, M. *Ann.,* ***1949***, *562*, 187; (b) Wittig, G.; Geissler, G. *Ibid,* ***1953***, *580*, 44; (c) Wittig, G.; Schöllkopf, U. *Chem. Ber.,* ***1954***, *87*, 1318; (d) Gensler, W.J. *Chem. Rev.,* ***1957***, *57*, 191 (see p. 218).

4 Kato, K.; Saino, T.; Nishizawa, R.; Takita, T.; Umezawa, H. *J. Chem. Soc., Perkin Trans. I,* ***1980***, 1618.

5 Igarashi, M.; Tamura, M.; Yanagi, M. *Bull.Chem.Soc., Jpn.,* ***1971***, *44*, 3468.

6 Szlompek-Nesteruk, D. *Prezemysl. Chem.,* ***1965***, *44*, 85 [*Chem. Abstr.,* ***1965***, *62*:16366f].

7 Mekhtiev, S.I.; Safarov, Yu.D.; Makhmedov, R.; Tagiev, R.B. *Azeb. Khim. Zh.,* ***1981***, 21 [*Chem. Abstr.,* ***1981***, *95*: 62618c].

8 Mita, R.; Higuchi, C.; Yamaguchi, A.; Murakami, H.; Kato, T. *Eur. Pat. Appl.* 18,177 [*Chem. Abstr.,* ***1981***, *94*; P175548c].

9 Jolidon, S.; Meul, T. *Eur. Patent App.* EP 144,980 [*Chem. Abstr.,* ***1986***, *105*:43325p].

10 Zharov, A.A.; Adadurov, G.A.; Kazakevich, A.G.; Zhulin, V.M.; Yakovleva, I.I. *Izv. Akad. Nauk. SSSR, Ser. Khim.,* ***1984***, 1199 (Engl. p. 1104).

11 Weisel, C.A.; Taylor, R.B.; Mosher, H.S.; Whitmore, F.C. *J. Am. Chem. Soc.,* ***1945***, *67*, 1071.

12 Robinson, J.B.; Thomas, J. *J. Chem. Soc.,* ***1965***, 2270.

13 Frankhauser, R.; Grob, C.A.; Krasnobajew, V. *Helv. Chim. Acta,* ***1966***, *49*, 690.

14 Simonova, N.I. *Tr. Leningr. Inst. Kinoinzhenerov,* ***1961***, 19 [*Chem. Abstr.,* ***1964***, *60*:4127b].

15 Röhnert, H. *Archiv. Pharm.,* ***1963***, *296*, 257.

16 Miyazawa, T.; Akita, E.; Ito, T. *Agr. Biol. Chem.,* ***1976***, *40*, 1651.

17 Corbett, W.M.; McKay, J.E.; Taylor, W. *J. Chem. Soc.,* ***1961***, 5041.

18 Habermehl, G.; Andres, H. *J.L. Ann. Chem.,* ***1977***, 800.

19 Pacheco, H.; Dreux, M.; Beauvillain, A. *Bull. Soc. Chim., Fr.,* ***1962***, 1379.

20 Teotino, U.M. *J. Org. Chem.,* ***1962***, *27*, 1906.

21 (a) Posner, T. *J.L. Ann. Chem.,* ***1912***, *389*, 1 (see p. 86; (b) also see

Basheeruddin, K.; Siddiqui, A.A.; Khan, N.H.; Saleha, S. *Synth. Commun.*, ***1979***, *9*, 705.

22 Perlmutter, P.; Tabone, M. *Tetrahedron Lett.*, ***1988***, *29*,949.

23 Vessiere, R. *Ann. Fac. Sci. Univ. Clermont Chim.*, ***1960***, 1 [*Chem. Abstr.*, ***1963***, *59*: 13811g].

24 (a) Chauvin, A.; Fabron, J.; Ait Yahia, M.O.; Pastor, R.; Cambon, R. *Tetrahedron*, ***1990***, *46*, 6705; (b) Haddach, M.; Pastor, R.; Riess, J.G. *Tetrahedron Lett.*, ***1990***, *31*, 1989.

25 Rich, D.H.; Sun, E.; Singh, J. *Biochem. Biophys. Res. Commun.*, ***1977***, *74*, 762.

26 Furukawa, M.; Okawara, T.; Terawaki, Y. *Chem. Pharm. Bull.* ***1977***, *25*, 1319.

27 Baldwin, J.E.; Harwood,L.M.; Lombard, M.J. *Tetrahedron*, ***1984***, *40*, 4363.

28 Nikishin, G.I.; Mustafaev, R.I.; Petrov, A.D. *Dokl. Akad. Nauk. SSSR*, ***1963***, *152*, 879 (Engl. p. 784).

29 Pfau, M.; Dulou, R. *Bull. Soc. Chim., Fr.*, ***1967***, 3336.

30 Fabrochnyi, B.P.; Shalavina, I.F.; Zurabyan, S.E.; Gol'dfarb, Ya.L.; Kostrova, S.M. *Zh. Org. Khim.*, ***1968***, *4*, 680 (Engl. p. 663).

31 Andruszkiewicz, R.; Silverman, R.B. *Synthesis*, ***1989***, 953.

32 Bethelot, P.; Vaccher, C.; Flouquet, N.; Debaert, M.; Luyckx, M.; Brunet, C. *J. Med. Chem.*, ***1991***, *34*, 2557.

33 Meerwein, H. *Org. Synth. Collect. Vol. 5*, ***1973***, 1080.

34 Sato, H.; Kusumi, T.; Imaye, K.; Kakisawa, H. *Chem. Lett.*, ***1975***, 965.

35 Oine, T.; Kugita, H.; Takeda, M. *Chem. Pharm. Bull.*, ***1963***, *11*, 541.

36 Masuo, F.; Yamamoto, K. *Jpn.* 23,161 ('63) [*Chem. Abstr.*, ***1964***, *60*: P2800g].

37 Smirnova, A.A.; Perekalin, V.V.; Shcherbakov, V.A. *Zh. Org. Khim.*, ***1968***, *4*, 2245 (Engl., p. 2166).

4

Condensations

An important general strategy for the preparation of amino acids involves generating a carbanion from an acid derivative (an enolate anion) and then reacting it with another carbonyl derivative. Such condensation reactions are well known and, when appropriate functionality is present, constitute an important route to non-α-amino acids. The sections in this chapter reflect different reaction types, including a few classical named reactions.

4.1. Knoevenagel and Doebner Condensations

4.1.A. Knoevenagel Condensations

Knoevenagel type[1] condensation reactions generally include the condensation of α-cyano-esters (and some other malonate analogs) with an aldehyde or ketone. An

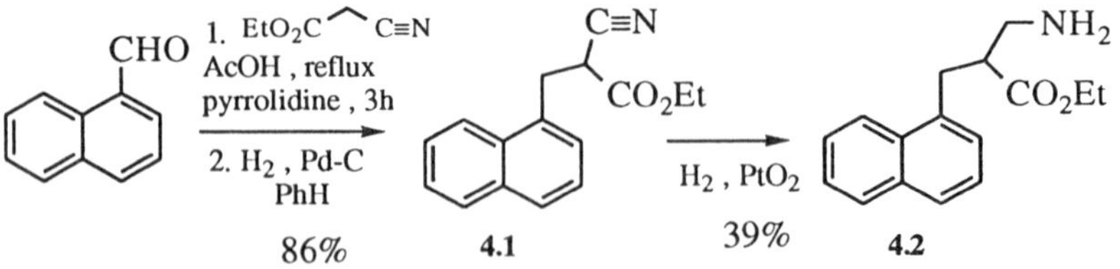

4.1 **4.2**

example is the condensation of naphthalene-1-carboxaldehyde with ethyl cyanoacetate to give ***4.1***.[2] Subsequent catalytic hydrogenation of the nitrile moiety gave ethyl 3-amino-2-(1-naphthylmethyl)propanoate, ***4.2***.

Enolate anions derived from cyano-esters can also be reacted with alkyl halides, as in the conversion of ***4.3*** to ***4.4*** via reaction with sodium hydroxide and 1-bromobutane.[3] Catalytic hydrogenation gave ethyl 2-butyl-2-methyl-3-aminopropanoate, ***4.5***.[3] Similarly prepared were 2,2-dimethyl-3-aminopropanoate, 2-methyl-2-ethyl-3-aminopropanoate, and 2-methyl-3-propyl-3-aminobutanoate in 64%, 71%, and 69% yield, respectively.[3]

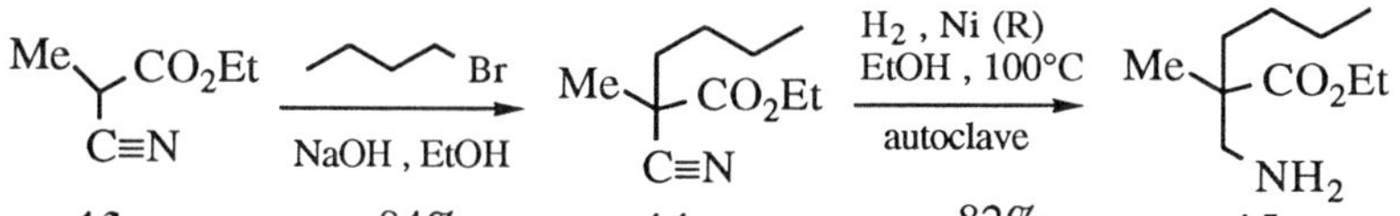

4.3 84% **4.4** 82% **4.5**

Cyano-esters can also be condensed with other nitriles, as in the reaction of methyl α-cyano acetate and α-amino acetonitrile to give ***4.6***.[4] Initial condensation of the cyano-ester enolate with the "other" nitrile moiety gave an iminium salt. Subsequent reaction with aqueous acid led to an imine, which isomerizes to the

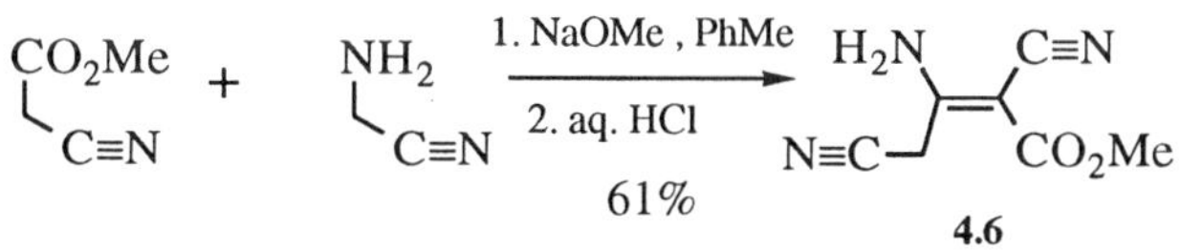

4.6

enamine form (N-C=C-) seen in ***4.6***. A related route reacted methyl α-cyano acetate with 2,2-dichloropropanenitrile.[5] In another variation, methyl α-cyanoacetate was

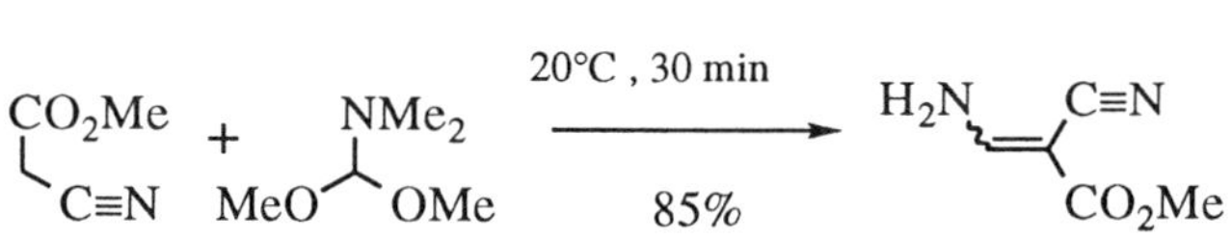

4.7

treated with a protected formamide derivative (the orthoamide shown) and displacement of methoxy was followed by elimination of methanol to give methyl 3-amino-2-cyanoprop-2-enoate, ***4.7***.[6]

4.1.B. Doebner Condensation

When malonic acid is used as the enolate partner, with pyridine as the base, the condensation reaction with aldehydes is known as the Doebner condensation.[7] This is an important route to non-α-amino acids. An example of this condensation when

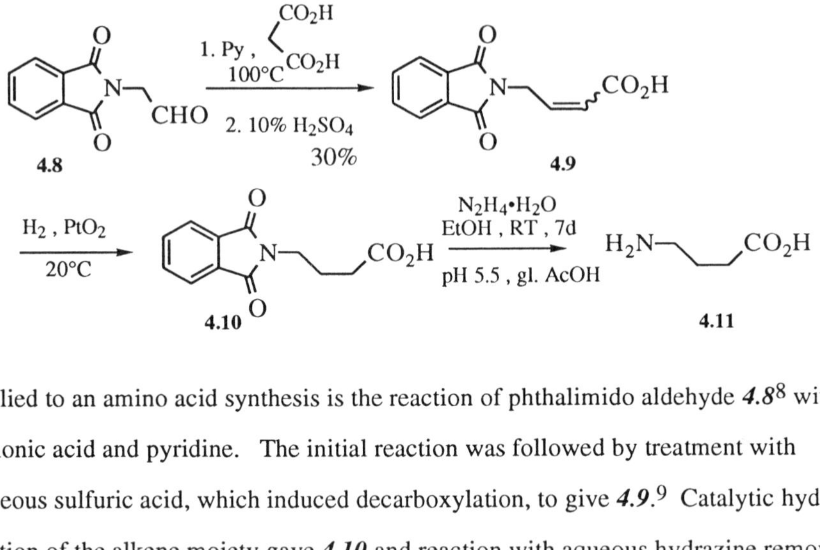

applied to an amino acid synthesis is the reaction of phthalimido aldehyde ***4.8***[8] with malonic acid and pyridine. The initial reaction was followed by treatment with aqueous sulfuric acid, which induced decarboxylation, to give ***4.9***.[9] Catalytic hydrogenation of the alkene moiety gave ***4.10*** and reaction with aqueous hydrazine removed

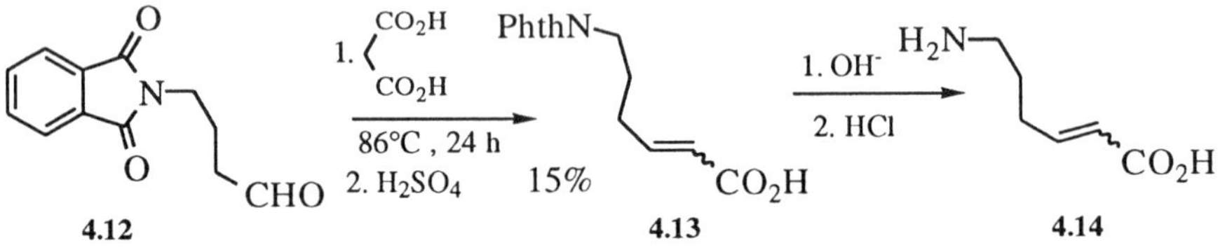

the phthalimidoyl group to give 4-aminobutanoic acid (***4.11***).[9] It is noted that hydrolysis of ***4.9*** followed by treatment with hydrazine gave 4-amino-2-buten-oic acid, but in only 7% yield.[9] The carbon chain of the phthalimidoyl aldehyde can be extended, as in ***4.12***,[10] which gave ***4.13*** in poor yield via Doebner condensation. This was converted to 6-aminohex-2-enoic acid, ***4.14***.[11]

Both aldehydes or ketones can be condensed with malonate derivatives, but ammonia can be used as a base rather than pyridine.[12] Reaction of butanal with ammonia in the presence of the half-ester of malonic acid, for example, gave ethyl

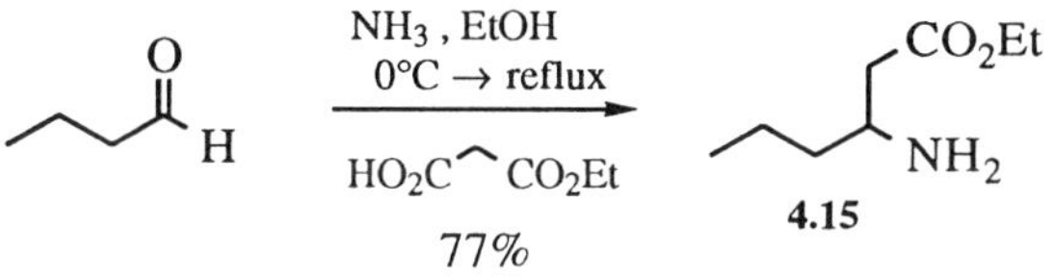

4.15

3-aminohexanoate (***4.15***).[13] 2,2-Disubstituted β-amino acids were also prepared by this method.[12-14] Ketones also react, but the yields are sometimes diminished relative to the analogous reaction with aldehydes. When 3-pentanone was treated

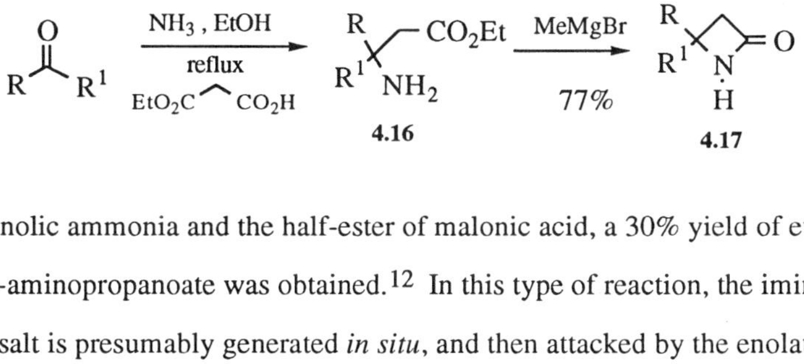

4.16 **4.17**

with ethanolic ammonia and the half-ester of malonic acid, a 30% yield of ethyl 3,3-diethyl-3-aminopropanoate was obtained.[12] In this type of reaction, the imine or iminium salt is presumably generated *in situ*, and then attacked by the enolate anion. When 3-amino-esters such as ***4.15*** are treated with base, a β-lactam can form. An

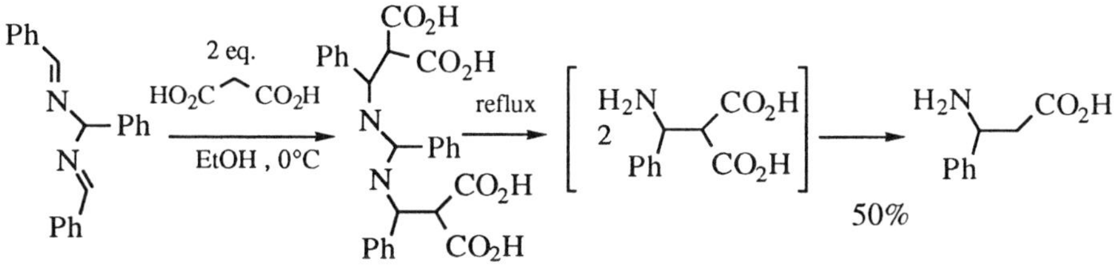

4.18 **4.19** **4.20** **4.21**

example is the reaction of ***4.16*** with methylmagnesium bromide to give β-lactam

4.17.[14] Hydrolysis regenerated to the β-alanine derivative (a 3-amino propanoic acid).

An earlier version of this reaction condensed ***4.18*** and malonic acid to give ***4.19***.[15] Refluxing ***4.19*** led to the monomer (***4.20***), which decarboxylated under the reaction conditions to give 3-amino-3-phenylpropanoic acid, ***4.21***. Doebner condensation with malonic acid itself sometimes leads to very poor yields of product, as seen above. An improvement in the reaction used nitro-ester ***4.22*** rather than malonic acid, and ***4.23*** [ethyl 3-(N-phenylamino)-2-nitroprop-2E-enoate] was produced in good yield.[16]

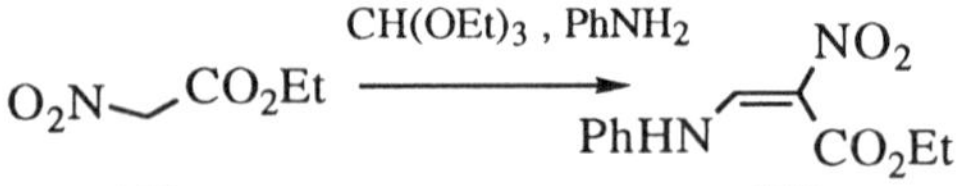

4.22 **4.23**

4.2. Mukaiyama Type Condensations

Although no examples are presented in this chapter where amino acids are prepared via an Aldol condensation, "Aldol-like" reactions have been used. The Mukaiyama Aldol reaction (or just the Mukaiyama reaction)[17] involves the condensation of a silyl enol ether with an aldehyde, catalyzed by metal salts such as $TiCl_4$. This approach has been reported several times for the synthesis of amino acids. This section will show the basic approach, along with some variations that have proved to be useful.

Silyl enol ethers react with imines under Mukaiyama conditions[17] to give amino-esters. Reaction of ***4.24*** with imine ***4.25***, for example, gave a mixture of diastereomeric methyl 3-(N-phenylamino)-2,3-diphenylpropanoates, ***4.26*** and ***4.27*** in a

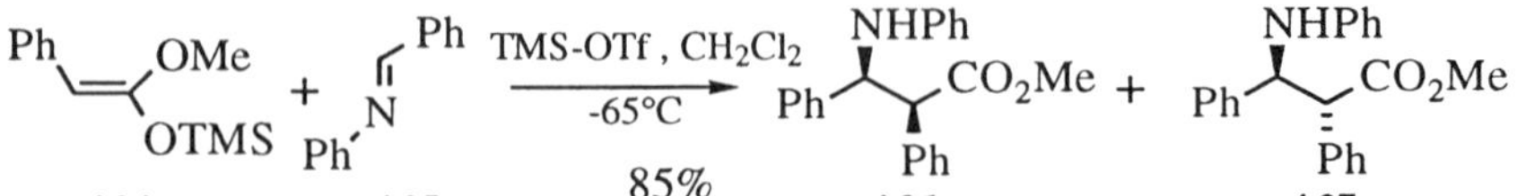

4.24 **4.25** **4.26** **4.27**

ratio of 14:86.[18] The identical reaction of CH_2=C(OMe)OTMS (***4.149***) led to an 85% yield of the *anti*-β-amino acid with none of the *syn*-diastereomer observed. Both 2- and 3-aryl-β-alanine derivatives have been prepared by the Mukaiyama reaction of imines when catalyzed by titanium tetrachloride.[19] The reaction of ketene silyl acetals with O-alkyl imines is another source of aryl-substituted β-alanines.[20]

O-Substituted hydroxylamines (an "O-imine") have been used in this type of reaction. When ***4.28*** reacted with O-silylketene acetal ***4.29***, in the presence of

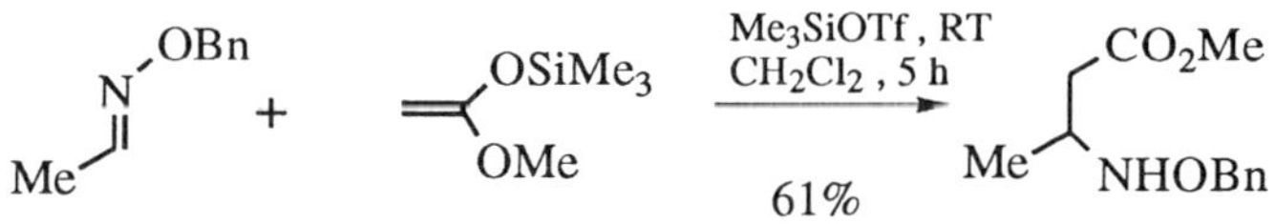

4.28 **4.29** **4.30**

trimethylsilyl triflate, methyl 3-(N-benzyloxy)butanoate, ***4.30*** was obtained.[20] Both 3-amino-2-methylpropanoic acid derivatives and 3-amino-2,2-dimethylbutanoic acid derivatives were prepared by this method. It is noted that in this synthesis, the N-OBn moiety was converted to the amine via acid hydrolysis.

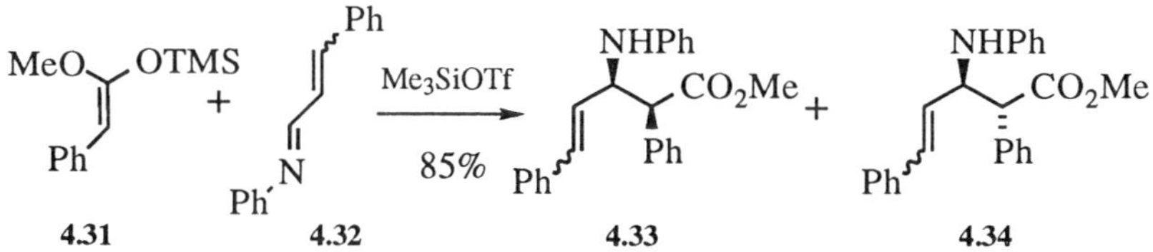

4.31 **4.32** **4.33** **4.34**

Conjugated imines such as ***4.31*** react with silyl enol ethers, in this case catalyzed by trimethylsilyl triflate, to give alkenyl amino acids. The reaction of ***4.31*** and ***4.32***

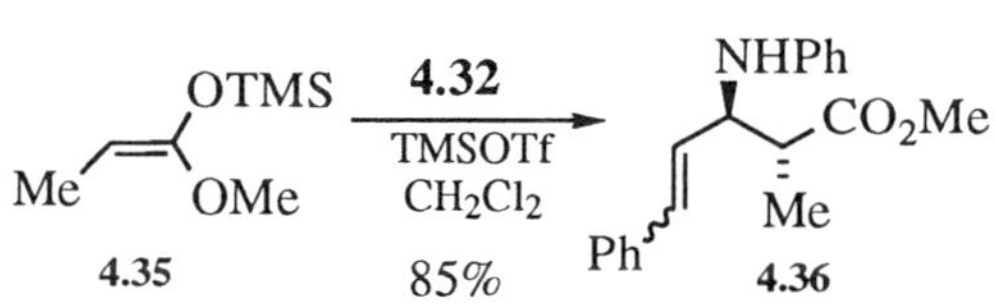

4.35 **4.36**

gave a 14:86 mixture of ***4.33:4.34***.[18] In some cases, the diastereoselectivity (see chapter five) is even higher than in the formation of ***4.33*** and ***4.34***. The reaction of ***4.35*** with ***4.32***, for example, generated diastereomer ***4.36*** as the only product.[18]

An O-trimethylsilyl derivative such as ***4.37*** (derived from methyl 2-methylpropanoate) can react with other amine surrogates, including $(Me_3Si)_2N$-CH_2OMe to give ***4.38***.[21] Removal of the N-TMS group gave methyl-3-amino-2,2-dimethylpropanoate, ***4.39***. The use of methyl propanoate in this sequence gave 2-methyl-3-aminopropanoic acid and cyclopentane carboxylic acid led to 1-amino-1-carbomethoxycyclopentane.

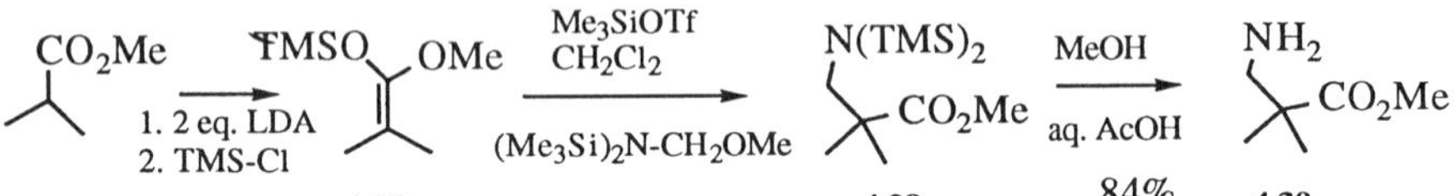

4.37 **4.38** **4.39**

Another amine surrogate is perhydro 1,3,5-triazine ***4.40***, which reacted with ***4.37*** to give methyl 2,2-dimethyl-N-benzyl-3-aminopropanoate (***4.41***).[22] The analogous N-isopropyl, N-ethyl, and N-phenyl derivatives were also prepared, as were the 2-methyl propanoate derivatives.[21]

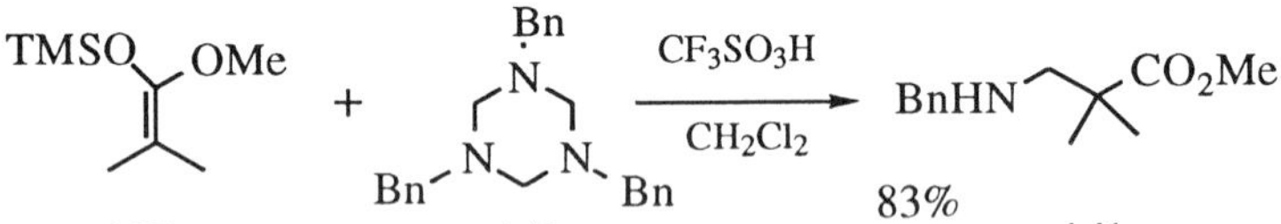

4.37 **4.40** **4.41**

4.3. Enolate Anion Reactions

The reactions discussed in section 4.1 obviously describe enolate anion reactions. The reactions in this section involve malonate derivatives that react with bases such as sodium hydride or lithium dialkylamides to generate the malonate anion, a highly stabilized enolate. This section also includes reactions of enolate anions derived from mono-esters and other acid derivatives.

4.3.A. Malonate Derivatives

Rather than pyridine or ammonia, bases such as sodium hydride or sodium ethoxide can be used to generate a malonate enolate anion. Such enolates are easier to control, from a synthetic viewpoint, and allow a wider range of reactions. For this reason, they are collected into this section. An example is the reaction of the sodium enolate of diethyl 2-methylmalonate with the bromine moiety in phthalimide derivative ***4.42***. This displacement reaction was followed by removal of the phthalimidoyl group, hydrolysis of the esters and decarboxylation to give 2-methyl-6-aminohexanoic acid (***4.43***).[23] Phthalimide ***4.42*** was prepared by reaction of 1,4-dibromobutane with potassium phthalimide.[23] The length of the carbon chain in the

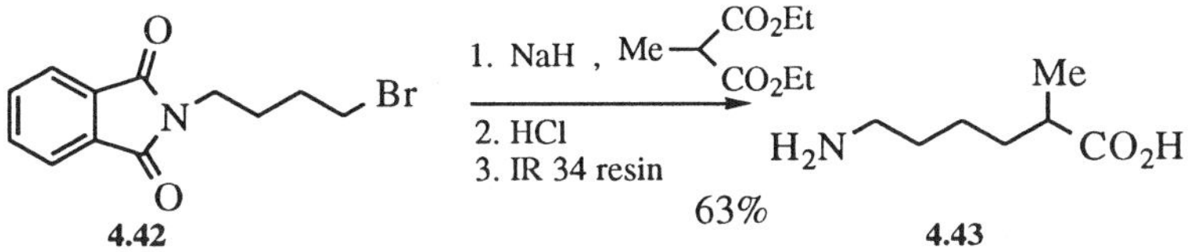

4.42 **4.43**

final amino acid is dependent on the halo-phthalimide precursors. N-chloromethyl phthalimide ***4.44***,[24] for example, reacted with the sodium enolate of diethyl 2-ethylmalonate to give ***4.45***.[25] Acid hydrolysis and accompanying decarboxylation gave a

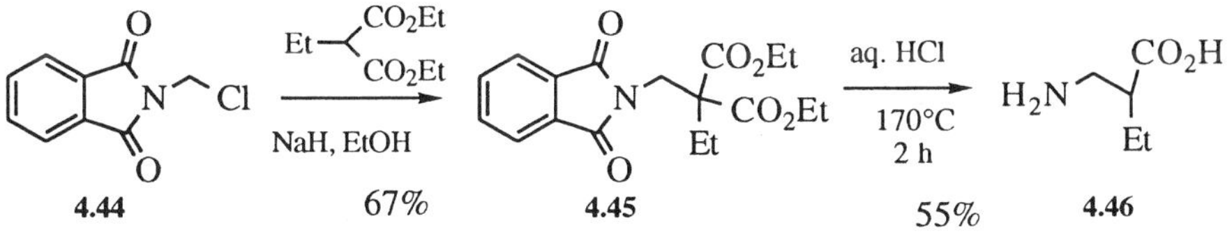

4.44 **4.45** **4.46**

55% yield of 3-amino-2-ethyl-propanoic acid, ***4.46***.[25] 3-Amino-2-methylpropanoic acid was prepared in an identical manner by using diethyl 2-methylmalonate as a precursor.[25]

Dibenzyl malonate was methylated via reaction of its sodium enolate with

iodomethane, and then condensed with α-bromo-ketone ***4.47*** to give ***4.48***.[26] Catalytic hydrogenation removed the benzyl group from the esters moieties and heating of the resulting dicarboxylic acid led to decarboxylation and formation of 6-(N-Boc amino)-3,8-dimethyl-5-oxononanoic acid, ***4.48.***

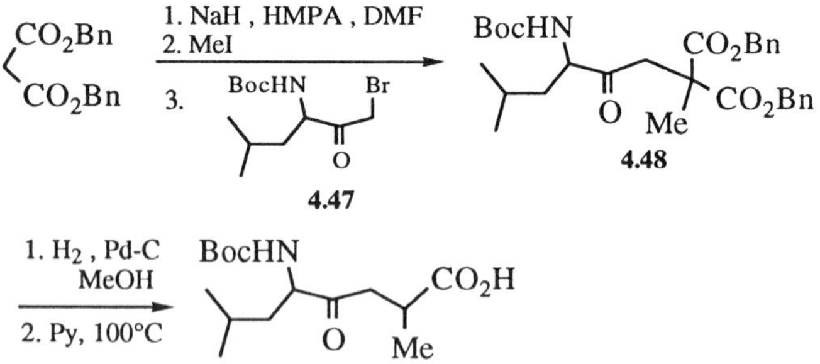

4.47 **4.48** **4.49**

Fluorine can be introduced into an amino acid via this strategy. Condensation of ***4.50*** with chlorodifluoromethane, for example, gave ***4.51***.[27] Bromination facilitated incorporation of nitrogen via reaction with phthalimide (see chapter one, section 1.1.B), and subsequent deprotection and decarboxylation gave 5-(N-phthalimidoyl)-2-(difluoromethyl)pent-3-enoic acid, ***4.52***.[27] This latter sequence was also used in a synthesis of 4-amino-2-fluorobut-2-enoic acid.[28]

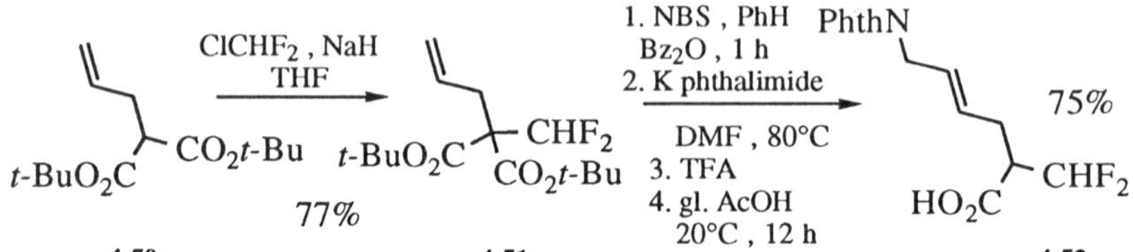

4.50 **4.51** **4.52**

Selectivity for one functional group in preference to another can be used to good advantage in reactions with malonate anions. One example is the reaction of diethyl 2-phenyl diethyl malonate and 2-chloroacetonitrile under basic conditions to give ***4.53***.[29] In this case, the malonate anion displaced chlorine rather than reacting with the cyano moiety. Catalytic hydrogenation of the cyano group and hydrolysis of the ester was followed by decarboxylation. The final product was 2-phenyl-4-aminobutanoic acid, ***4.54***. A similar sequence with 3-chloropropanenitrile led to 2-phenyl-5-

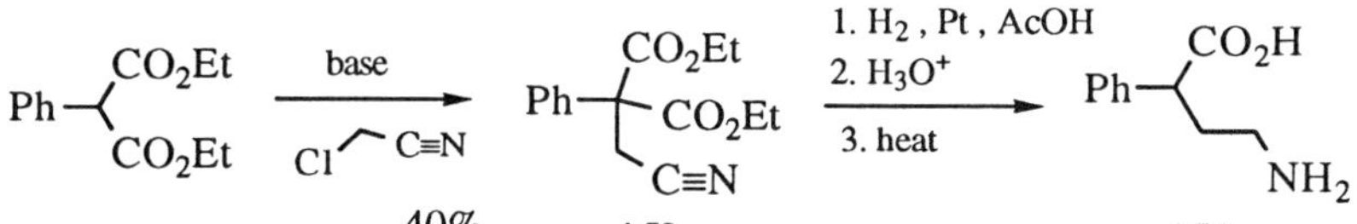

4.53 **4.54**

aminopentanoic acid.[29] In a related example, malonate anion reacted with trifluoroacetonitrile at the cyano moiety rather than at the less labile C-F moiety. Condensation of dimethyl malonate and trifluoroacetonitrile gave an ***4.55***, for example, via initial attack at the nitrile.[30] Subsequent treatment with chlorine gave methyl 4,4,4-trifluoro-3-aminobut-2-enoate (***4.56***).

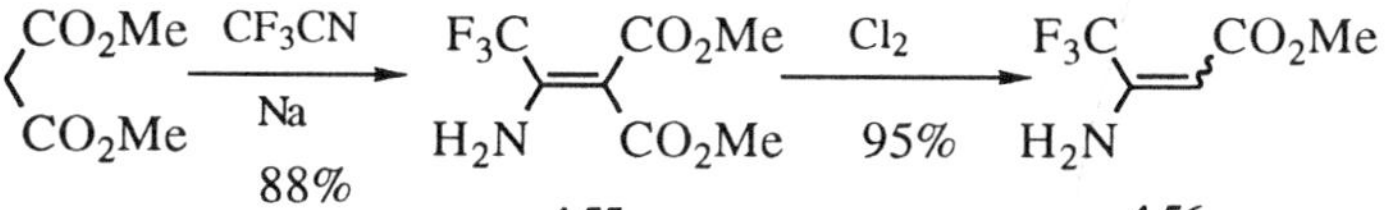

4.55 **4.56**

Several electrophilic groups can be opened by malonate anions. In one interesting example, aziridine derivatives were opened, as in the condensation of ***4.57*** with diethyl 2-fluoromalonate to give ***4.58*** . Subsequent treatment with aqueous HCl gave 4-amino-2-fluorobutanoic acid, ***4.59***.[31] Another strategy using a different

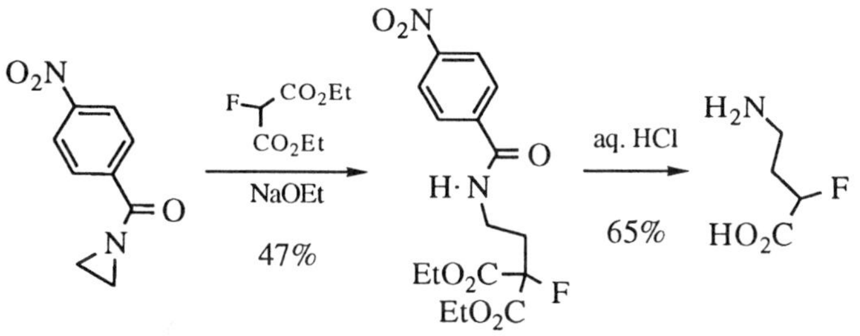

4.57 **4.58** **4.59**

electrophilic group for the preparation of fluorinated amino acids was shown in this same work. The conjugate addition of the sodium enolate of diethyl 2-fluoromalonate to 2-nitro-1-butene generated ***4.60***.[31] Acid hydrolysis followed by reduction of the nitro group with iron metal in HCl gave 4-amino-2-fluorohexanoic acid (***4.61***).[31]

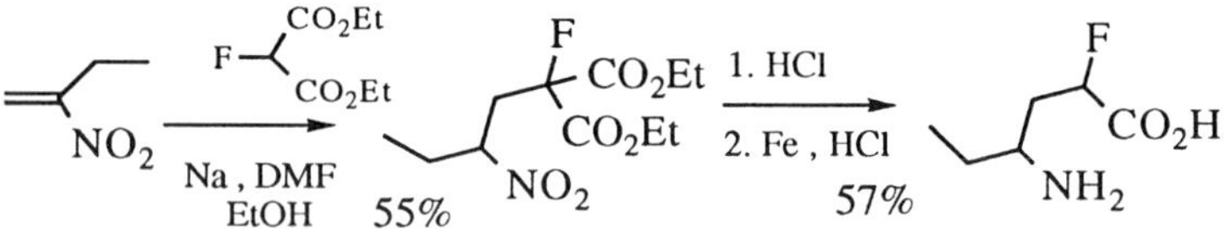

4.60 **4.61**

Another conjugate addition strategy used the sodium enolate of ***4.62*** in a reaction with ketene ***4.63***[32] to give ***4.64***.[33] In this particular example, both the amino and the carboxyl moieties of the final amino acid were present in the starting material (***4.62***). Aqueous acid hydrolysis of the ester groups and the amide was accompanied by decarboxylation to give 2-amino-3-methylenebutanedioic acid, ***4.65***.[33]

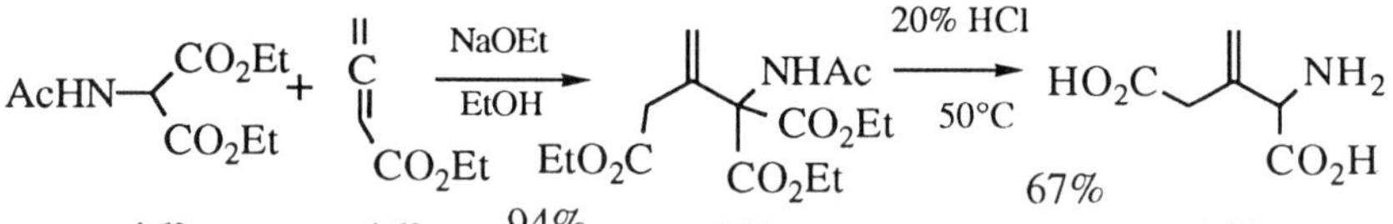

4.62 **4.63** **4.64** **4.65**

Clearly related to malonate derivatives are acetoacetic acid esters. In general, β-keto-esters are convenient sources of oxo-alkenyl amino acids (as described in

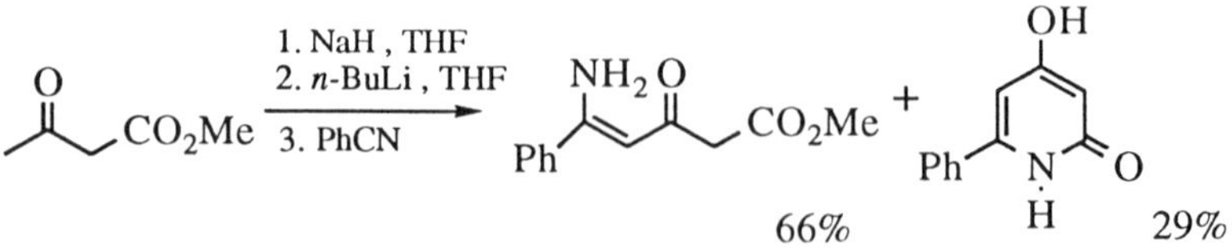

4.66 **4.67**

chapter one, section 1.1.C.ii). An example is presented here for comparison with the malonate anion reactions. Formation of the *bis*-anion of methyl acetoacetate allowed condensation with benzonitrile (at the methyl carbon) to give ***4.66*** as the major product, along with some of the cyclized product, pyridone ***4.67***.[34]

4.3.B. Mono-Ester Enolates

This section will describe reactions of ester enolates that are generated using "modern" enolate anion technology. The use of lithium diisopropylamide and related "non-nucleophilic" bases allows a measure of control in ester condensation reactions that is unmatched by the more traditional techniques described in previous sections.

4.3.B.i. Reaction with Halides

As applied to an amino acid synthesis, the ester enolate must react with another molecule that contains a nitrogen moiety. In one example, methyl 2-methylpropanoate was treated with lithium diisopropylamide and then with 4-bromobutanenitrile to give ***4.68***.[35] Catalytic hydrogenation of the cyano group gave methyl 6-amino-2,2-dimethylhexanoate (***4.69***).[35a] In this case, the nitrile was the amine surrogate and the ester was the acid precursor.

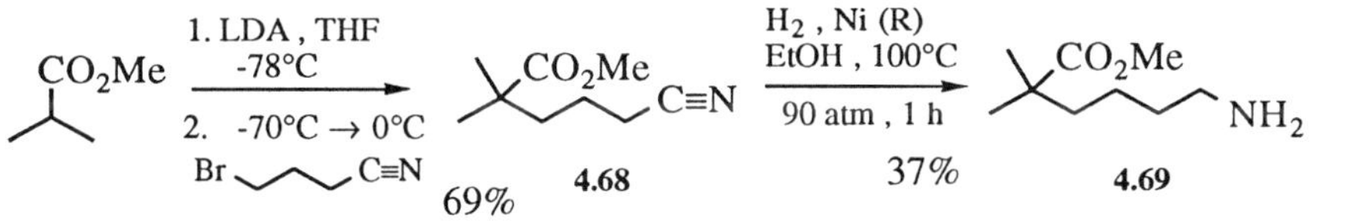

4.3.B.ii. Reaction with Imine-type Reagents

In one approach, an imine moiety was generated *in situ* by reaction of a methoxy carbamate with lithium diisopropylamide. This imine was then condensed with an ester enolate to give the amino-ester, protected as a carbamate. Reaction of ***4.70*** with lithium diisopropylamide gave imine ***4.71***, for example. Subsequent reaction with the metal enolate shown (M = Li, Na) gave a mixture of diastereomeric ethyl 3-(N-methylcarbamoyl)-2-ethyl-4-methylpentanoates, ***4.72*** and ***4.73*** in a 7:3 ratio.[21] Substituted butanoic acid derivatives were also prepared in this manner, as well as

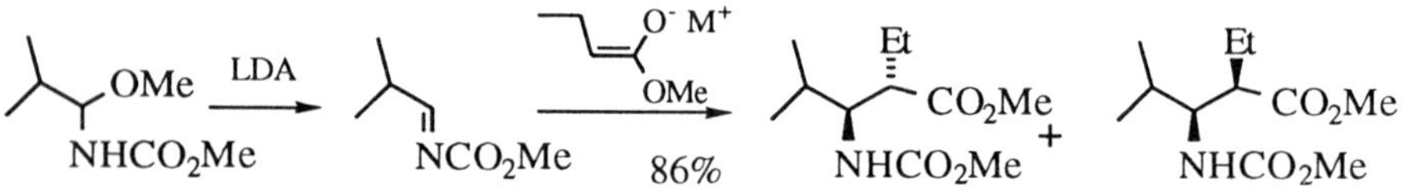

4.70 **4.71** **4.72** **4.73**

2-ethyl-, 2-isopropyl-, and 2-phenyl derivatives.[35,36] The enolate of *t*-butyl acetate reacted with ***4.71*** to give the analogous amino acid. If the methoxy carbamate also contains an alkenyl group, alkenyl amino acids can be prepared by this method. The condensation reaction of ***4.74*** with the lithium enolate of ethyl acetate, for example, gave methyl 3-(N-methylcarbamoyl)hex-5-enoate, ***4.75***.[37]

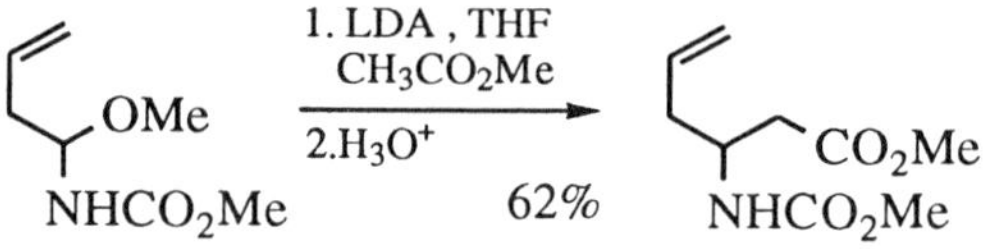

4.74 **4.75**

Schiff bases are readily prepared by reaction of an aldehyde with an amine. The imines can then be condensed with an ester enolate to generate an amino-ester

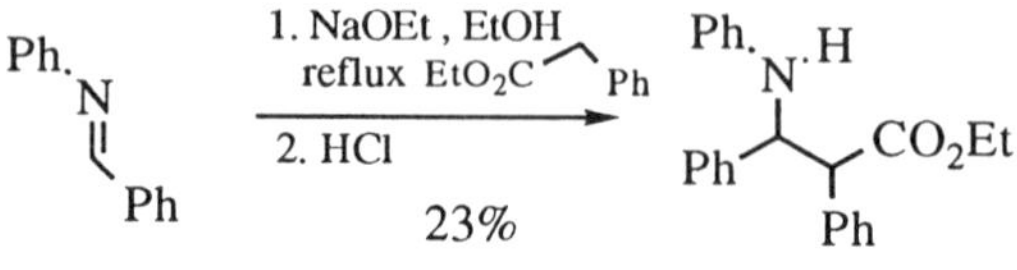

4.76 **4.77**

directly. An example is the condensation of Schiff base ***4.76*** with ethyl phenyl-acetate to give ethyl 3-(N-phenylamino)-2,3-diphenyl-3-aminopropanoate (***4.77***).[38] The N-(4-chlorophenyl) derivative was also prepared by this method.

This fundamental approach can be "turned around" so the acid moiety is incorporated into the amine-bearing substrate, by using a fluorinated amine such as ***4.78***. Conversion to imine ***4.79*** and condensation with the lithium enolate of methyl 2-methylpropanoate, for example, led to methyl 3-(N-phenylthioamino)-4,4,4-trifluoro-2,2-dimethylbutanoate, ***4.80***.[39]

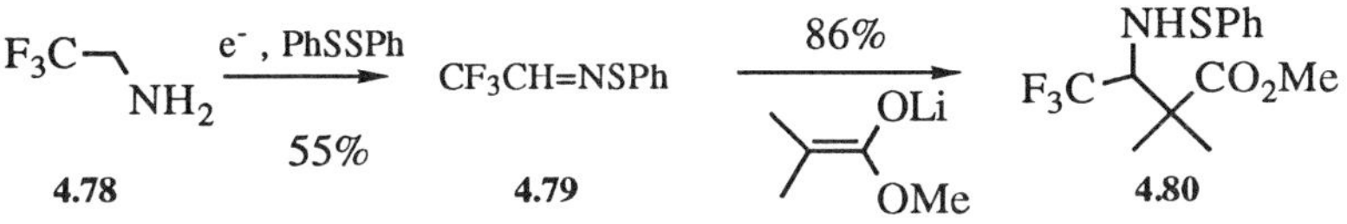

4.78 **4.79** **4.80**

4.3.B.iii. Reaction with Nitriles

The condensation reaction of ester enolates with nitriles are an important general source of alkenyl-β-amino acids. In a simple example, the reaction of *t*-butyl acetate with magnesium diisopropylamide and then propanenitrile led to *t*-butyl 3-amino-hex-2-enoate, ***4.81***.[40] In addition, *t*-butyl 3-aminobut-2-enoate (66%), 3-amino-4-methylpent-2-enoate (74%), 3-amino-4,4-dimethylpent-2-enoate (43%), 3-amino-4-phenylpent-2-enoate (25%) and other 4-aryl and 2-alkyl-substituted alkenylamino

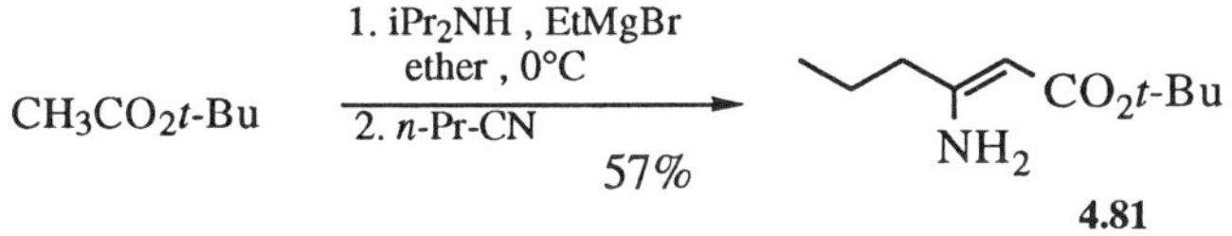

4.81

acids were prepared. Aromatic nitriles can be condensed with ester enolates to form aryl substituted alkenyl amino acids. The enolate of *t*-butyl acetate reacted with benzonitrile, for example, to give *t*-butyl 3-amino-3-phenylprop-2-enoate, ***4.82***.[41]

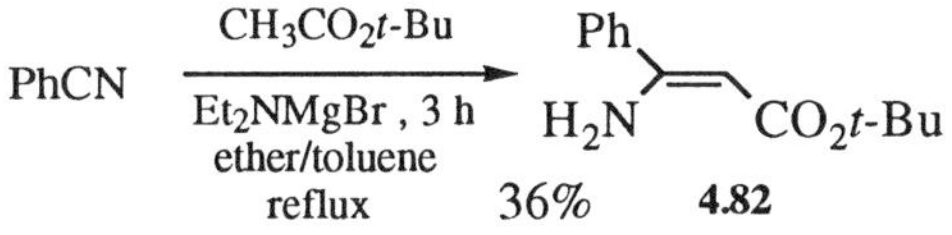

4.82

More "sophisticated" nitriles can be used to prepare highly functionalized amino acids. The reaction of ***4.83*** with enolate ***4.84***, for example, gave ***4.85***.[40c] This particular alkenyl amino acid was used to synthesize 4-amino-2-(5H)-furanones.

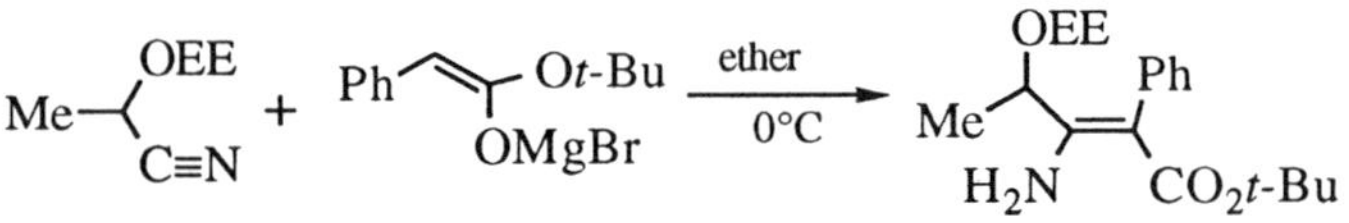

4.83 **4.84** **4.85**

4.3.C. Carbanions of Carboxylic Acids and Acid Surrogates

In previous sections, ester enolates were used as or carboxyl synthon. Modern techniques allow generation of both mono- and di-anions of carboxylic acids (see the formation of ***4.66*** in section 4.3.A). These enolate anions undergo C-alkylation and C-condensation reactions. Such compounds are also carboxyl surrogates that have proved to be useful. This section gives examples of these reactions when applied to the synthesis of amino acids.

4.3.C.i. Carboxylic Acid Enolates

The dianion of diphenylacetic acid (or 2-phenylpropanoic acid) was generated by reaction of sodium naphthalenide. This C-nucleophile opened the three-membered ring of N-benzylaziridine to give 4-(N-benzoylamino)-2,2-diphenylbutanoic acid (or 4-(N-benzoylamino)-2-methyl-2-phenylbutanoic acid after hydrolysis (***4.86***, R = Ph, Me respectively).[42]

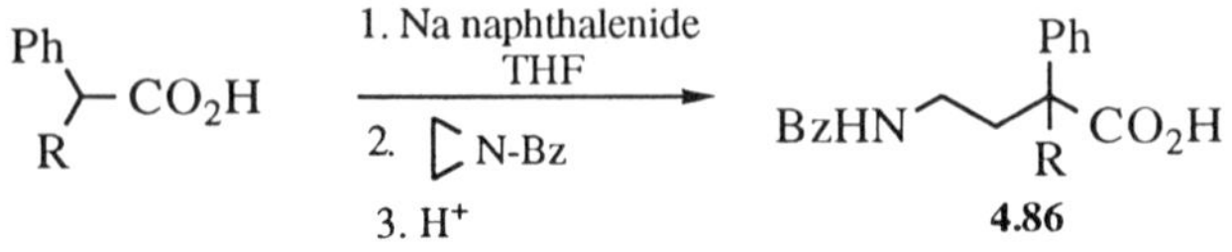

4.86

Although malonate derivatives were described previously, one example is included here to demonstrate their use as a carboxyl enolate. Reaction of ***4.87*** with carbonyl diimidazole (CDI, see chapter two, section 2.4) gave the acyl imidazole (a very highly activated acyl derivative), which reacted with the magnesium enolate of mono-ethyl malonate to give ***4.88***.[43] In this instance, the starting material happened to have a protected alcohol moiety incorporated into its structure. Subsequent treatment of the β-keto-ester moiety (see chapter one, section 1.1.C.ii) with ammonium acetate gave the protected hydroxy derivative [ethyl 5-O-(B-4,5-diphenyl-2-oxazoloyl)-3-aminopent-2-enoate, ***4.89***].[43]

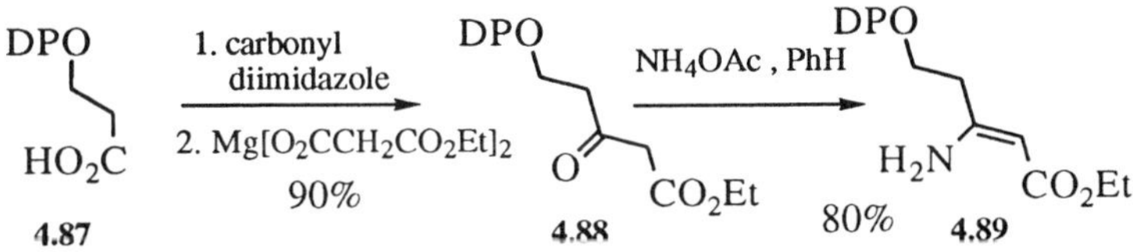

DP = B-4,5-diphenyl-2-oxazoloyl

4.3.C.ii. Anions of Carboxyl Surrogates

An interesting acid surrogate is oxazolidine ***4.90*** (developed by Meyers),[44] which reacted with methoxy carbamate ***4.70*** (see section 4.3.B.ii) in the presence of lithium diisopropylamide and $TiCl(OiPr)_3$ to give ***4.91***.[21] Acid hydrolysis converted ***4.91*** to an acid and esterification gave methyl 3-(N-methylcarbamoyl)-4-methylpentanoate, ***4.92*** (90% ee, R; see chapter five for other enantioselective reactions).[21] In addition, the 3-aminobutanoate derivative was prepared by this method, as was the 3-amino-hex-5-enoic acid derivative.

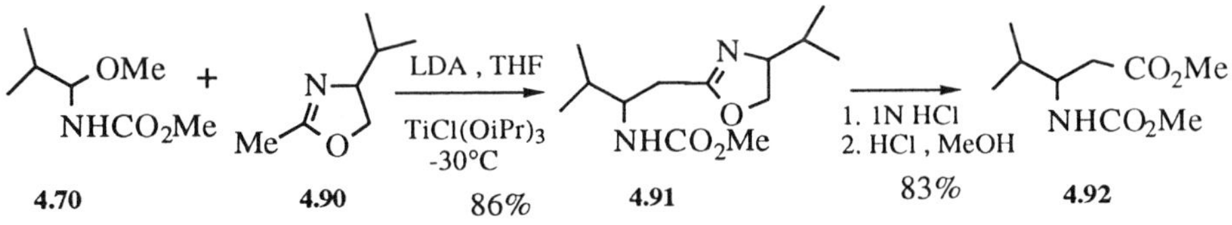

4.3.D. Nitrile Enolates

Another type of carboxylate surrogate is the nitrile moiety, although it can also be converted to an aminomethyl group. This section will include amino acids that are prepared by methods using the enolate of alkyl nitriles in alkylation reactions or condensation reactions.

The sodium salt of succinimide reacts with α,ω-dibromides such as ***4.93*** to give ***4.94***.[45] In a second step, the enolate of α-phenyl nitrile displaced the bromide moiety in ***4.94*** to give ***4.95***. Acid hydrolysis converted the nitrile to an acid and the imide to an amine, leading to ***4.96***.[45a] Reaction of ***4.94*** (n = 4) with the indicated nitrile (R = H) led to a 57% yield of 2-phenyl-6-aminohexanoic acid (***4.96a***).[45] Similarly prepared were 2-phenyl-7-aminoheptanoic acid (***4.96b***) in 68% yield; 2-phenyl-8-aminooctanoic acid (***4.96c***) in 68% yield; 2-phenyl-10-aminodecanoic acid (***4.96d***) in 44% yield; and, 2-phenyl-12-aminododecanoic acid (***4.96e***) in 26% yield, all with R = H.[45] Many other nitriles were used in this sequence to give a wide variety of phenyl

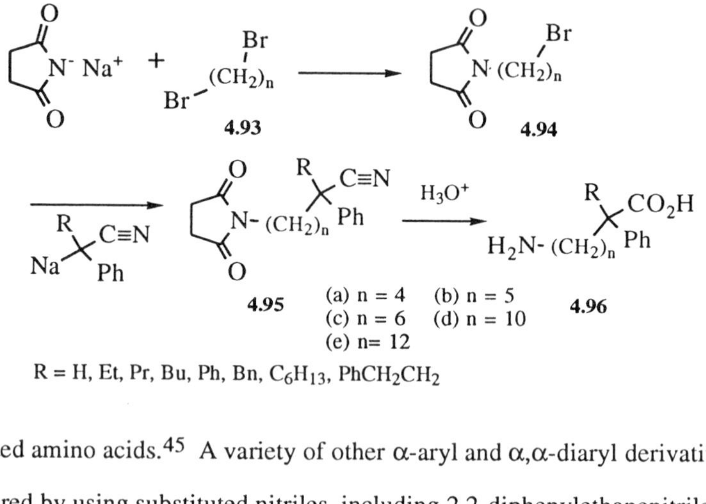

4.93 **4.94** **4.95** **4.96**

(a) n = 4 (b) n = 5
(c) n = 6 (d) n = 10
(e) n= 12

R = H, Et, Pr, Bu, Ph, Bn, C_6H_{13}, $PhCH_2CH_2$

substituted amino acids.[45] A variety of other α-aryl and α,α-diaryl derivatives can be prepared by using substituted nitriles, including 2,2-diphenylethanenitrile, 2-phenylbutanenitrile, 2-phenylpentanenitrile, 2,3-diphenylpropanenitrile, etc.

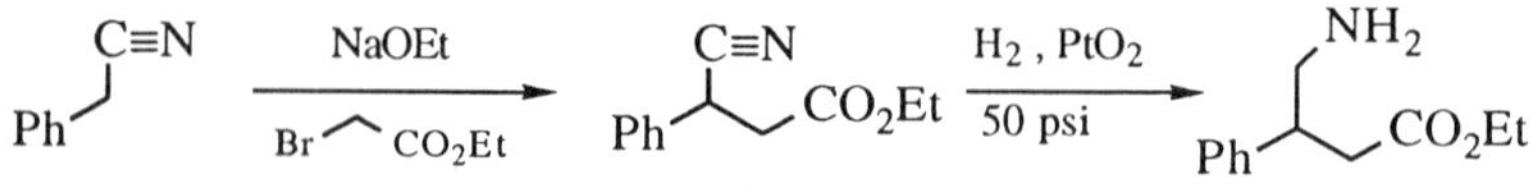

4.97 **4.98**

A nitrile enolate can react with a carboxyl-bearing electrophile, allowing the nitrile to be an amine surrogate. Phenyl acetonitrile reacted with ethyl bromoacetate after conversion to the enolate by reaction with sodium ethoxide, for example, to give ***4.97***, Catalytic hydrogenation of the cyano group gave ethyl 4-amino-3-phenyl-butanoate, ***4.98***.[46]

Nitrile enolates can also react with carbonyl derivatives such as oxalate, which serves as a carboxyl surrogate. Reaction of 2-phenylacetonitrile and diethyl oxalate, in the presence of sodium amide, gave ***4.99***.[47] Catalytic hydrogenation of the cyano group and hydrolysis led to ethyl 2-phenyl-3-aminopropanoate, ***4.100***. In this case also, the nitrile moiety was an amine surrogate. Other α-aryl acetonitrile derivatives

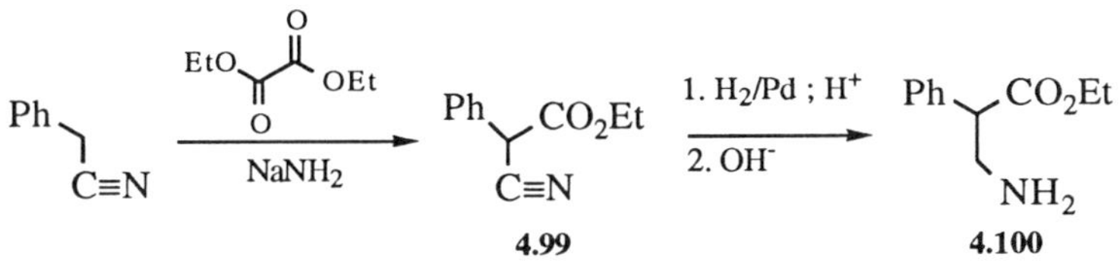

4.99 **4.100**

were similarly converted to the following β-alanine derivatives: 2-(3-chlorophenyl)-; 2-(4-chlorophenyl); 2-(2-methylphenyl)-; 2-(4-methylphenyl)-; and, 2-(3-methyl-phenyl)-3-aminopropanoates.[48]

A specialized reaction involves condensation of diphenylacetonitrile (***4.101***) with formaldehyde, and then reaction with ***4.102 in situ*** to give ***4.103***.[49] Acid hydrolysis gave 4-(N,N-dimethylamino)-2,2-diphenylpentanoic acid, ***4.104***.

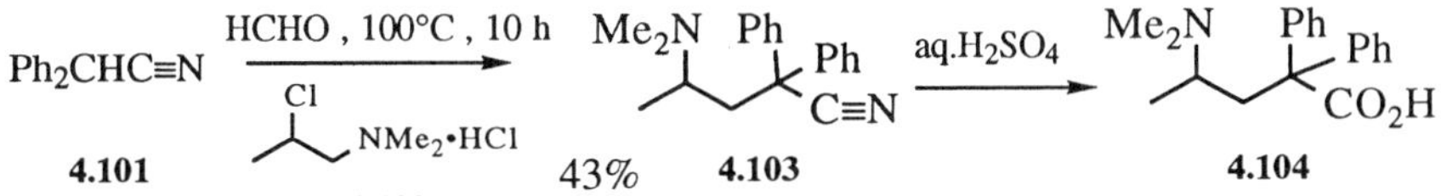

4.101 **4.102** **4.103** **4.104**

4.3.E. Nitroalkane Enolates

This section will describe syntheses that use an enolate anion derived from nitro-alkyl derivatives. In general, these anions will react with a species bearing a carboxyl group or a carboxyl surrogate. Later in the reaction sequence, the nitro group will be reduced to an amine (see chapter 1, section 1.1.B.iii).

An example of this process is the reaction of nitromethane with sodium methoxide to give the nitro enolate. Subsequent reaction with methyl 3-bromopropanoate gave a nitro-ester, and catalytic hydrogenation led to 4-aminopentanoic acid (***4.105***) in 19% overall yield.[50] The enolate of nitroethane reacted similarly with methyl

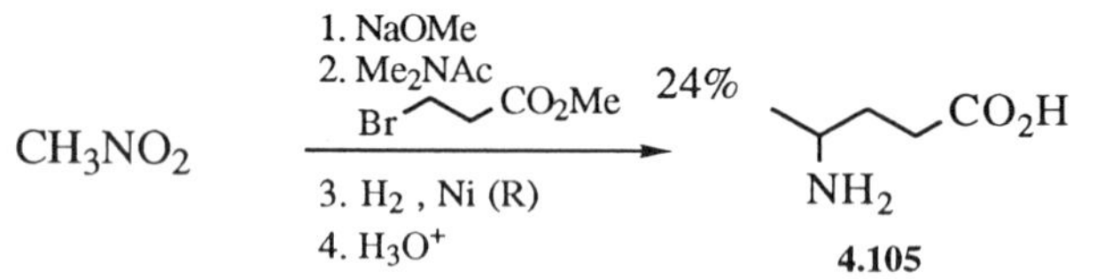

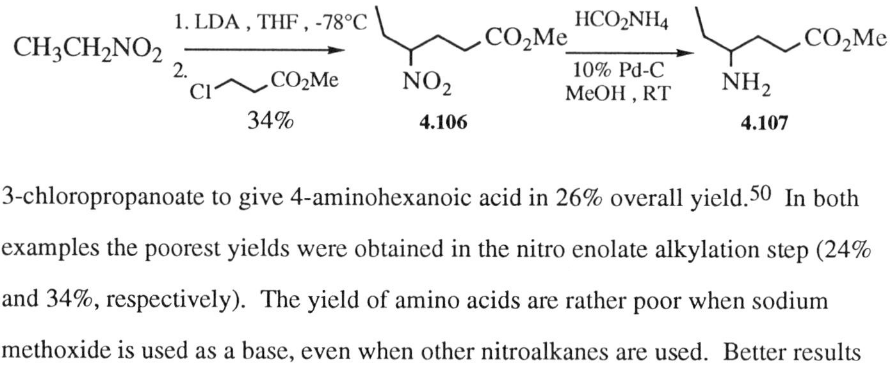

3-chloropropanoate to give 4-aminohexanoic acid in 26% overall yield.[50] In both examples the poorest yields were obtained in the nitro enolate alkylation step (24% and 34%, respectively). The yield of amino acids are rather poor when sodium methoxide is used as a base, even when other nitroalkanes are used. Better results were obtained when lithium diisopropylamide (LDA) was used to generate the nitro enolate. Reaction of nitroethane with LDA followed by quenching with methyl 3-chloropropanoate gave ***4.106*** and reduction with ammonium formate gave an 81%

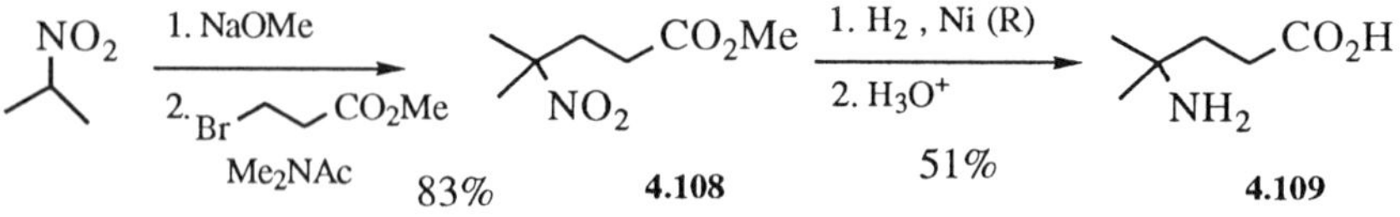

overall yield of ***4.107***.[50] Enolates derived from secondary nitro compounds can give better yields in alkylation reactions with unhindered halides than do primary nitro compounds. The enolate of 2-nitropropane, for example, reacted with methyl 3-bromopropanoate to give methyl 4-methyl-4-nitropentaoate, ***4.108***. Catalytic hydrogenation with Raney nickel led to 4-amino-4-methylpentanoic acid, ***4.109.*** Enolate anions of benzylic nitro compounds also tend to give better yields in alkylation reactions. Condensation of α-nitrotoluene with methyl bromoacetate and N,N-dimethyl-acetamide to give a 68% yield of methyl 2-hydroxy-3-nitro-3-phenylpropanoate. Hydrolysis and catalytic hydrogenation of the nitro group gave a 24% yield of 3-amino-2-hydroxy-3-phenylpropanoic acid.

Another related method uses the anion of nitroalkanes in a reaction with halo-esters. Nitromethane reacted with lithium diisopropylamide to form the nitro enolate and then with methyl 3-chloropentanoate to give ***4.110***.[51] Reduction of the nitro group with ammonium formate gave methyl 2-phenyl-3-aminopropanoate, ***4.111***.

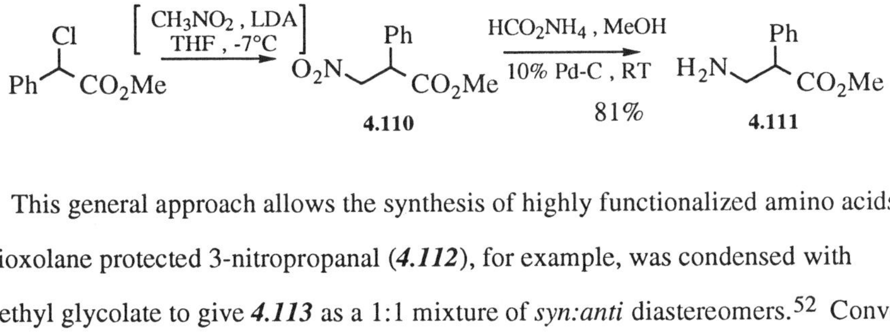

4.110 **4.111**

This general approach allows the synthesis of highly functionalized amino acids. Dioxolane protected 3-nitropropanal (***4.112***), for example, was condensed with methyl glycolate to give ***4.113*** as a 1:1 mixture of *syn:anti* diastereomers.[52] Conversion of the alcohol to its dimethyl-*t*-butylsilyl derivative was followed by hydrogenation of the nitro group to an amino group, giving methyl 3-amino-2-hydroxybutanoate 4-carboxaldehyde (protected as a dioxolane, ***4.114***).

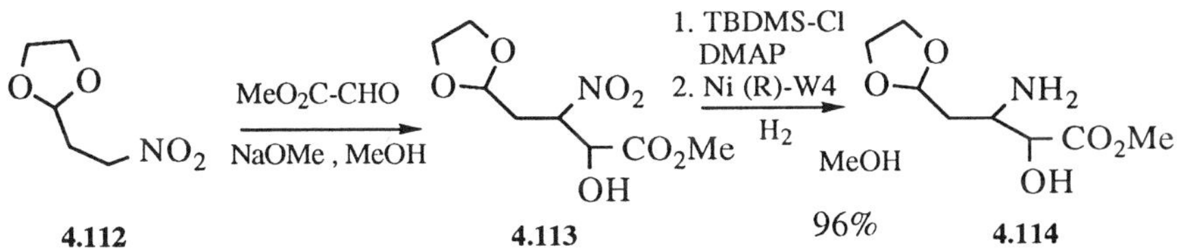

4.112 **4.113** **4.114**

In a very different condensation reaction, nitro enolates derived from nitro acetates reacted with orthoesters, in the presence of an amine, to give aminopropenoic acid derivatives. Ethyl α-nitroacetate, for example, was converted to ethyl 3-(N-phenylamino)-2-nitroprop-2-enoate, ***4.115***, in good yield.[16]

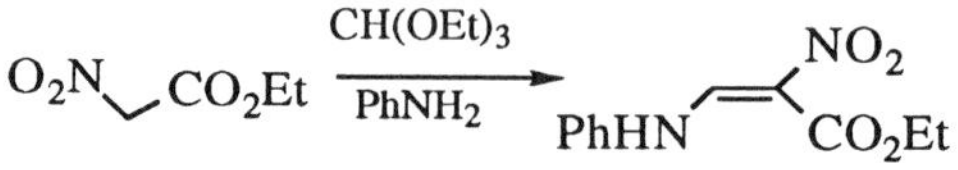

4.115

4.4. Condensation Reactions Involving Organozinc Intermediates

There are several condensation reactions that use zinc enolates rather than sodium, magnesium, or lithium enolates. These are usually categorized as a classical Reformatsky reaction or as a modified Blaise reaction.

4.4.A. Reformatsky Reactions

The condensation reaction of an imine derived from an aldehyde with ethyl bromoacetate under Reformatsky[53] conditions led to a mixture of amino-esters and β-lactams. When acetaldehyde was condensed with methylamine, Schiff base ***4.116*** was the product. A Reformatsky reaction with ethyl bromoacetate gave a mixture of the β-lactam (***4.117***) and the amino-ester (***4.118***). Acid hydrolysis and hydrogenation led to the β-amino acid N-methyl-3-aminobutanoic acid (***4.119***). When benzaldehyde was the aldehyde starting material, 3-(N-methylamino)-3-phenylpropanoic acid (***4.120***) was the final product.[54]

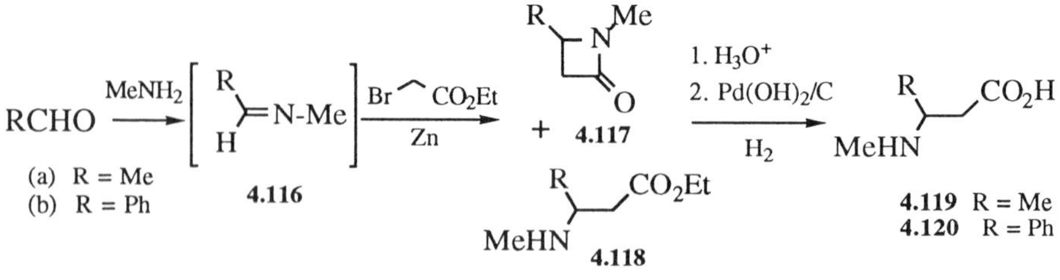

4.116 **4.117** **4.118** **4.119** R = Me **4.120** R = Ph

An alternative approach used phthalimido-ketone ***4.121***, which was converted to an N-phthalimidoyl alcohol under Reformatsky conditions,[54] but removal of the phthalimidoyl group led to formation of lactam ***4.122***.[55] Treatment with acid led to conjugated lactam ***4.123*** rather than the amino acid. If ***4.121*** was reacted with bromo *t*-butyl acetate and zinc, alcohol ***4.124*** was obtained after hydrolysis of the ester.[55] Removal of the phthalimidoyl group with hydrazine then led to 5-amino-3-hydroxy-3-methylpentanoic acid, ***4.125***. 2-Aryl-3-hydroxy derivatives such as

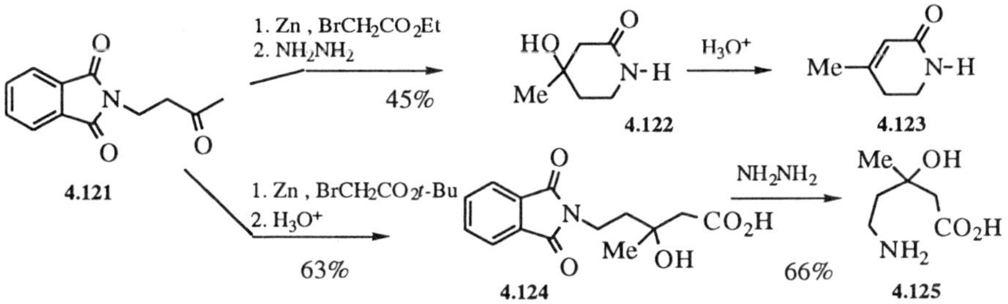

5-amino-3-hydroxy-3-(3-chlorophenyl)pentanoic acid were also prepared by this latter approach (in 86% and 59% yield for the two steps).

4.4.B. Blaise Reactions

The Blaise reaction[56] involves the condensation of an α-bromoester and a nitrile, in the presence of zinc metal, to give β-keto esters. The fundamental reaction is illustrated by the reaction of ***4.126*** to give ***4.128***, via a zinc iminium salt (***4.127***).

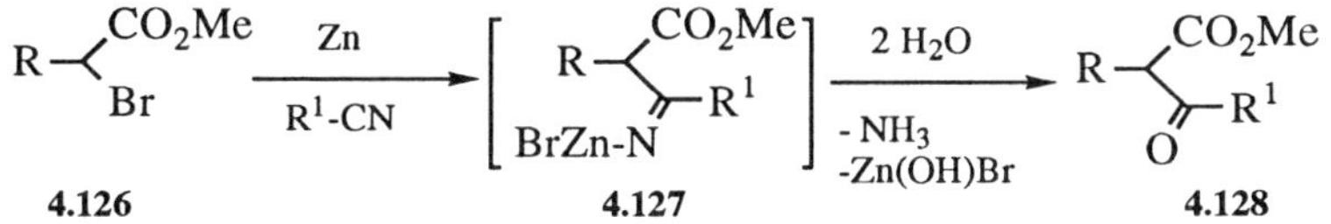

A modified Blaise reaction (the imine intermediate is converted to an enamine moiety rather than hydrolyzed to a carbonyl) can be used for the preparation of

amino acids. When ***4.129*** was reacted with benzonitrile and zinc, methyl 3-amino-2-methyl-3-phenylprop-2-enoate, ***4.130***, was obtained.[57] Several alkenyl amino acids were prepared, including 3-amino-7-chlorohept-2-enoate (in 95% yield) using

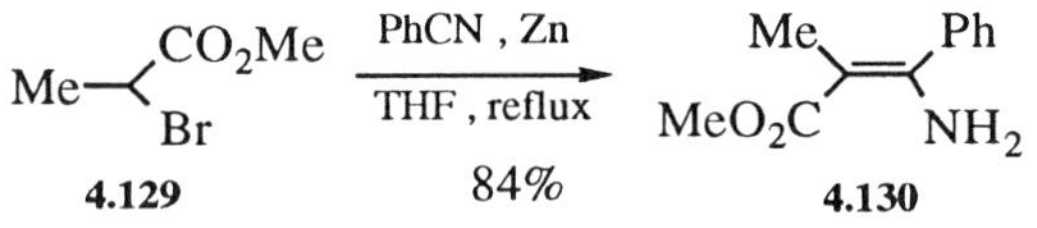

5-chloropentanenitrile as a precursor. A variation in this process reacted benzonitrile with zinc and cupric acetate. Subsequent treatment with a mixture of iodine and methyl iodoacetate gave methyl 3-amino-3-phenylprop-2Z-enoate, ***4.131***.[58]

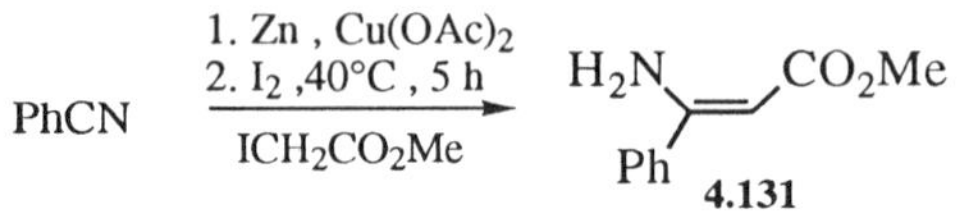

4.5. Specialized Condensation Reactions

This section includes a few "condensation-type" reactions that are not readily categorized in previous sections. They are presented to show the variations that are possible with this fundamental strategy.

One interesting condensation was reported between 2-methylmalonic acid and 1-aminoethanol (***4.132***), but the yields were poor.[59] Refluxing in ethanol followed by reaction with ammonia gave 2-methyl-3-aminobutanoic acid, ***4.133***.[59]

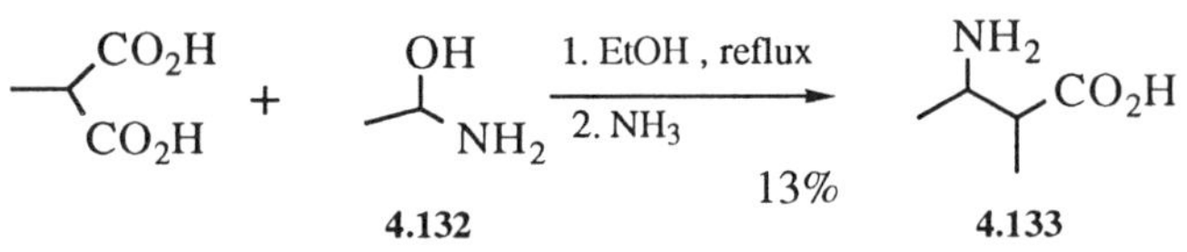

Another variation reacted ethyl 3,3-diphenylpropenoate (***4.134***) with potassium metal in ammonia to give an intermediate bearing a carbanionic carbon β- to the ester moiety. This intermediate was trapped with 2-chloro-N,N-dimethylamino-ethane to give a 30% overall yield of ethyl N,N-dimethyl-3,3-diphenyl-5-amino-pentanoate, ***4.135***.[60]

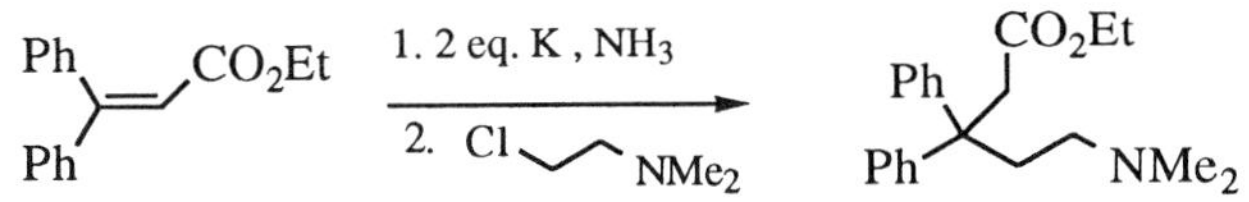

4.134 **4.135**

A synthesis of aryl substituted amino acids involved the reaction of maleic acid with aryl aldehydes, in the presence of ammonium acetate, to give β-aryl-β-amino acids.[61] Specifically, maleic acid reacted with 3-nitrobenzaldehyde (***4.136***) to generate 3-amino-3-(3-nitrophenyl)propanoic acid, ***4.137***.[61]

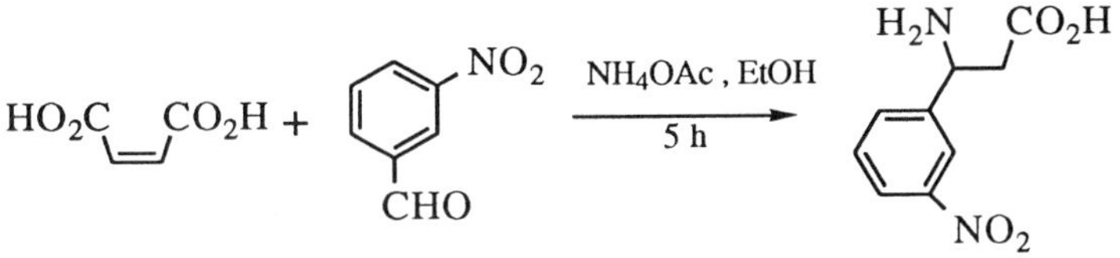

4.136 **4.137**

Functionalized amino acids can be prepared by manipulation of specialized precursors. Reaction of diethyl 2-chloromalonate with sodium phenoxide gave

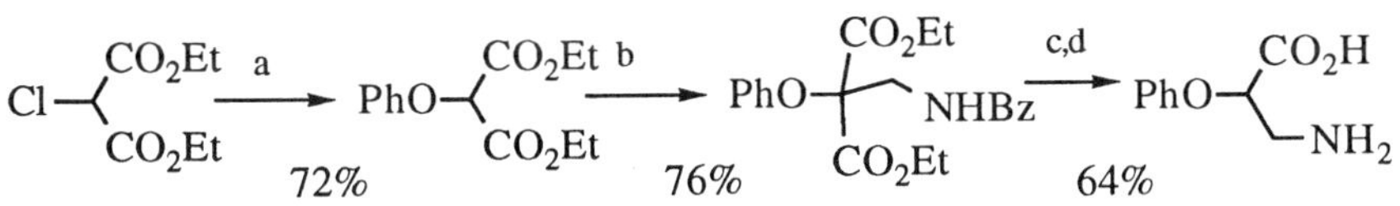

4.138 **4.139** **4.140**

(a) Na , PhOH , EtOH (b) *Chem. Ber., 1957, 90,* 1357 (c) 20% HCl , reflux
(d) ion exchange resin KU-2 (H^+)

2-phenoxy malonate ***4.138***.[62] Incorporation of the amino group by known procedures led to ***4.139***,[63] and this was converted to 2-phenoxy-3-aminopropanoic acid (***4.140***). Both the 2-(4-chlorophenoxy) and the 2-(2,4,5-trichlorophenoxy) derivatives were prepared by this method.[63]

REFERENCES

1 (a) Japp, F.R.; Streatfeild, F.W. *J. Chem. Soc.*, ***1883***, *43*, 27; (b) Knoevenagel, F. *Ber.*, ***1896***, *29*, 172; (c) *Ibid*, ***1898***, *31*, 730; (d) Jones, G. *Org. React.*, ***1967***, *15*, 204; (e) Mundy, B.F.; Ellerd, M.G. *Name Reactions and Reagents in Organic Synthesis*, Wiley, New York, ***1988***, pp. 120-121.

2 Harada, H.; Iizuka, K.; Kamijo, T.; Akahane, K.; Yamamoto, R.; Nakano, Y.; Tsubaki, A.; Kubota, T.; Shimaoka, I.; Umeyama, H.; Kiso, Y. *Chem. Pharm. Bull.*, ***1990***, *38*, 3042.

3 Eisenbach, C.D.; Lenz, R.W. *Macromolecules*, ***1976***, *9*, 227.

4 Junek, H.; Wibmer, P.; Thierrichter, B. *Synthesis*, ***1977***, 560.

5 Bodnarchuk, N.D.; Kolomoets, D.P.; Gavrilenko, B.B.; Derkach, G.I. *Fiziol. Aktiv. Vehchestva*, ***1972***, 13 [*Chem. Abstr.*, ***1973***,*79*: 4886x].

6 Bredereck, H.; Effenberger, F.; Botsch, H. *Chem. Ber.*, ***1964***, *97*, 3397.

7 Doebner, O. *Ber.*, ***1900***, *33*, 2140.

8 Radde, E. *Ber.*, ***1922***, *55*, 3174.

9 Balenović, K.; Jambrešić, I.; Urbas, B. *J. Org. Chem.*, ***1954***, *19*, 1589.

10 Holmes, H.L. *Org. React.*, ***1948***, *4*, 60

11 Stammer, C.H.; Webb, R.G. *J. Org. Chem.*, ***1969***, *34*, 2306.

12 Rodionow, W.M.; Federowa, A.M. *Arch. Pharmaz. Ber. dtsch. Pharmaz. Ges.*, ***1928***, *266*, 116.

13 Testa, E.; Fontanella, L.; Aresi, V. *J.L. Ann. Chem.*, ***1964***, *673*, 60.

14 Chernova, N.G.; Rybkina, E.I.; Berlin, A.Ya. *Zh. Obshch. Khim*, ***1964***, *34*, 2129 (Engl., p. 2142).

15 Johnson, T.B.; Livak, J.E. *J. Am. Chem. Soc.*, ***1936***, *58*, 299.

16 Knippel, E.; Knippel, M.; Michalik, M.; Kelling, H.; Kristen, H. *Z. Chem.*, ***1975***, *15*,446 [*Chem. Abstr.*, ***1976***, *84*: 121068w].

17 (a) Mukaiyama, T. *Org. React.*, ***1982***, *28*, 203; (b) Mukaiyama, T. *Angew. Chem. Int. Ed., Engl.*, ***1977***, *16*, 817.

18 Guanti, G.; Narisano, E.; Banfi, L. *Tetrahedron Lett.*, ***1987***, *28*, 4331, 4335.

19 Ojima, I.; Inaba, S.; Yoshida, K. *Tetrahedron Lett.*, ***1977***, 3643.

20 Ikeda, K.; Achiwa, K.; Sekiya, M. *Tetrahedron Lett.*, ***1983***, *24*, 4707.

21 Shono, T.; Tsubata, K.; Okinaga, N. *J. Org. Chem.*, ***1984***, *49*, 1056.

22 Ikeda, K.; Achiwa, K.; Sekiya, M. *Tetrahedron Lett.*, ***1983***, *24*, 913.

23 Overberger, C.G.; Parker, G.M. *J. Polym. Sci., A-1*, ***1968***, *6*, 513.

24 Sachs, F. *Ber. dtsch. Chem. Ges.*, ***1898***, *31*, 1225 (see p. 1232).
25 Böhme, H.; Broese, R.; Eiden, F. *Chem. Ber.*, ***1959***, *92*, 1258.
26 Harbeson, S.L.; Rich, D.H. *J. Med. Chem*, ***1989***, *32*, 1378.
27 Bey, P.; Gerhart, F.; Van Dorsselaer, V; Danzin, C. *J. Med. Chem.*, ***1983***, *26*, 1551.
28 Tolman,V.; Vereš, K. *Tetrahedron Lett.*, ***1964***, 1967.
29 Leonard, F.; Wajngurt, A.; Klein, M.; Meyer, H. *J. Med. Chem.*, ***1963***, *6*, 539.
30 Bodnarchuk, N.D.; Momot, V.V.; Gavrilenko, B.B. *Zh. Obshch. Khim.*, ***1975***, *45*, 873 (see p. 856).
31 Buchanan, R.L.; Pattison, F.L.M. *Can. J. Chem.*, ***1965***, *43*, 3466.
32 (a) Lang, R.W.; Hansen, H.J. *Org. Synth. Coll. Vol. 7*, ***1990***, 232; (b) Rosenthal, R.W.; Schwartzman, L.H. *J. Org. Chem.*, ***1959***, *24*, 836.
33 Paik, Y.H.; Dowd, P. *J. Org. Chem.*, ***1986***, *51*, 2910.
34 Huckin, S.N.; Weiler, L. *Can. J. Chem.*, ***1974***, *52*,1343.
35 (a) Čefelín, P.; Lochman, L.; Stehlíček, J. *Coll. Czech. Chem. Commun.*, ***1973***, *38*, 1339; (b) Čefelín, P.; Lochman, L. *Czech.* 161,459 [*Chem. Abstr.*, ***1976***, *85*: P143778a].
36 Shono, T.; Tsubata, K.; Okinaga, N. *J. Org. Chem.*, ***1984***, *49*, 1056.
37 Shono, T.; Kise, N.; Sanda, F.; Ohi, S.; Tsubata, K. *Tetrahedron Lett.*, ***1988***, *29*, 231.
38 Simova, E.; Kurtev, B. *Monatsh. Chem.*, ***1965***, *96*, 722.
39 Fuchigami, T.; Ichikawa, S.; Konno, A. *Chem. Lett.*, ***1992***, 2405.
40 (a) Hiyama, T.; Kobayashi, K. *Tetrahedron Lett.*, ***1982***, *23*, 1597; (b) Hiyama, T.; Kobayashi, K.; Nishide, K. *Bull. Chem. Soc., Jpn.*, ***1987***, *60*, 2127; (c) Hiyama, T.; Oishi, H.; Suetsugu, Y.; Nishide, K.; Saimoto, H. *Ibid*, ***1987***, *60*, 2139.
41 Sosnovskikh, V.Ya.; Ovsyannikov, I.S.; Nikol'ski, A.L. *Zh. Org. Khim.*, ***1986***, *22*, 1775 (Engl., p. 1595).
42 Stamm, H.; Weiss, R. *Synthesis*, ***1986***, 395.
43 Neyer, G.; Achatz, J.; Danzer, B.; Ugi, I. *Heterocycles (Special Ed.)*, ***1990***, *30*, 863.
44 (a) Meyers, A.I.; Nabeya, A.; Adickes, H.W.; Politzer, I.R. *J. Am. Chem. Soc.*, ***1969***, *91*, 763; (b) Meyers, A.I.; Nabeya, A.; Adickes, H.W.; Politzer, I.R.; Malone, G.R.; Kovelesky, A.C.; Nolen, R.L.; Portnoy, R.C. *J. Org. Chem.*,

1973, *38*, 36.

45 (a) Salmon-Legagneur, F.; Neveu, C. *Bull. Soc. Chim., Fr.*, ***1965***, 2270; (b) Idem, *C.R. Acad. Sci.*, ***1964***, *259*, 1878; (c) Idem, *Ibid*, ***1963***, *256*, 187.

46 Buu, N.T.; Van Gelder, N.M. *Br. J. Pharmac.*, ***1974***, *52*, 401.

47 Cignarella, G.; Mariani, L.; Testa, E. *Gazz. Chim. Ital.*, ***1965***, *95*, 831.

48 (a) Leonard, F.; Wajngurt, A.; Klein, M.; Smith, C.M. *J. Org. Chem.*, ***1961***, *26*, 4062; (b) Leonard, F. *US Patent* 3,125,583 [*Chem. Abstr.*, ***1964***, *60*: P13193d].

49 Marquina, J.G.; Millet, J.M.; Sistare, J. *Farm. Nueva (Madrid)*, ***1961***, *26*, 381 [*Chem. Abstr.*, ***1963***, *58*: 3386a].

50 Kaji, E.; Igarashi, A.; Zen, S. *Bull. Chem. Soc., Jpn.*, ***1976***, *49*, 3181.

51 Ram, S.; Ehrenkaufer, R.E. *Synthesis*, ***1986***, 133.

52 Williams, T.M.; Crumbie, R.; Mosher, H.S. *J. Org.Chem.*, ***1985***, *50*, 91.

53 (a) Reformatsky, S. *Ber.*, ***1887***, *20*, 1210; (b) Rathke, M.W. *Org. React.*,***1975***, *22*, 423; (c) Diaper, D.G.M.; Kuksis, A. *Chem. Rev.*, ***1959***, *59*, 89.

54 Furukawa, M.; Okawara, T.; Noguchi, Y.; Terawaki, Y. *Chem. Pharm. Bull.*, ***1978***, *26*, 260.

55 Carganico, G.; Cangiano, G.; Cozzi, P.; Lovisolo, P.P. *Framaco Ed. Sci.*, ***1987***, *42*, 277.

56 (a) Blaise, E.E. *C.R. Hebd. Sci., Acad. Sci.*, ***1901***, *132*, 478; (b) Cason, J.; Rinehart, K.L.; Thornton, S.D. *J. Org. Chem.*, ***1953***, *18*, 1594.

57 Hannick, S.M.; Kishi, Y. *J. Org. Chem.*, ***1983***, *48*, 3833.

58 Cassar, L.; Iqbal, A.; Rochat, A.C. *Eur. Pat. Appl.* EP 102,318 [*Chem. Abstr.*, ***1984***, *101*: P55530g].

59 Kurtev, P.; Pozharliev, I. *Compt. Rend. Acad. Bulgare Sci.*, ***1963***, *16*, 65 [*Chem. Abstr.*, ***1963***, *59*: 15173d].

60 Patchett, A.A.; Giarrusso, F.F. *J. Med. Pharm. Chem.*, ***1961***, *4*, 403.

61 Profft, E.; Becker, F.J. *J. Prakt. Chem.*, ***1965***, *30*, 18.

62 Mamaev, V.P.; Mikhaleva, M.A. *Izv. Sibirsk. otd. Akad. Nauk. SSSR*, ***1962***, 145 [*Chem. Abstr.*, ***1963***, *59*: 5051h].

63 Hellmann, H.; Haas, G. *Chem. Ber.*, ***1957***, *90*, 1357.

5

Diastereoselective and Enantioselective Syntheses

Most of the syntheses presented in chapters 1-4 generated achiral amino acids or chiral, racemic amino acids. A few chiral, nonracemic amino acids were prepared, but only to illustrate some specific point of a synthesis. The presence of a functional group on an amino acid chain usually means that its synthesis will generate diastereomers and/or enantiomers. Preparation of these compounds using methods that proceed with high diastereoselectivity and/or enantioselectivity are very important since any biological activity usually resides in a single diastereomer, if not a single enantiomer (see chapter six for biologically active amino acids). This chapter will focus attention on the diastereoselectivity of synthetic methods that give amino acids. It will also discuss the enantioselectivity of certain methods, or will discuss approaches where high enantioselectivity is observed in the formation of the amino acid. The enantioselective methods include the use of chiral auxiliaries or chiral templates (as defined by Hanessian).[1]

5.1. Amino Acid Refunctionalization

In this first section, chiral, nonracemic α-amino acids (also a few non-α-amino acids) will be synthetically manipulated to produce chiral, nonracemic non-α-amino acids. Several of the synthetic methods discussed in previous chapters will be used here, but their use in diastereoselective and/or enantioselective preparations will be highlighted.

5.1.A. Alanine and Derivatives

A condensation route to amino acids involves the use of a hydropyrimidin-2-one (*5.1*, derived from β-alanine), as shown below. Reaction with lithium diisopropylamide, followed by reaction with iodobutane, gave a substituted heterocycle (*5.2*) with high diastereoselectivity.[2] Acid hydrolysis cleaved the ring to give 2-butyl-3-aminopropanoic acid (*5.3*). In a similar manner, alkylation with iodomethane led to 53% yield of 2-methyl-3-aminopropanoic acid; benzyl bromide gave 2-benzyl-3-aminopropanoic acid in 33% yield; and, iodohexane gave 2-hexyl-3-aminopropanoic acid in 47% yield.[2] Juaristi also used *S*-asparagine(see next section) as a starting material to prepare 2-methyl-3-aminopropanoic acid.[3]

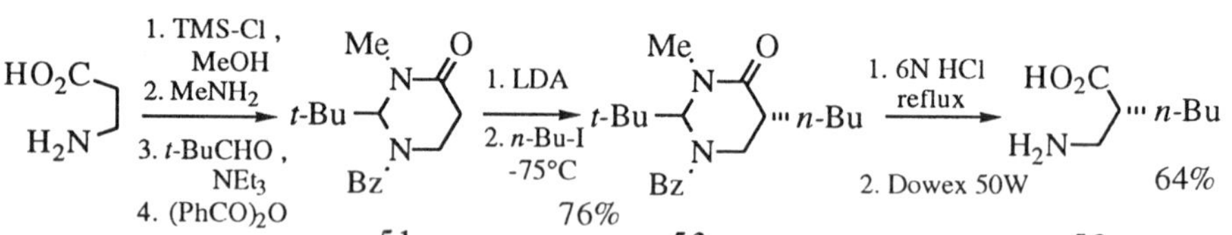

5.1 5.2 5.3

In many cases, the carboxyl portion of the amino acid starting material is converted to an aldehyde, allowing several reactions that are not available to carboxylic acids. An example of this approach is the conversion of N-Boc alanine (*5.4*) to N-Boc alinal (*5.5*)[4] via conversion to the ester and reduction with lithium borohydride. The next step will be a common feature of this and succeeding chapters. Oxidation of the alcohol moiety in *5.5* to an aldehyde, in this case using SO_3•pyridine in DMSO, allowed condensation with sodium cyanide to give a cyanohydrin. Hydrolysis of the nitrile moiety to an acid and led to 2-hydroxy-3-aminobutanoic acid (*5.6*). Similarly, Boc-glycine led to ethyl 2-hydroxy-3-aminobutanoate in about 84% yield.[4]

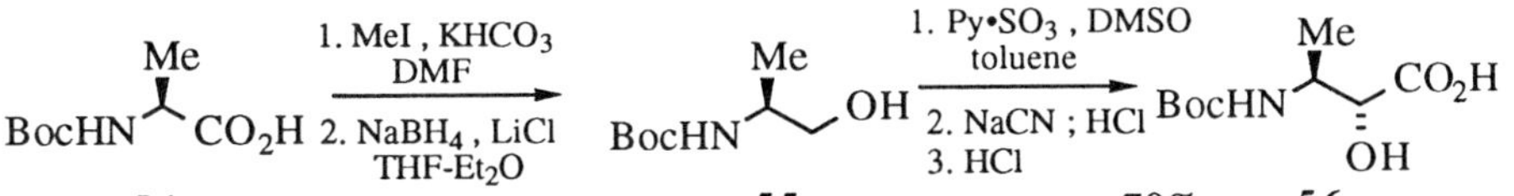

5.4 **5.5** **5.6**

Another important reaction made possible by the presence of the aldehyde moiety is an olefination reaction that extends the chain and adds a new carboxyl moiety. A Horner-Wadsworth-Emmons olefination[5] of N-Cbz alinal (*5.7*) gave methyl 4-(N-Cbz amino)pent-2-enoate (*5.8*) as a mixture of *cis:trans* isomers (>30:1).[6] A number of other derivatives were prepared by starting with different amino acids.

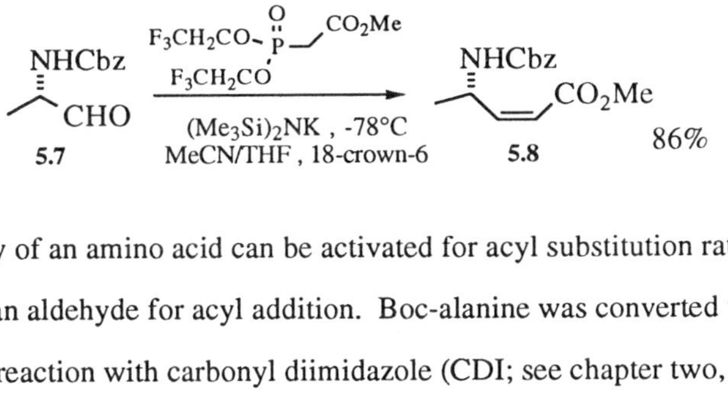

5.7 **5.8**

The acid moiety of an amino acid can be activated for acyl substitution rather than converted to an aldehyde for acyl addition. Boc-alanine was converted to an acyl imidazole by reaction with carbonyl diimidazole (CDI; see chapter two, section 2.4), and then condensed with the magnesium enolate of the mono ethyl ester of malonic acid to give keto-ester *5.9*. Subsequent catalytic hydrogenation of the ketone moiety gave ethyl 3-hydroxy-5-(N-Boc amino)pentanoate, ***5.10***.[7] Once the

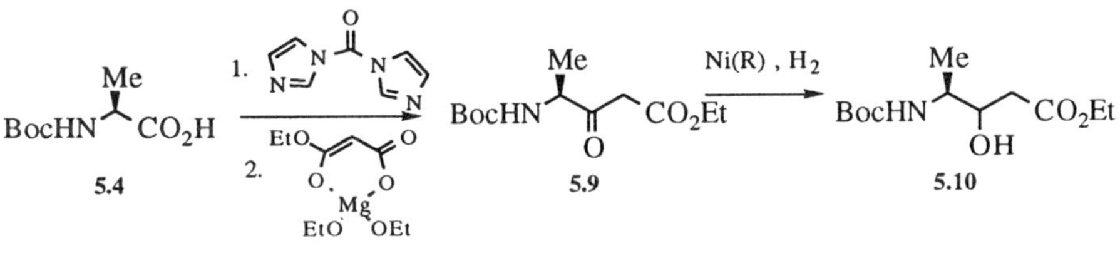

5.4 **5.9** **5.10**

oxo moiety is incorporated in an amino acid, reduction of that group is a common method for the preparation of hydroxy amino acids. There are many methods for the reduction of a ketone moiety to an alcohol.

A different synthetic route converted alanine to the diasteromeric mixture of oxazolidin-2-ones, ***5.11a*** and ***5.11b***.[8] After separation of these diastereomers, oxidative cleavage of the alkene moiety in ***5.11a*** (with $RuCl_3$ and $NaIO_4$) led to ***5.12a*** and basic hydrolysis gave *D*-isothreonine (***5.13a***).[9] *L*-allo-Threonine (***5.13b***) was formed in a similar manner from diastereomer ***5.11b***, via ***5.12b***. Note that a variety of amino acids were prepared from alanine. These include both 2-amino, 3-amino, and 4-amino acid derivatives.

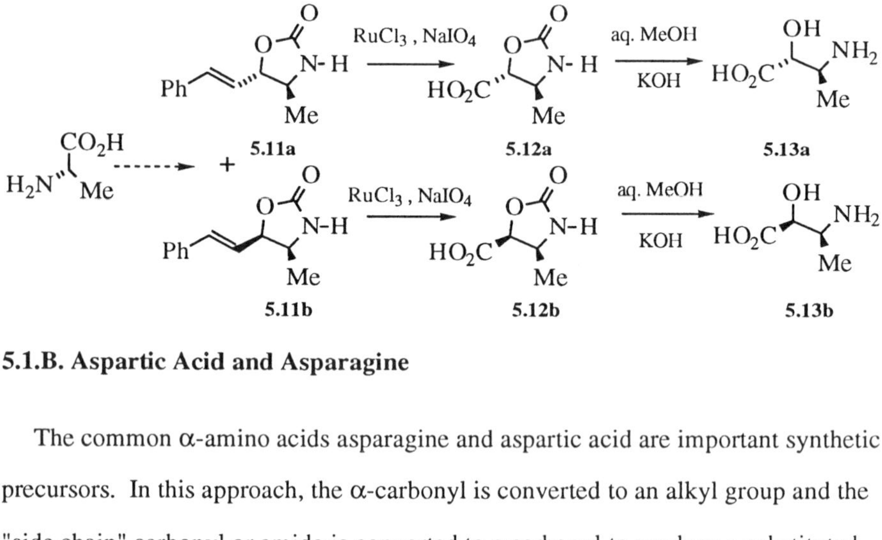

5.1.B. Aspartic Acid and Asparagine

The common α-amino acids asparagine and aspartic acid are important synthetic precursors. In this approach, the α-carbonyl is converted to an alkyl group and the "side chain" carbonyl or amide is converted to a carbonyl to produce a substituted β-amino acid. In all cases shown here, either a 3-amino- or a 4-amino acid is prepared.

5.1.B.i. Aspartic Acid Precursors

A chiral, nonracemic lactone intermediate is prominent in the conversion of aspartic acid (***5.14***) first to anhydride ***5.15***, and then to ***5.16*** via borohydride reduction.[10] Opening the lactone with iodotrimethylsilane gave ethyl 3-(N-tosylamino)-4-iodobutanoate (***5.17***), which was converted to 3-aminononanoic acid, ***5.18***. The use of the appropriate organocuprate in the final sequence allowed the synthesis of to 3-aminopentanoic acid, 3-aminooctanoic acid, as well as several alkyl substituted amino acids.

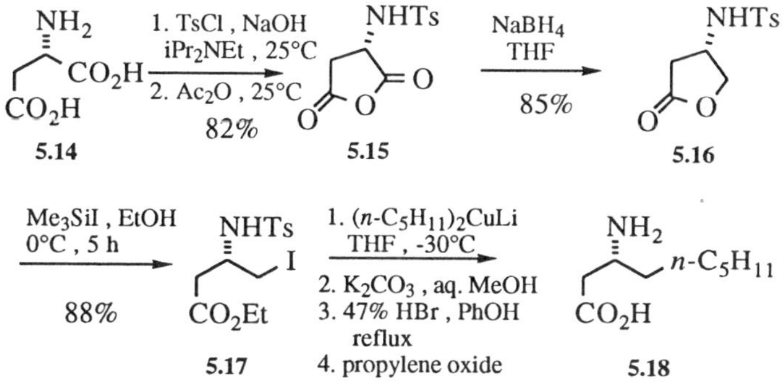

This synthesis is related to an earlier sequence that produced lactone ***5.20*** by reduction of anhydride ***5.19***,[11] which was formed by treating N-Cbz aspartic acid

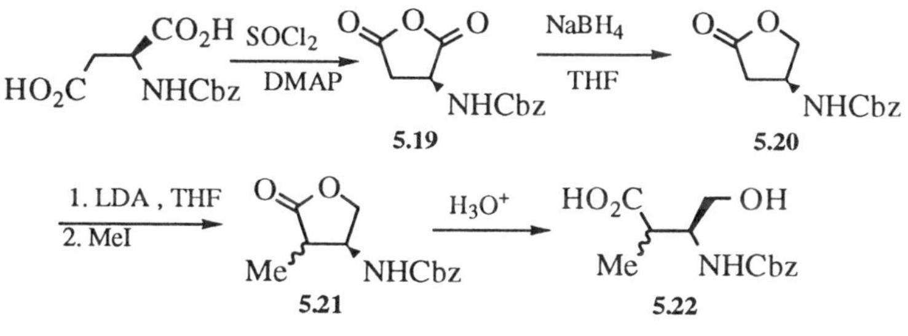

with thionyl chloride and N,N-dimethylaminopyridine (DMAP). In this case, methylation of the lactone enoate gave ***5.21*** and hydrolysis of the lactone ring gave 3-(N-Cbz amino)-4-hydroxy-2-methylbutanoic acid, ***5.22***.[11]

The amino group in *L*-aspartic acid (***5.14***) was converted to an alcohol moiety in the mono-amide of *L*-2-hydroxysuccinic acid, ***5.23***.[12] Dehydration of the amide moiety led to nitrile ***5.24***, which was to function as an amine surrogate. Reduction of the nitrile and hydrolysis of the acetate gave 2-hydroxy-4-aminobutanoic acid (***5.25***).[13,12]

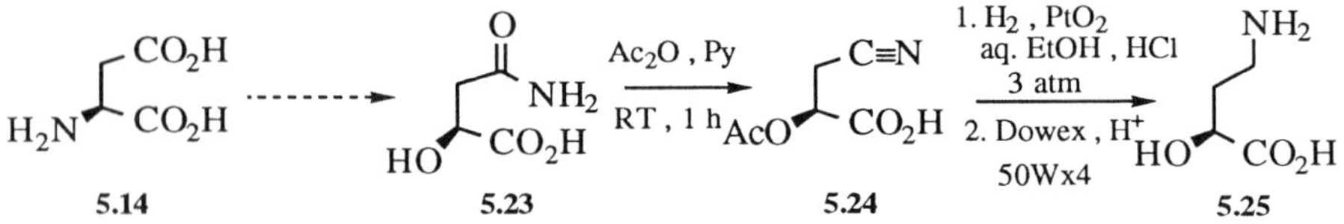

A different approach reduced both the acid moieties in aspartic acid to give an amino-diol, which was cyclized to an oxazolidinone by treatment with sodium hydride.[14] Oxidation of the remaining alcohol moiety led to aldehyde ***5.26***. The aldehyde was protected as the dithiolane, the heterocyclic ring was hydrolyzed, and the amine group protected to give ***5.27***. Reaction of the alcohol moiety in ***5.27*** with phosgene generated a primary chloride, which reacted with the lithium enolate of ethyl acetate to give a 1.5:1 mixture of 3S,4S:3R,4S ethyl 4-(N-Boc amino)-3-hydroxy-pentanoate 5-carboxaldehyde protected as the dithiolane (***5.28***).[14]

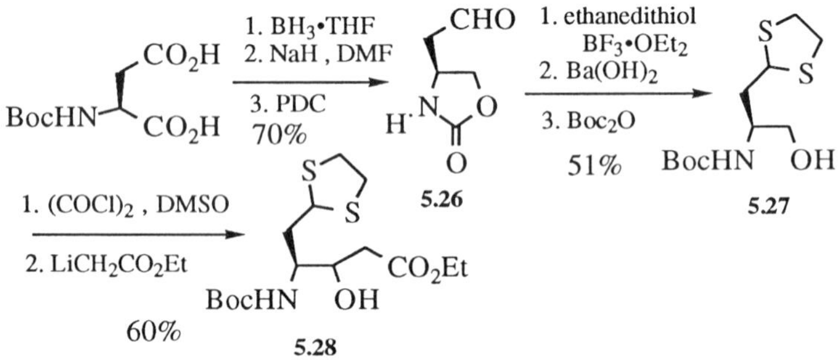

5.1.B.ii. Asparagine Precursors

Asparagine is obviously related to aspartic acid, since it is the mono-amide of aspartic acid. When asparagine (***5.29***) was first converted to the N,N-dibenzyl derivative[15] and then to the benzyl ester (***5.30***), treatment with tosyl chloride and pyridine gave the corresponding nitrile (***5.31***) in 65% overall yield.[16] Reduction of the ester with $LiAlH_4$ and conversion to the mesylate (91% from ***5.31***) allowed displacement by organocuprates to give ***5.32*** (**a**, R = Me, 65%; **b**, R = *n*-Bu, 72%; **c**, R = Ph, 48% from the mesylate).[16] Hydrolysis with aqueous hydrochloric acid gave the protected amino acid in about 90% yield and removal of the N-benzyl groups gave the free amino acid (3-aminopentanoic acid, ***5.33a***, R = Me;[17] 3-aminooctanoic acid, ***5.33b***, R = *n*-Bu; 3-amino-4-phenylbutanoic acid, ***5.33c***, R = Ph) in 90-95% yield.[16] In a similar manner, aspartic acid was converted to asparagine and then to ***5.33b*** and also to 3-amino-5-phenylpentanoic acid.[18]

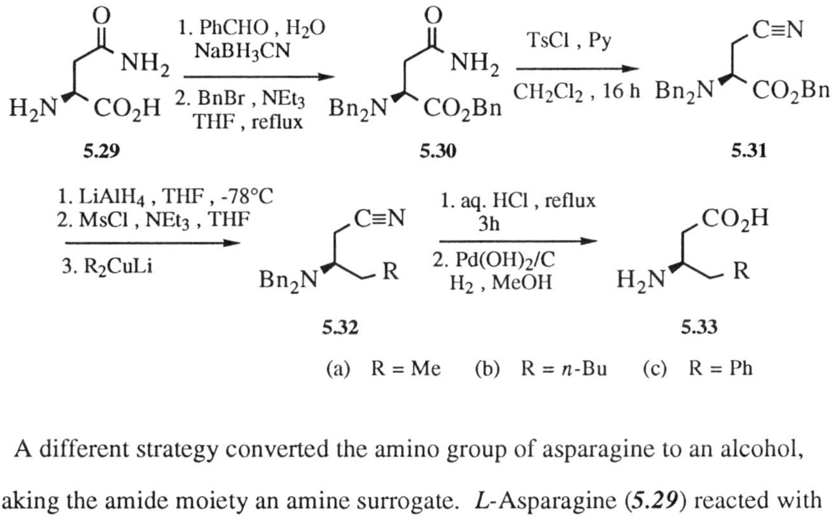

5.29 **5.30** **5.31** **5.32** **5.33**

(a) R = Me (b) R = *n*-Bu (c) R = Ph

A different strategy converted the amino group of asparagine to an alcohol, making the amide moiety an amine surrogate. *L*-Asparagine (***5.29***) reacted with sodium nitrite and acetic acid (HONO) to give malamidic acid (***5.34***),[19] analogous to the conversion of the amino group of aspartic to an alcohol moiety previously noted in this section. Treatment with NaOCl led to 2-hydroxy-3-aminopropanoic acid (isoserine, ***5.35***). In a similar manner, glutamic acid (see next section) was converted to 4-amino-2-hydroxybutanoic acid.[20]

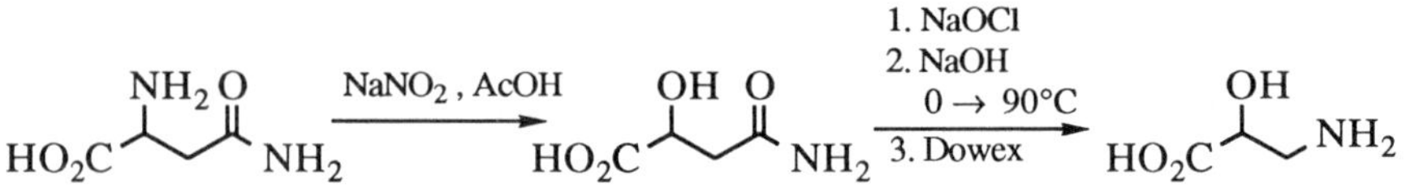

5.29 5.34 5.35

5.1.C. Glutamic Acid

Glutamic acid is important for the preparation of 4-amino acid derivatives. There are two important strategies. The first involves conversion of glutamic acid to a chiral, nonracemic lactone. The second involves conversion of glutamic acid to a chiral nonracemic lactam (pyroglutamate).

5.1.C.i. Conversion to Lactone Intermediates

L-Glutamic acid (***5.36***) is a readily available starting material and treatment with nitrous acid (formed from $NaNO_2$ and HCl) led to formation of the chiral, non-racemic γ-butyrolactone 4-carboxylic acid (***5.37***).[21] This is analogous to the conversion of the amino group in both aspartic acid and asparagine to an alcohol moiety. A difference in this system, however, is that the hydroxy acid spontaneously cyclizes to form a five-membered ring lactone (***5.37***). Selective reduction of the carboxyl

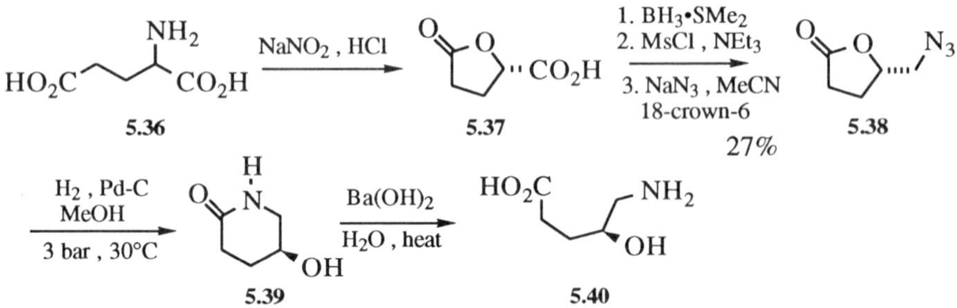

5.36 5.37 5.38 5.39 5.40

group in ***5.37*** with borane, followed by mesylation of the resulting alcohol, set up a displacement reaction with azide to give azido-lactone ***5.38***. Catalytic hydrogenation of the azide generated the amine, which attacked the lactone to form lactam

5.39.[21] Basic hydrolysis of the lactam (see chapter one, section 1.3) gave 4-hydroxy-5-aminopentanoic acid, *5.40*.[21] It is noted that lactone *5.2* has been used in several synthetic transformations.[22]

5.1.C.ii. Conversion to Pyroglutamate Derivatives

Treatment of glutamic acid with thionyl chloride and ethanol gave an 84% yield of 5S-carboethoxy-2-pyrrolidinone (ethyl pyroglutamate, *5.41*.)[23] Reduction of the ester moiety with lithium borohydride gave *5.42*. Subsequent acid hydrolysis and isolation via passage through a Dowex ion exchange column gave 4-amino-5-hydroxypentanoic acid (*5.43*).[23] It is important to note that in this sequence, the chiral center in *5.41* did not racemize during this synthetic sequence. Another

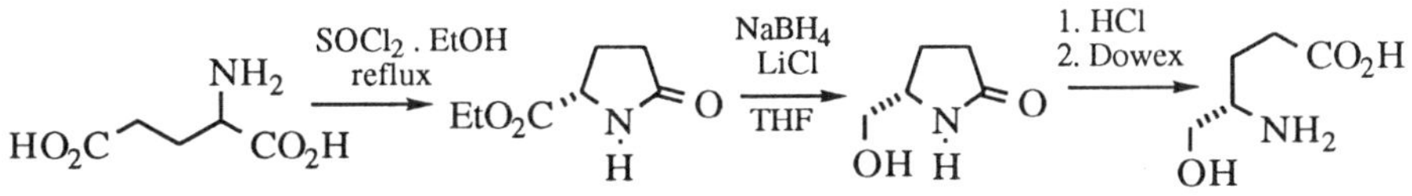

synthesis of *5.43* began with *5.36* but protected the nitrogen and then converted the carbonyl group distal to nitrogen into an ester.[24] This allowed selective reduction of the carbonyl group proximal to the α-nitrogen and deprotection gave *5.43*. Amino acid *5.43* was also prepared from N-Cbz glutamine (the amide form of glutamic acid).[25] Reduction of the acid moiety followed by deprotection afforded *5.43* in 39% overall yield.[24]

Silverman prepared 5S-hydroxymethyl-2-pyrrolidinone (*5.42*)[23] by essentially the same route as shown above. Conversion of the hydroxymethyl group to a mesylate was followed by its reduction to a methyl group with zinc and sodium iodide, giving 5R-methyl-2-pyrrolidinone (*5.44*).[26] The 5S-methyl derivative was prepared in an identical manner from *D*-glutamic acid. When *5.44* was heated with 6N HCl, ethyl 4R-aminopentanoic acid (***5.45***) was obtained.[26] It is noted that Smith recently prepared ***5.44*** by a slightly different route and reacted it with formaldehyde and chlorotrimethylsilane to form 5R-methyl-N-(chloromethyl)-2-pyrrolidinone.[27] This reagent was shown to react with alcohols, and the resulting N-alkoxymethyl derivative could be analyzed by proton NMR to determine the enantiomeric excess if the alcohol reactant contained a chiral center.[27]

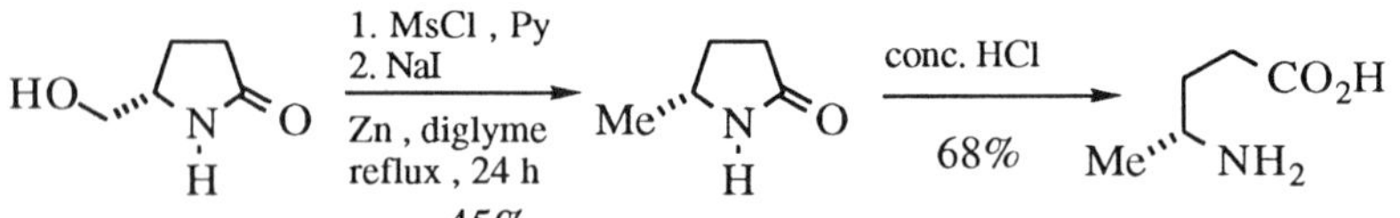

5.42 **5.44** **5.45**

Holmes used a strategy similar to that of Silverman, just discussed, to produce longer chain amino acids. Reaction of ***5.42*** with tosyl chloride was followed by reaction with di-*n*-butyl cuprate to give ***5.46***. This was hydrolyzed with 2M HCl to give a quantitative yield of 4S-aminononanoic acid, ***5.47***.[28]

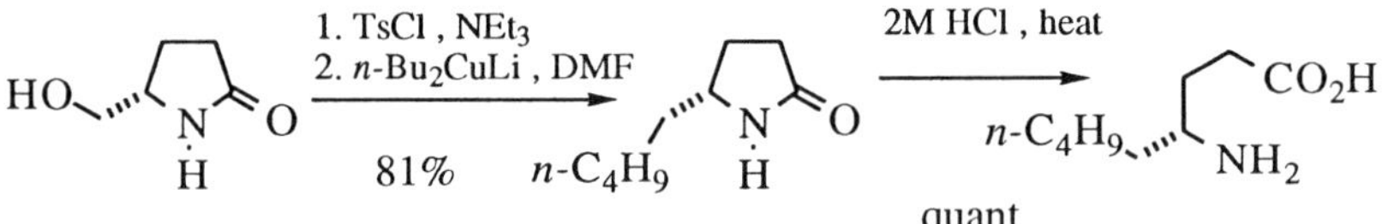

5.42 **5.46** **5.47**

A number of functionalized chiral, nonracemic amino acids can be prepared using ethyl pyroglutamate as a chiral template. Silverman reduced ethyl pyroglutamate (***5.41***) to the alcohol (***5.42***) with lithium borohydride.[23] This was followed by treatment with bromine and triphenylphosphine to give 5-bromomethyl-2-pyrrolidinone, ***5.48***. Similarly prepared were 5-chloromethyl-2-pyrrolidinone (***5.50***); 5-fluoro-methyl-2-pyrrolidinone (***5.52***); and, 5-cyanomethyl-2-pyrrolidinone.[23] In all cases, reaction with HCl and isolation via ion exchange chromatography gave the 5-halo-4-aminopentanoic acid [5-bromo-4-aminopentanoic acid (***5.49***); 5-chloro-

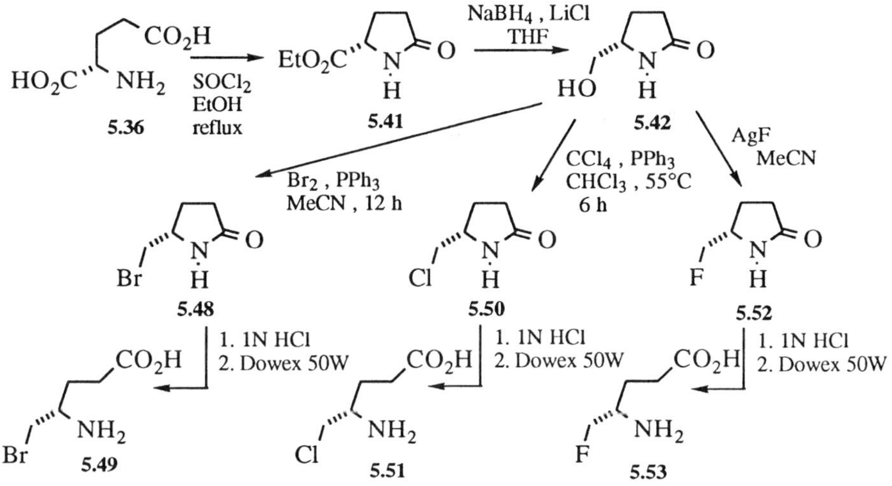

4-aminopentanoic acid (*5.51*); 5-fluoro-4-aminopentanoic acid (*5.53*)].[23] Ethyl pyroglutamate (*5.41*) can also be converted to 4-amino-5-cyano-pentanoic acid (*5.55*) via 5-cyanomethyl-2-pyrrolidinone (*5.54*).[23]

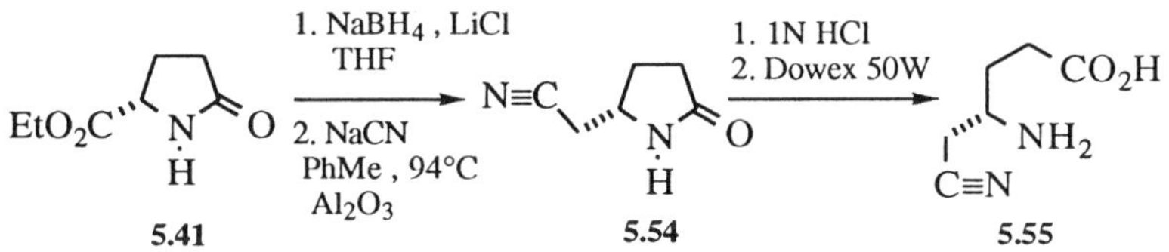

5.1.C.iii. Other Glutamic Acid Derivatives

An alternative route to functionalized amino acids begins with *5.56* (1-azabicyclo[3.3.0]-6-oxo-7-phenyloctan-2-one) derived from glutamic acid by reaction of *5.42* (see above) with benzaldehyde. The double bond was introduced via selenation, oxidation, and *syn*-elimination of the resulting selenoxide to give *5.57*.[29] Osmylation of the alkene moiety and protection of the resulting diol as the acetonide gave *5.58*. Deprotection gave *5.59* and conversion of the hydroxy-methyl group to a nitrile was followed by protection of the nitrogen. Basic hydrolysis of the lactam ring followed, to give the acetonide (*5.60*) of 3-(N-Boc amino)-5S-cyano-2S,3S-dihydroxypentanoic acid.

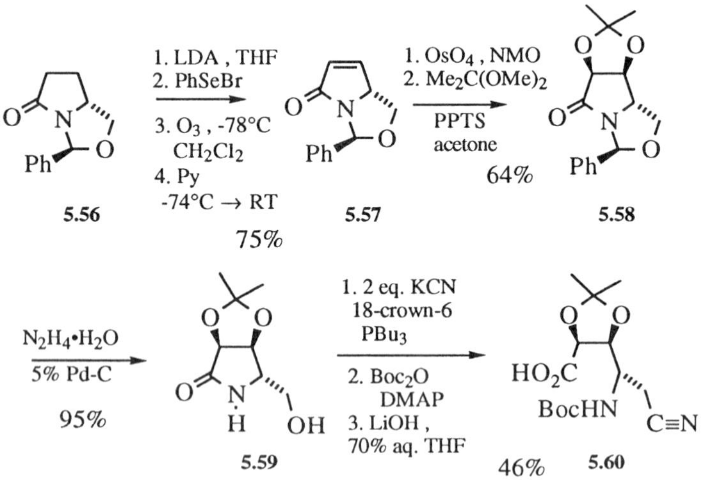

5.56 5.57 5.58 5.59 5.60

5.1.D. Leucine and Derivatives

Leucine is an important synthetic starting material for the preparation of statine and its derivative (see chapter six, section 6.3) due to the biological importance of those compounds. The synthesis of other derivatives, some clearly related to statine, will be presented in this section. The reactions noted here are largely presented as a preview for the synthetic strategies used to prepare statine.

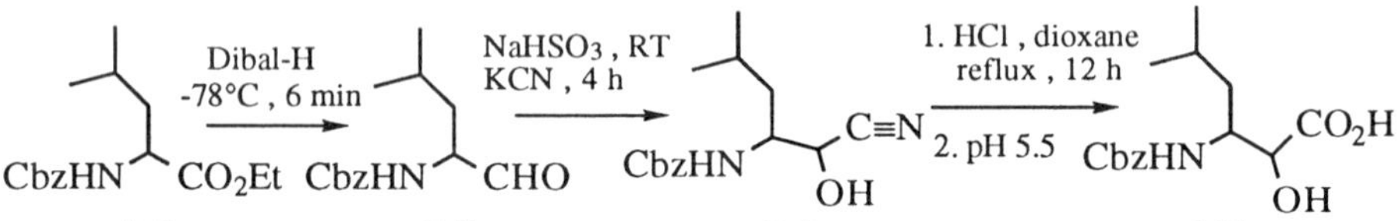

5.61 5.62 5.63 5.64

Reduction of the ethyl ester of N-Cbz leucine (*5.61*) with diisobutylaluminum hydride gave the corresponding aldehyde (*5.62*), as seen when other amino acid precursors were used.[30] Aldehyde *5.62* reacted with potassium cyanide and sodium bisulfite to give cyanohydrin ***5.63***. Hydrolysis and deprotection gave 3-amino-2-hydroxy-5-methylhexanoic acid ***5.64***. The diastereoselectivity of the condensation reaction of ***5.62*** to form ***5.63*** is not outstanding, and the diastereomeric mixture must be chromatographically separated. A similar strategy formed ***5.61*** by reducing

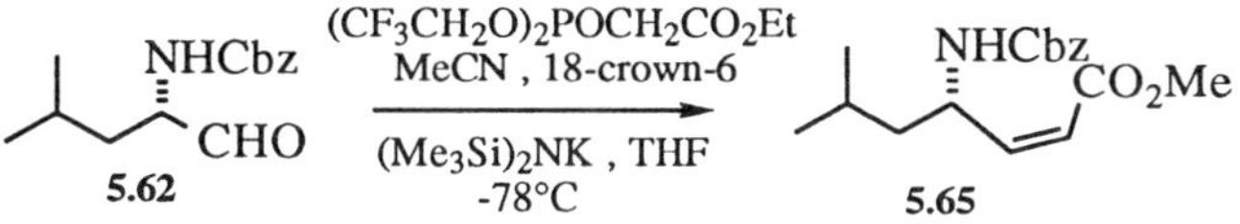

leucine and then oxidizing the resultant alcohol with sulfur trioxide-pyridine-DMSO.[4] A related route used trimethylsilyl cyanide and zinc iodide to produce an O-trimethylsilyl cyanohydrin intermediate, which was converted to ***5.64*** in 75% yield.[31]

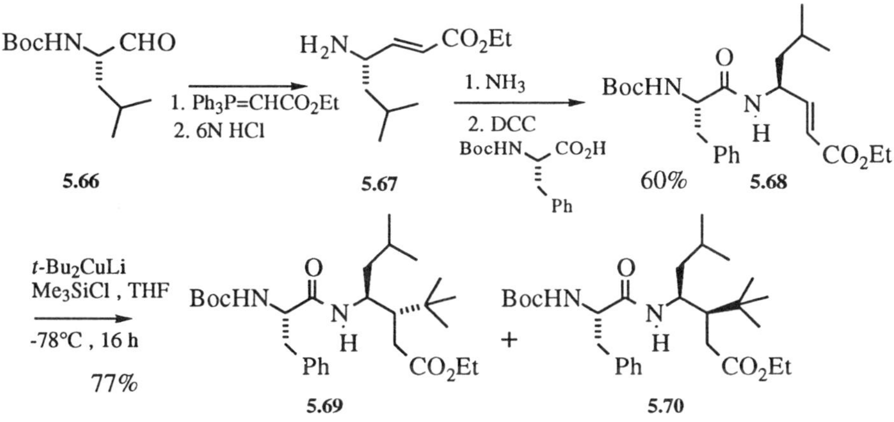

Once an aldehyde such as ***5.62*** is formed, other reactions can be done, as seen in previous sections. One is an olefination reaction (here using a Horner-Wadsworth-Emmons reaction;[5] see ***5.7*** in section 5.1.A) with ***5.62*** to generate methyl 4S-(N-Cbz amino)-6-methylhept-2-enoate, ***5.65*** .[32] Using a similar strategy, a Wittig reaction[33] with ***5.66*** gave ***5.67*** and this was converted to dipeptide ***5.68***.[34] Subsequent reaction with di-*t*-butylcuprate gave an 80:20 mixture of ***5.69:5.70***. Organocuprates also added to ***5.67***, but better selectivity was obtained in reactions with dipeptides such as ***5.68***.[33]

5.1.E. Phenylalanine and Derivatives

Phenylalanine is another important amino acid precursor, probably because 4-amino-3-hydroxy-5-phenylpentanoic acid is considered to be an analog of statine (see chapter six, section 6.3). In one example, N-Boc phenylalanine (***5.71***) was converted to ***5.72*** via activation of the acid with imidazole dicarbonyl and then condensation with the magnesium salt of mono ethyl malonate.[35] Catalytic hydrogenation (using a chiral ruthenium catalyst) reduced the ketone moiety to a hydroxyl group with high diastereoselectivity, giving enantiopure 4-(N-Boc amino)-3-hydroxy-5-phenylbutanoic acid, ***5.73***. In this specific example, the phenyl group was later reduced to give the corresponding cyclohexyl derivative. In another

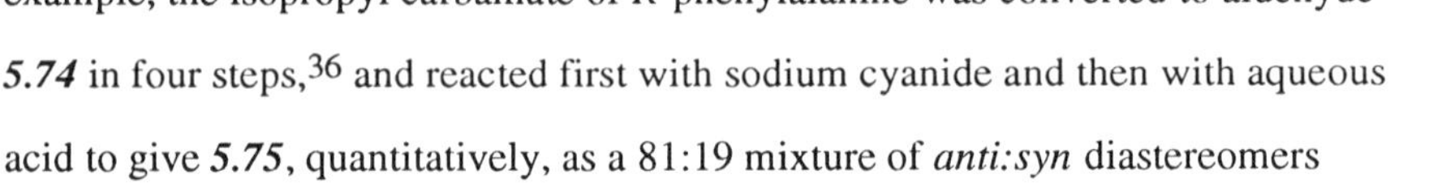

5.71 **5.72** **5.73**

example, the isopropyl carbamate of *R*-phenylalanine was converted to aldehyde ***5.74*** in four steps,[36] and reacted first with sodium cyanide and then with aqueous acid to give ***5.75***, quantitatively, as a 81:19 mixture of *anti:syn* diastereomers ***5.75a:5.75b***.[37]

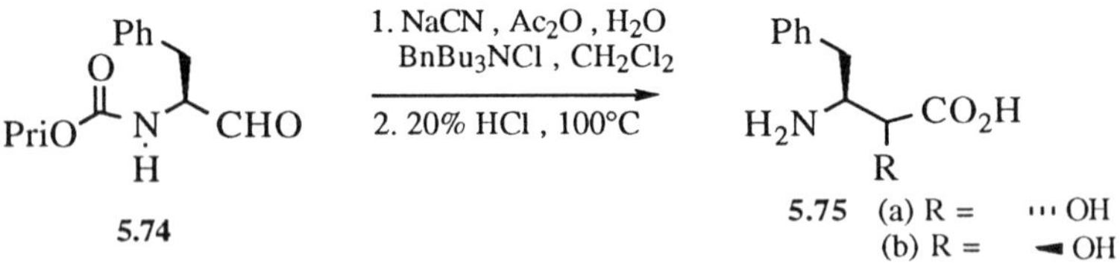

5.74

5.75

There are other synthetic transformations of phenylalanine, including conversion of the acid moiety to a halide (*5.78*), that allow a chain extension reaction by coupling with the anion of dimethyl malonate.[38] In this case, the amino group was incorporated in the starting material rather than using an amine surrogate. This

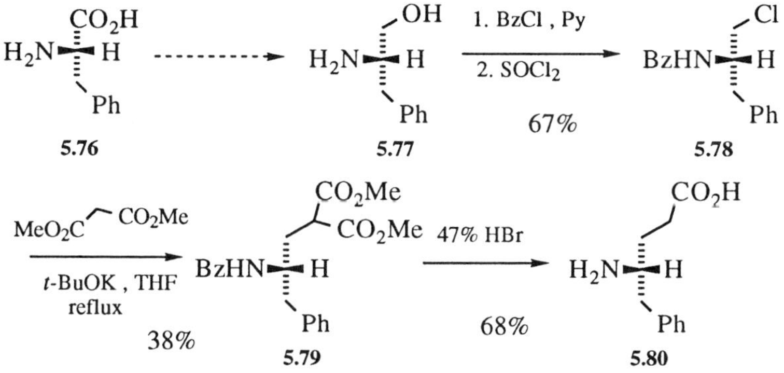

particular sequence began by conversion of phenylalanine to *5.77*, by a known route.[39] Protection of the amine and conversion of the alcohol moiety to chloride *5.78*, via reaction with thionyl chloride, was followed by reaction with the enolate anion of dimethyl malonate to give methyl 4-(N-benzylamino)-2-carbomethoxy-5-phenylpentanoate, *5.79*. Treatment of this malonate moiety with 47% HBr led to hydrolysis and decarboxylation, giving 4-amino-5-phenylpentanoic acid (*5.80*) as the final product.[38]

Phenylalinal derivatives can also be reacted with potassium cyanide to generate hydroxy amino acids after further manipulation of the product. Conversion of N-Boc phenylalanal (*5.81*) to 3S-amino-2-hydroxy-4-phenylbutanoic acid (*5.82a*)[38] and conversion of *5.83* to 3R-amino-2-hydroxy-4-phenylbutanoic acid (*5.82b*) as an 81:19 *anti:syn* mixture illustrates this synthetic strategy.[39,40] Note that modifying the synthetic sequence and also the carbamate protecting group allowed a measure of control in the stereochemistry of the final amino acid. Another condensation

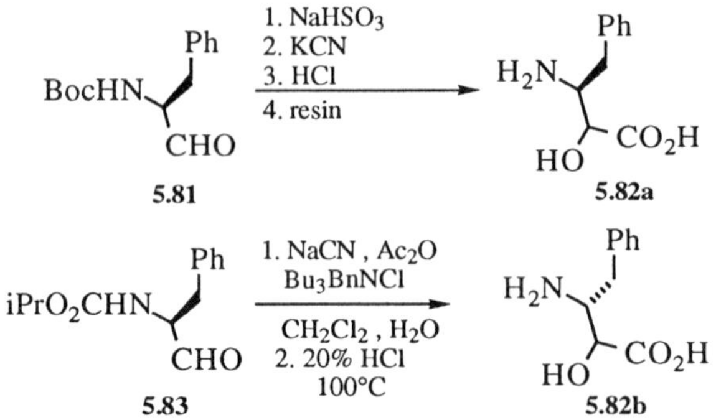

reaction used titanium reagent ***5.84*** in a reaction with ***5.81*** to give ethyl 5-(N-Boc amino)-4-hydroxy-6-phenylhexanoate, ***5.85***, in a highly stereocontrolled synthesis.[41]

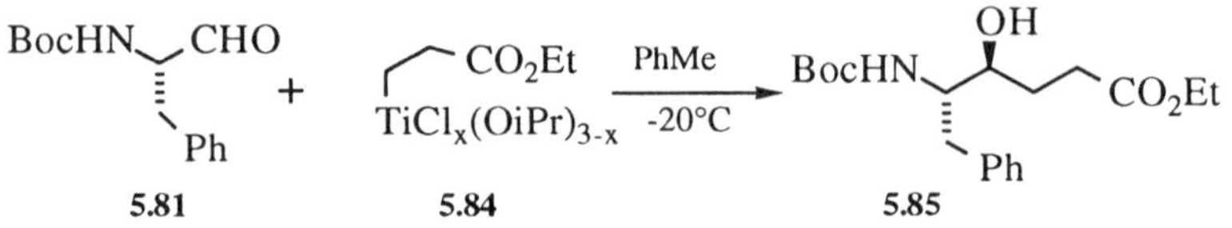

The reaction of an aldehyde with a phosphorus ylid was noted several times in previous sections. Sulfur ylids such as dimethylsulfonium methylid used here also react with aldehydes, in this case with ***5.81***, to give an epoxide. Subsequent opening

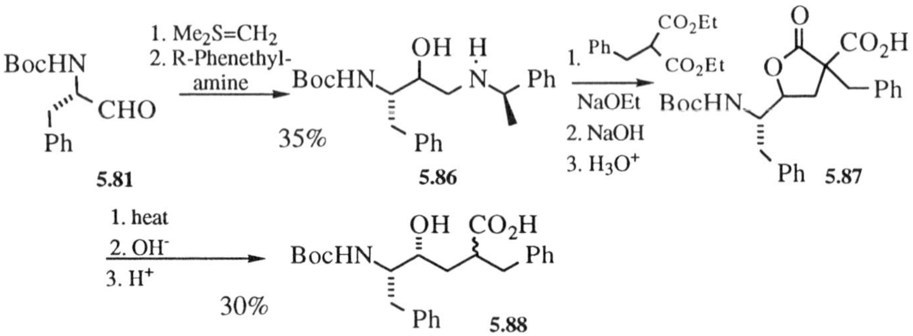

of the epoxide ring by the nitrogen of *R*-phenethylamine gave 2R,3S,5R ***5.86***.[42] In this case, the phenethyl group is both a nitrogen protecting group and also a chiral auxiliary. Condensation of ***5.86*** with the sodium enolate of with ethyl 2-benzylmalonate led to lactone ***5.87***, which was opened and decarboxylated to give N-Boc 5-amino-4-hydroxy-6-phenyl-2-benzylhexanoic acid (***5.88***).[42]

Phenylalanine is an important source of another class of amino acids, those that contain a cyclohexyl ring. Clearly, a phenyl ring can be reduced to a cyclohexane derivative, usually by catalytic hydrogenation or by treatment with an alkali metal in liquid ammonia. Such derivatives are often used as isosteres of statine (see chapter six, section 6.3), but several fundamental reaction types are shown here. Treatment of ***5.89***, for example, with hydrogen and a rhodium catalyst reduced the phenyl ring to a cyclohexane ring. Subsequent oxidation with sulfur trioxide•pyridine and DMSO gave the corresponding aldehyde, ***5.90***.[43] This aldehyde was then treated with HCN to give 3-amino-4-cyclohexyl-2-hydroxybutanoic acid, ***5.91***, as a mixture of diastereomers (7:3 2R,3S:2S,3S) in 60% overall yield.[43] In a continuation of

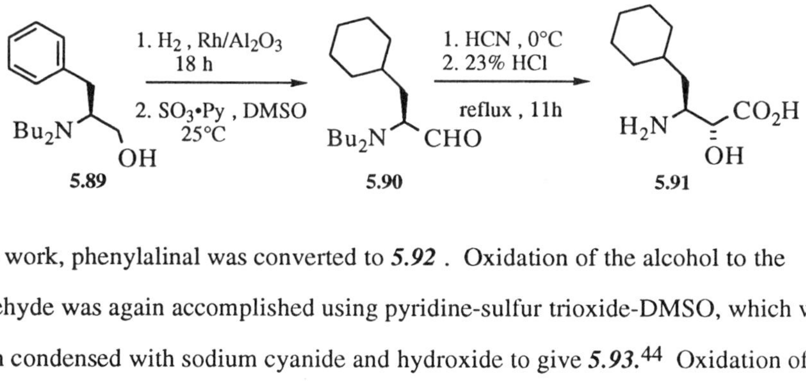

5.89 **5.90** **5.91**

this work, phenylalinal was converted to ***5.92*** . Oxidation of the alcohol to the aldehyde was again accomplished using pyridine-sulfur trioxide-DMSO, which was then condensed with sodium cyanide and hydroxide to give ***5.93***.[44] Oxidation of the N-benzyl group led to the *syn*- derivative rather than the *anti*- derivative of ***5.93***.

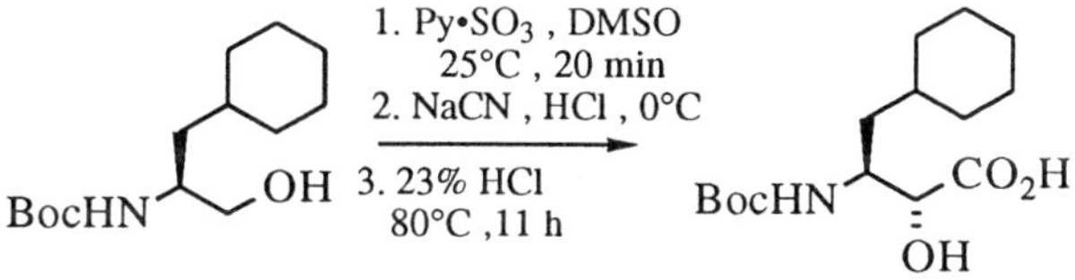

5.92 **5.93**

A similar sequence to those described above converted phenylalanine (***5.76***) to N-tritylcyclohexyl derivative ***5.94***.[45] This reaction sequence took a novel turn when ***5.94*** was converted to a phosphite ester (***5.95***). Treatment with sodium hydride gave a very novel, highly functionalized ylid that reacted with the keto-ester shown to give ***5.96***. Basic hydrolysis led to 5-(N-tritylamino)-6-cyclohexyl-2-dimethyl-methylene-4-oxohexanoic acid, ***5.97***.

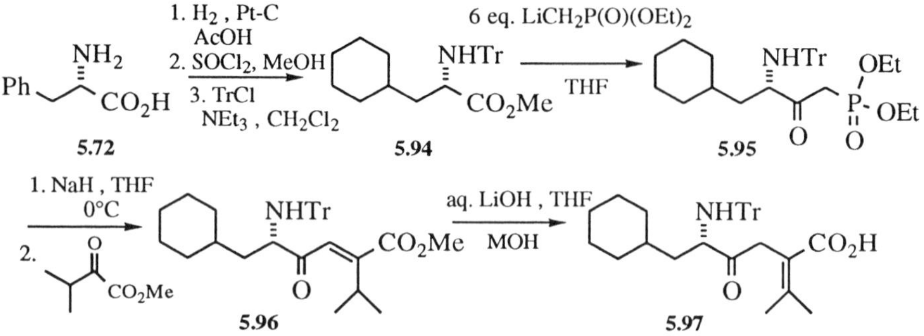

5.1.F. Serine and Derivatives

L-Serine (***5.98***) incorporates an hydroxyl group in the amino acid structure. As with other amino acids seen in this chapter, the amino group can be transformed into an hydroxyl group and this leads to an interesting transformation. The amino group of ***5.98*** was converted to a bromine, with complete retention of configuration, by reaction with NOBr to give *S*-***5.99***.[46] Upon treatment with ammonium hydroxide,

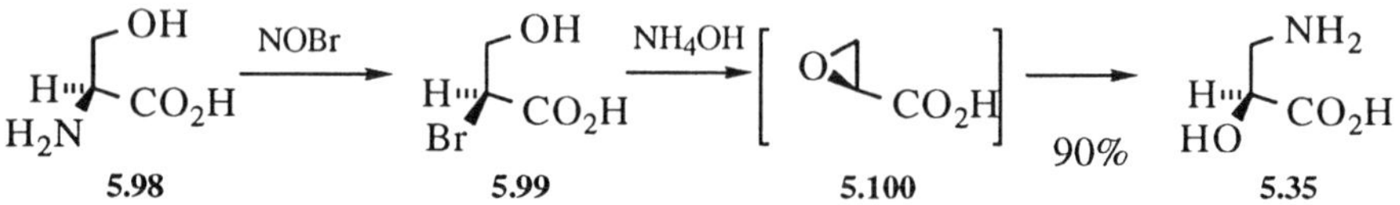

epoxide ***5.100*** was generated *in situ* and then opened by ammonia to give 3-amino-2-hydroxypropanoic acid (*D*-isoserine, ***5.35***).[46] The net result is that the CH_2OH moiety was transformed into a $-CH_2NH_2$ moiety and the $-NH_2$ in ***5.98*** was converted to -OH in ***5.35***. Similarly, *L*-threonine was converted to 2R,3S-2-hydroxy-3-amino-butanoic acid (isothreonine).[47] The isolation and direct conversion of ***5.100*** to ***5.35*** has been reported via reaction with aqueous ammonia.[48]

In another sequence, the methyl ester of N-Boc-*D*-serine (***5.101***) was converted to isoxazolidine carboxaldehyde ***5.102***.[49] Garner used a Diels-Alder reaction between the aldehyde moiety and Danishefsky's diene[50] to give ***5.103***, as a 3:1 β:α-hydrogen mixture. The pyranose moiety was converted to an hydroxy-ester and deprotection of the isoxazolidine group gave methyl 4-(N-Boc amino)-3,5-dihydroxypentanoate, ***5.104***.[49]

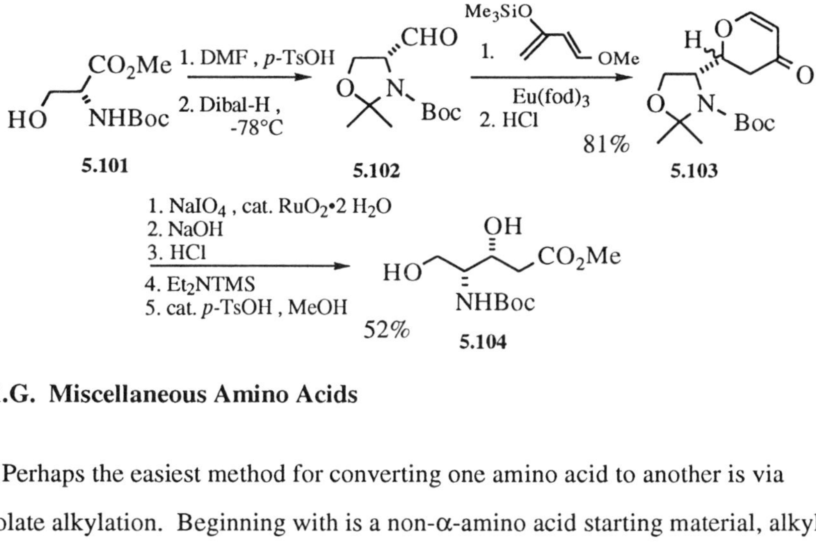

5.101 **5.102** **5.103** **5.104**

5.1.G. Miscellaneous Amino Acids

Perhaps the easiest method for converting one amino acid to another is via enolate alkylation. Beginning with is a non-α-amino acid starting material, alkylation of its enolate anion can generate a new amino acid, sometimes with high

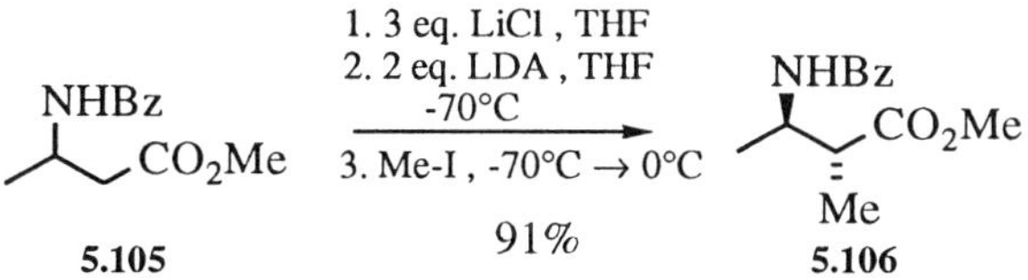

5.105 **5.106**

diastereoselectivity. Addition of three equivalents of lithium chloride to the lithium enolate of ***5.105*** enhanced the diastereoselectivity (>99% de) for the *anti*-diastereomer [methyl 3-(N-benzoylamino)-2-methylbutanoate, ***5.106***] upon reaction with iodomethane.[51]

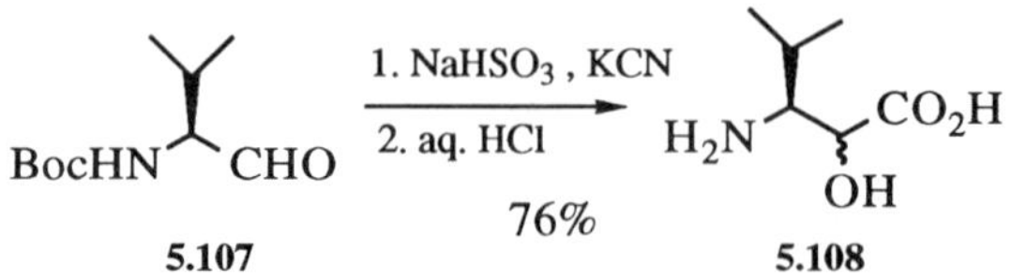

As in previous sections, α-amino acids can be used to prepare non-α-amino acids. The methyl ester of *L*-valine, for example, was converted to N-Boc-valinal (***5.107***) in 82% yield.[52] Subsequent treatment with sodium bisulfite and potassium cyanide,[18b]

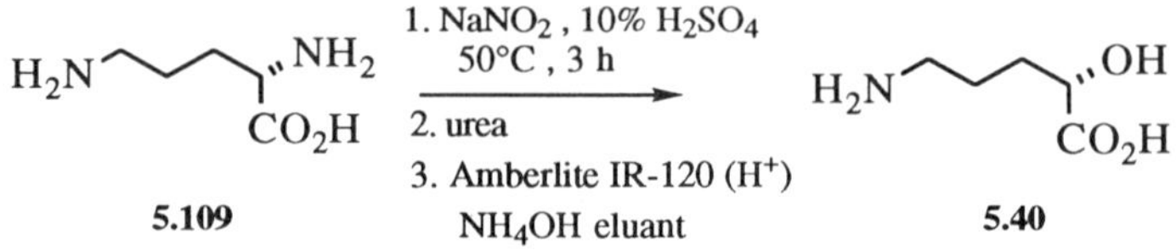

followed by acid hydrolysis gave 3-amino-2-hydroxy-4-methylpentanoic acid (***5.108***), as a mixture of diastereomers.[18a,30b] Another readily available α-amino acid is *L*-ornithine (***5.109***, see reference 22 for other synthetic transformations of ornithine),

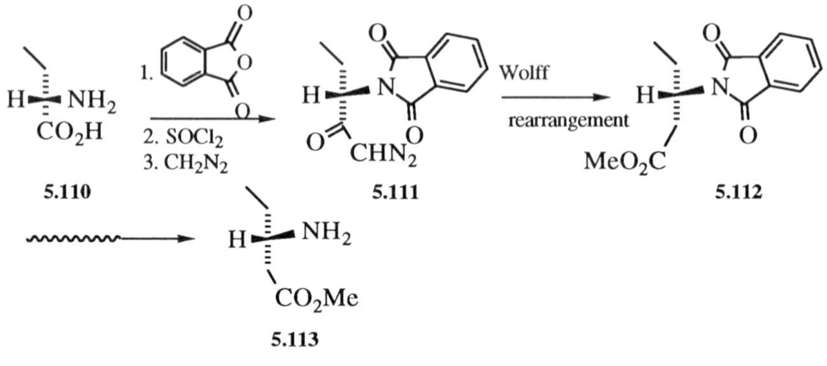

which reacted with $NaNO_2$ in sulfuric acid (forming HONO). This reagent converted the α-amino group to a hydroxyl moiety (as seen in previous examples in this chapter), and hydrolysis generated 2-hydroxy-5-aminopentanoic acid (*5.40*) in good yield.[21]

The Wolff rearrangement[53] is a classical method for the chain extension of acids via rearrangement of an α-diazoketone intermediate. This technique has been used to conversion α-amino acids to β-amino acids. When S-2-aminobutanoic acid (*5.110*) was first converted to the phthalimido derivative, the acid moiety was treated with thionyl chloride and then diazomethane to give diazoketone ***5.111***.[54] Wolff rearrangement and esterification generated methyl ester ***5.112***, which was converted to 3-aminopentanoic acid, ***5.113***.[54]

A somewhat different synthetic approach began with amino lactone ***5.114*** (derived from homoserine). Reduction with diisobutylaluminum hydride gave hemi-acetal derivatives ***5.115:5.116*** as a 1:5 mixture.[55] Subsequent Wittig olefination[33] gave ***5.117*** in 80% overall yield from ***5.114***. Catalytic hydrogenation of the alkene moiety and cleavage of the trityl group gave 6-hydroxy-4-aminohexanoic acid, ***5.118***.[55]

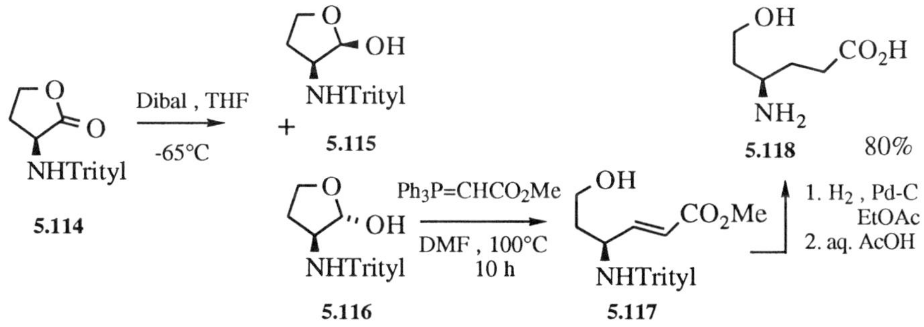

A more elaborate synthesis involved condensation of ***5.119*** and ***5.120*** to give ***5.121***.[56] Removal of the acyl protecting group and reduction gave ***5.122***. Treatment with camphorsulfonic acid induced a transfer of the "protecting group" (converted an oxazolidine ring to an acetonide) to give ***5.123***. Oxidation of the primary alcohol moiety to a carboxylic acid gave the acetonide of 5-(N-Boc amino)-6-cyclohexyl-2-isopropyl-3,4-dihydroxyhexanoic acid, ***5.124***.

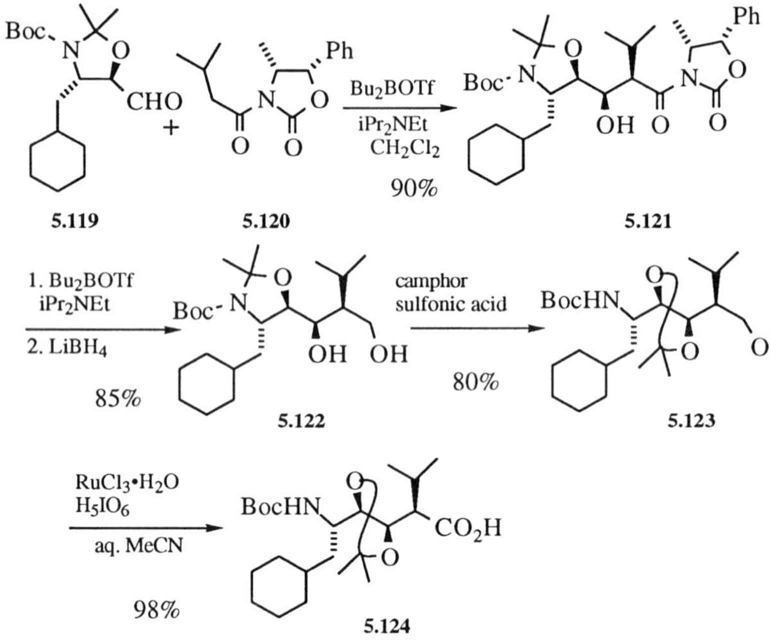

Aldehydes can also be generated by oxidative cleavage of alkenes. Acetylation of ***5.125*** was followed by oxidative cleavage of the alkene and conversion of the resultant aldehyde to a conjugated ester via Wittig olefination.[33] The final product was *t*-butyl-3-amino-6-carbobenzoxy-4-(O-acetyl)-7-phenylhept-5-enoate, ***5.126***.[57] Other amino acids were prepared by changing the ylid substituent from carbo-alkoxymethyl to alkyl, and by modifying the phenyl group to an alkyl substituent.

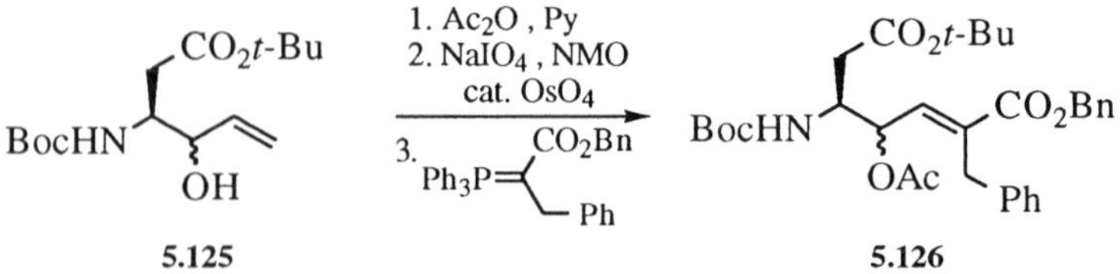

5.2. Diastereoselective and Enantioselective Reactions

A variety of reaction types have been used for the synthesis of amino acids in this chapter and the ones preceding it. Many reactions do not fit into a single category, as did the reactions in section 5.1. Those reactions that proceed with good diastereoselectivity or with reasonable enantioselectivity are collected into this section.

5.2.A. Diastereoselective Ester Enolate Reactions

There are two general categories in this section. Reactions that produce diastereomers but with no enantioselectivity comprise one category. The other will present examples that produce amino acids of good enantiopurity. A chiral auxiliary is used in some cases. Such auxiliaries are chiral, nonracemic molecular fragments that are chemically attached to another molecular fragment. The asymmetry of the auxiliary will induce asymmetry in subsequent reactions and is later chemically cleaved from the molecule. The resulting product will have some percentage of enantiopurity (usually reported as % ee, or % enantiomeric excess) due to asymmetric induction when the auxiliary was attached.

5.2.A.i. No Auxiliaries

In this section, no auxiliary will be attached to the substrates, but the reactions will proceed with some degree of diastereoselectivity and will generally produce chiral, racemic products, as seen in section 5.1.G. Seebach used an ester enolate

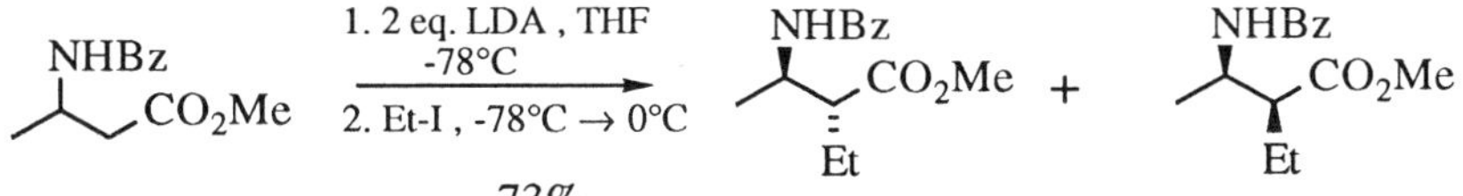

5.105 **5.127a** **5.127b**

strategy to produce alkyl substituted aminoalkanoic acids, but used β-amino-ester derivatives as precursors. When ***5.105*** was treated with two equivalents of lithium diisopropylamide and then with iodoethane, racemic methyl 3-(N-benzoylamino)-2-ethylbutanoate was produced, as a 16:1 mixture of diastereomers ***5.127a:5.127b***.[58] Similar reaction with iodomethane gave a 50% yield of methyl N-benzoyl-3-amino-2-methylbutanoate (4:1 *anti:syn*). Reaction with allyl bromide gave a 90% yield of methyl N-benzoyl-2-allyl-3-aminobutanoate (31:1 *anti:syn*) and benzyl bromide led to a 90% yield of methyl 3-(N-benzoylamino)-2-benzylbutanoate (36:1 *anti:syn*).[58] When ***5.127*** was used in this sequence, with benzyl bromide, methyl 3-(N-benzoyl-amino)-2-benzyl-2-ethylbutanoate (***5.128***) was formed (99:1 *anti:syn*).[58]

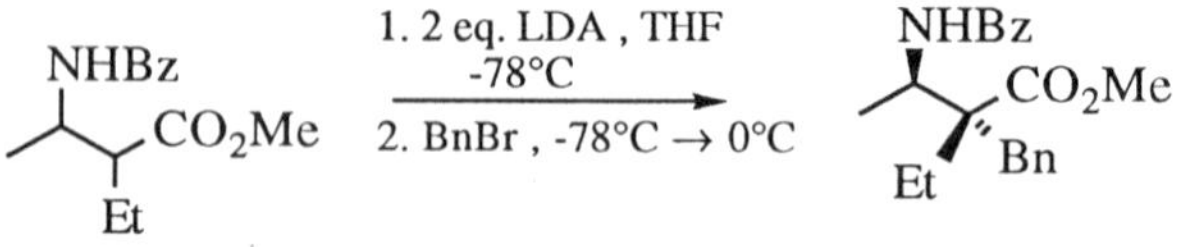

5.127 **5.128**

5.2.A.ii. With Chiral Auxiliaries or Using Chiral Templates

Two general methods can be used to induce asymmetry in a synthesis: using a chiral auxiliary, as defined above, or using a chiral, nonracemic starting material (a

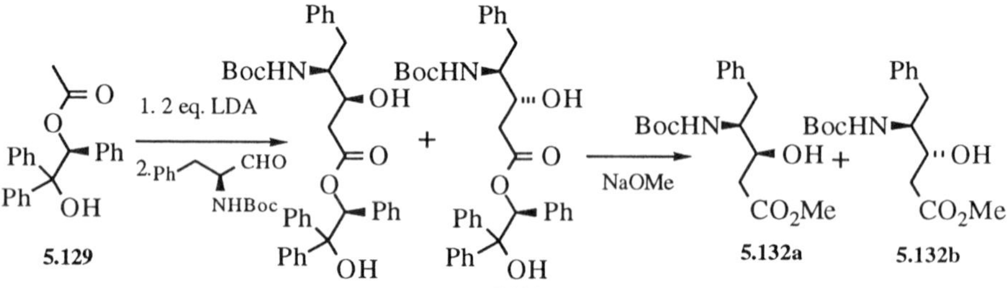

5.129 **5.130** **5.131** **5.132a** **5.132b**

chiral template).[1] Ester ***5.129*** contains a chiral auxiliary in the form of the alcohol portion of the ester. The enolate derived from ***5.129*** in a reaction with N-Boc phenyl-alinal to give a 9:1 mixture of ***5.130*** and ***5.131***.[59] Treatment of this mixture with sodium methoxide led to a 9:1 mixture of ***5.132a*** and ***5.132b*** in 60% overall yield.

Seebach used amino-esters such as ***5.133***[51] where the asymmetric center at C_3 (the amino-bearing carbon) makes this amino-ester a chiral template. In subsequent reactions, that chiral center will induce asymmetry into the new products. Ethyl N-benzyl-3-aminobutanoate (***5.133***) was treated with 2.2 equivalents of lithium diiso-propylamide to give dianion ***5.134***.[51] Quenching with iodomethane gave ***5.135a*** (80% de, where de refers to diastereomeric excess) and reaction with iodoethane gave ***5.135b*** (93%de).[51] The alkylation reaction was highly diastereoselective and each diastereomer was produced with good enantiopurity. The methyl ester gave similar results, although the yield of the alkylation product varied with the nature of the halide that was used. Dianion ***5.134*** was also condensed with aldehydes to give hydroxy amino acid derivatives.

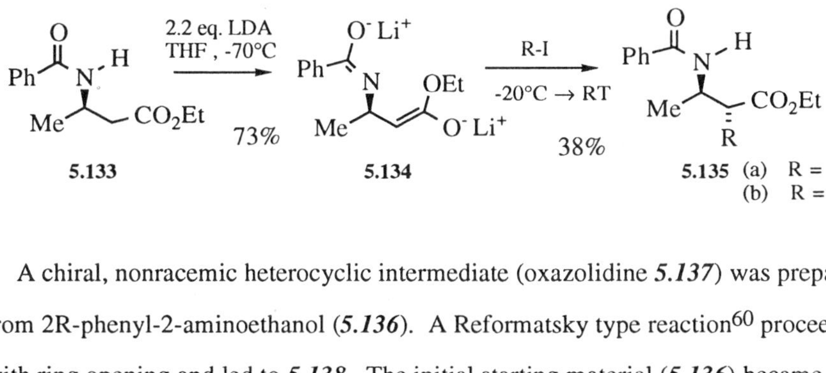

5.133 **5.134** **5.135** (a) R = Me
(b) R = Et

A chiral, nonracemic heterocyclic intermediate (oxazolidine ***5.137***) was prepared from 2R-phenyl-2-aminoethanol (***5.136***). A Reformatsky type reaction[60] proceeded with ring opening and led to ***5.138***. The initial starting material (***5.136***) became part

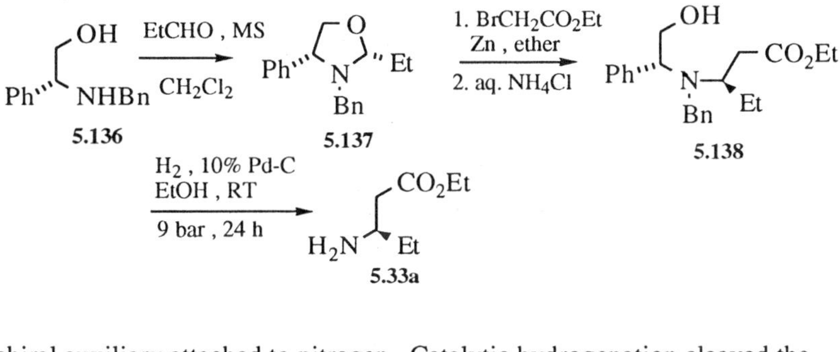

5.136 **5.137** **5.138** **5.33a**

of a chiral auxiliary attached to nitrogen. Catalytic hydrogenation cleaved the benzylic auxiliary to give ***5.33a*** (ethyl 3-aminopentanoate) in 55% overall yield (72% ee, R).[61] Both ethyl 3-aminobutanoate and ethyl 3-aminoheptanoate were prepared by this method.

A final example used a chiral, nonracemic amino aldehyde as a chiral template. Reaction of ***5.139*** with vinyl Grignard led to amino alcohol ***5.140*** as a 6:1 (*syn:anti*) mixture of diastereomers.[62] Reaction with 1-ethoxy-ethanol protected the alcohol and oxidative cleavage of the alkene led to 3-(N-Boc amino)-2-(1-ethoxyethoxy)-3-phenylpropanoic acid, ***5.141***.

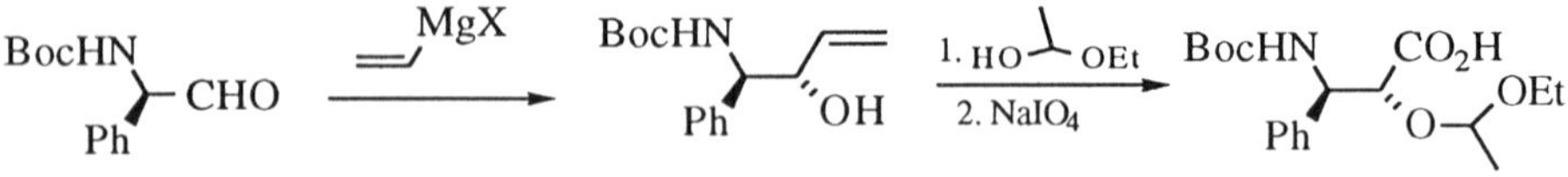

5.139 5.140 5.141

5.2.B. Diastereoselective Conjugate Additions

Several conjugate addition reactions have been used for the synthesis of amino acids, as seen in chapter three. It is easy to prepare conjugated esters and using chiral auxiliary or a chiral template, conjugate addition can lead to amino acids with high diastereoselectivity and enantioselectivity.

5.2.B.i. Diastereoselective Reactions With and Without Chiral Auxiliaries

Conjugate addition of the lithium salt of a chiral amine to a β-substituted α,β-unsaturated ester leads to formation of a chiral, nonracemic amino acid. Addition of the chiral, nonracemic lithium amide ***5.143*** (contains a phenethyl auxiliary) to ***5.142*** gave the amino-ester.[63] Catalytic hydrogenation removed both benzylic groups (the auxiliary and the benzyl group) and acid hydrolysis of the ester moiety led to 3-amino-3-(4-benzyloxyphenyl)-propanoic acid, ***5.144***.[69] The initial Michael adduct was formed with >99% dr (dr is diastereomeric ratio), leading to high enantioselectivity in ***5.144*** after removal of the auxiliary.

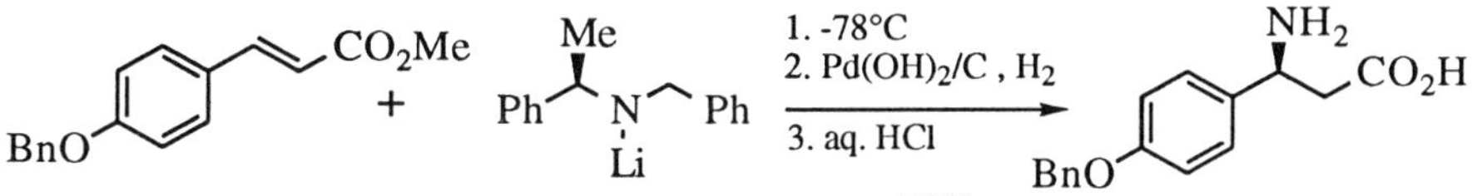

5.142 **5.143** **5.144**

Addition of amines to conjugated esters can be highly diastereoselective, even if a chiral auxiliary is not attached to the amine. Benzylamine, for example, added to ***5.145*** to give ***5.146a*** and ***5.146b*** with good selectivity for ***5.146a*** (the *threo* diastereomer).[64] Catalytic hydrogenation cleaved the N-benzyl group to give ethyl 3-amino-2-(3,4-dichlorophenyl)butanoate.

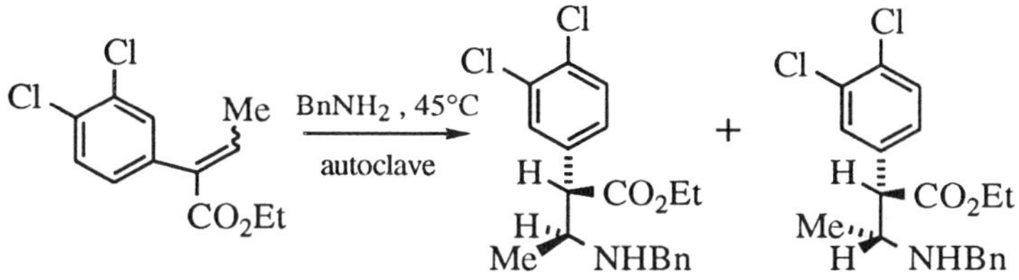

5.145 **5.146a** **5.146b**

A different approach is to attach a chiral auxiliary to the ester, and then react it with an achiral amine. This approach was used by d'Angelo when diphenylmethylamine was added to 2-(1-β-naphthyl-1-methylethyl)-5-methylcyclohexyl but-2E-enoate (***5.147***), a menthol analog. High pressure[65] Michael addition (15 kbar) led to

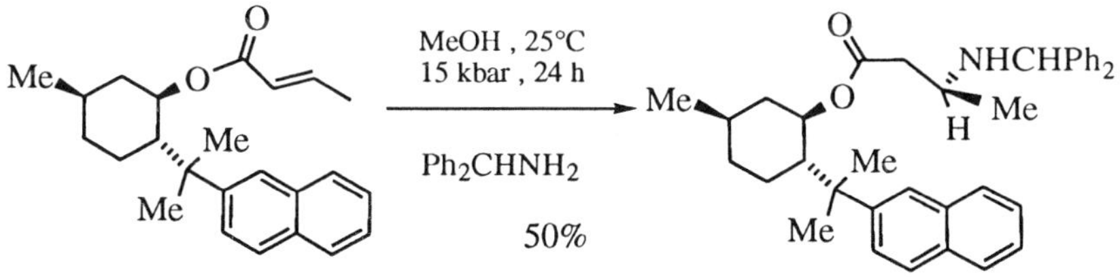

5.147 **5.148**

that same ester of 3S-(N-diphenylmethylamino)butanoic acid (***5.148***) in >99%ee.[66]

It is also possible to do "conjugate" additions with α,β-epoxy esters using amine nucleophiles. A problem with this approach is that the amine can attack epoxy carbons at both the α- and the β-position relative to the ester moiety. An example is the reaction of epoxy acid ***5.149***[67] with diethylamine, which opened the epoxide ring to give a 1:6 mixture of ***5.150a:5.150b***. The diastereoselectivity of this ring opening was improved to >20:1 (***5.150a:5.150b***) by addition of *tetrakis*-isopropoxy titanium (IV) to the initial reaction with diethylamine.[67]

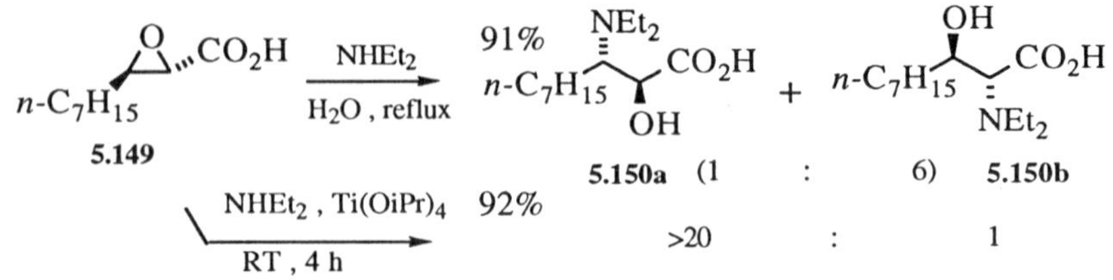

Another unusual conjugate addition involved the reaction of a conjugated ester with an isocyanate. α-Fluoro-ester ***5.151***, for example, was converted to aldehyde ***5.152*** and then Horner-Wadsworth-Emmons olefination gave ***5.153***.[68] Reaction of ***5.153*** with chlorosulfonyl isocyanate (CSI, see chapter two - section 2.4), followed by treatment with base, gave ***5.154***. Subsequent basic hydrolysis gave 3-amino-4-fluoro-5-hydroxyl-4-methylhexanoic acid (***5.155***).

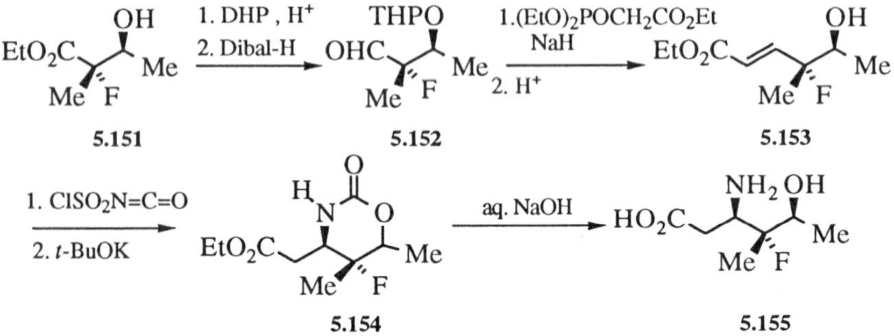

An isocyanate was used in another "conjugate" addition reaction. When ***5.156*** (which contains a chiral phenethyl group on nitrogen as a protecting group and as a chiral auxiliary), was condensed with tosyl isocyanate, pyrazolone ***5.157*** was formed.[69] Removal of the N-protecting group and opening the ring with lithium and ammonia was followed by acid hydrolysis to give 3,4-diaminobutanoic acid, ***5.158***.

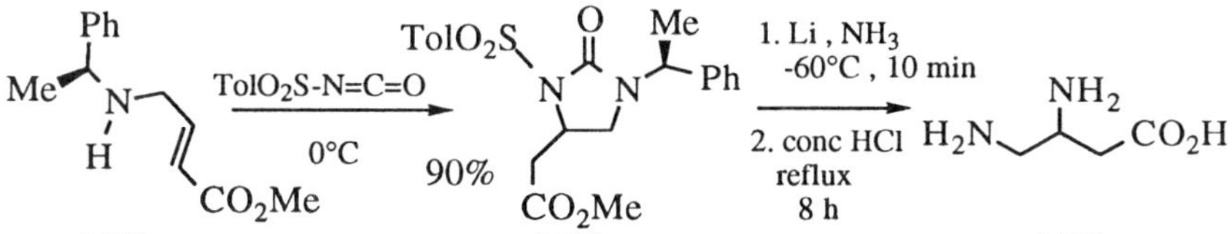

5.156 **5.157** **5.158**

5.2.B.ii. Using Chiral Templates

One example was reported that could be included in this section. The conjugate addition reaction was internal, where the attached carbamoyl nitrogen adds to the

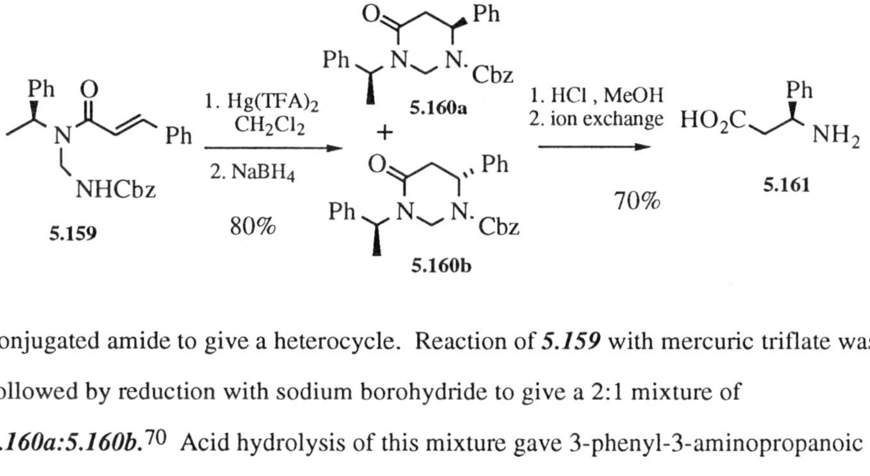

conjugated amide to give a heterocycle. Reaction of ***5.159*** with mercuric triflate was followed by reduction with sodium borohydride to give a 2:1 mixture of ***5.160a:5.160b***.[70] Acid hydrolysis of this mixture gave 3-phenyl-3-aminopropanoic acid (***5.161***, with 33%ee).

5.2.C. Enzymatic Transformations

The carbonyl moiety of oxo amino acids can be reduced to give hydroxy amino acids. Although sodium borohydride is a good reagent, it does not give highly diastereoselective reductions in the absence of steric effects or neighboring group effects. Enzymatic reductions, however, often proceed with excellent enantioselectivity and good-to-excellent diastereoselectivity. In one example, the organism *Saccharomyces carlsbergenssis* ATCC 2345 reduced the ketone moiety selectively in methyl 2-oxo-4-aminobutanoate (***5.162***) to give methyl S-4-amino-2-hydroxybutanoate, ***5.163***.[71]

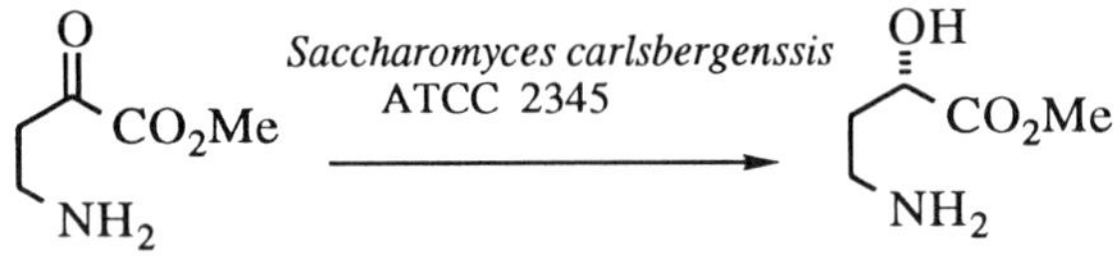

5.162 **5.163**

Specific enzymatic processes can introduce functionality in an enantioselective process. N-Benzyl-2-pyrrolidinone (***5.164***) was hydroxylated directly using an available enzymatic system, *Beauveria sulfurescens* (ATCC 7159).[72] The hydroxy-lactam (***5.165***) was then hydrolyzed and de-benzylated to give S-2-hydroxy-4-aminobutanoic acid, ***5.25***.

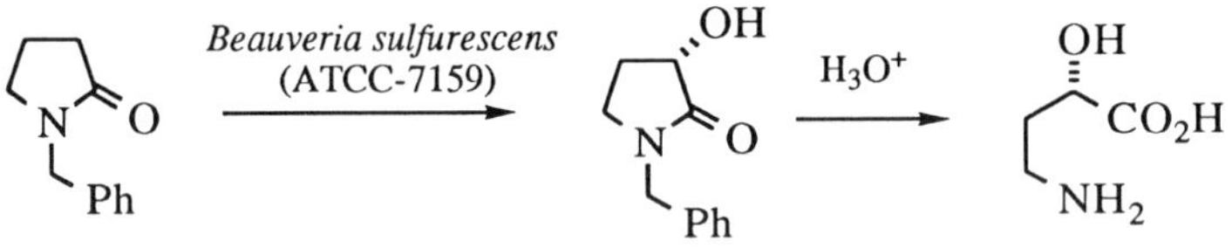

5.164 **5.165** **5.25**

One method for preparing chiral, nonracemic molecules is by resolution. If a reaction occurs with one enantiomer faster than with its antipode, that enantiomer forms a product and its antipode remains unchanged. Since these different molecules can be separated, resolution has occurred. When diester ***5.166*** was treated

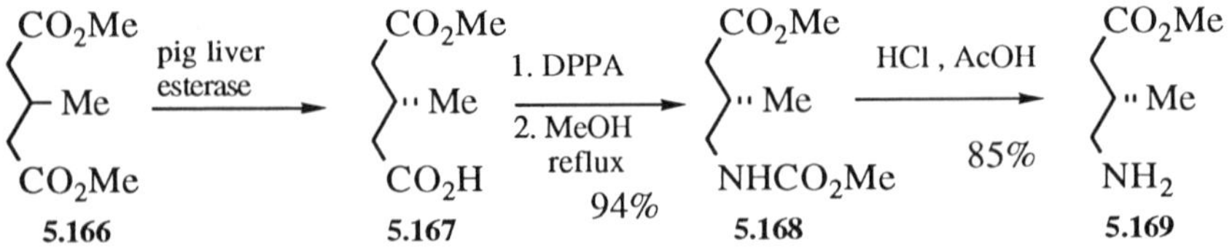

5.166 **5.167** **5.168** **5.169**

with pig liver esterase, a 46% yield of half-ester (*5.167*) was obtained (since only 50% of the ester could react, 50% represents 100% reaction and a 46% yield represents a 92% yield based on reactivity).[73] The unreacted diester was separated from the half-ester product of the enzymatic hydrolysis. Reaction of the acid moiety in *5.167* with diphenylphosphoryl azide (DPPA) followed by conversion of the resulting amine to its carbamate (with methanol) resulted in ***5.168.*** Acid hydrolysis cleaved the carbamoyl group to give ***5.169*** (3S-methyl 4-aminobutanoic acid) with high overall enantioselectivity.[73] It is noted that when ***5.167*** was reacted with ammonia and an iodoso compound, a 60% yield of 3R-methyl 4-aminobutanoic acid was obtained.[73]

5.2.D. Reductions

Reductive amination was used in chapter one (section 1.1.C.i) for the preparation of amino acids. This reaction can be highly diastereoselective and when a chiral template such as ***5.170*** was used, the reduction was highly enantioselective. Reaction of keto-ester ***5.170*** with hydroxylamine gave an oxime, which was treated with cyanoborohydride to reduce to imine moiety and give hydroxylamine ***5.171.*** This was reduced by hydrogenation over Raney nickel to give ***5.172***.[74] Acid hydrolysis gave 2-methyl-3-aminobutanoic acid, ***5.173***.

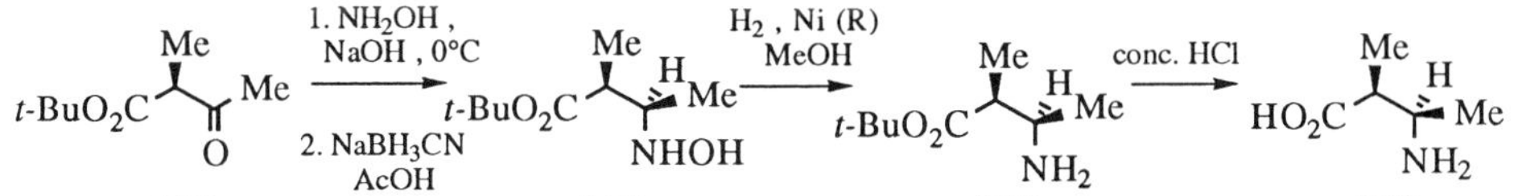

5.170 5.171 5.172 5.173

5.2.E. Transformations of Malic Acid Derivatives as Chiral Templates

Malic acid can be obtained in optically pure form and it serves as a useful chiral template for the synthesis of amino acids. *S*-Malic acid (***5.174***) was easily converted to ***5.175*** via reaction with formaldehyde.[75] This allowed the preparation of diazoketone ***5.176*** and a Curtius rearrangement[76] led to 2-hydroxy-3-aminopropanoic acid (isoserine, ***5.35***) after hydrolysis. In this work, isoserine was cited as having important biological activity[77] as well as being a constituent of the antibiotic edeine.[78] *S*-Malic acid was also converted to a chiral, nonracemic imide that was an important

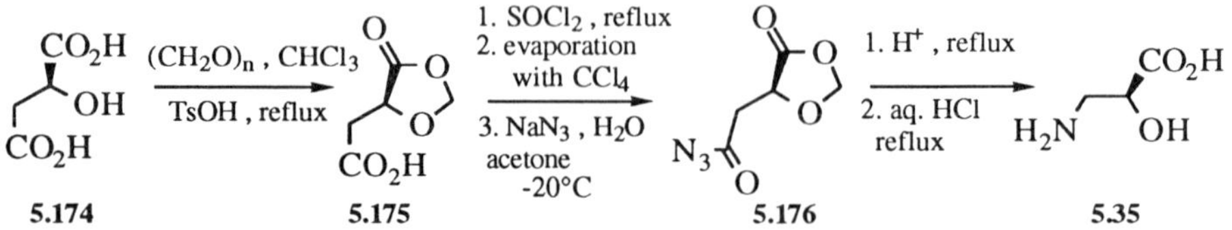

synthetic intermediate. Reaction of ***5.174*** with benzylamine gave the chiral, nonracemic imide, ***5.177***.[79] Reduction of one imide carbonyl with sodium borohydride, protection of the resulting alcohol, and coupling with tri-*n*-butylallyl tin led to selective formation of ***5.178*** (as a 3:1 *trans:cis* mixture). Removal of both the O- and N-benzylic groups gave ***5.179***. A new set of protecting groups (see ***5.180***) was necessary to allow separation of diastereomers before final ring opening of the lactam[80] led to 4S-amino-3S-hydroxyheptanoic acid, ***5.181***.[80]

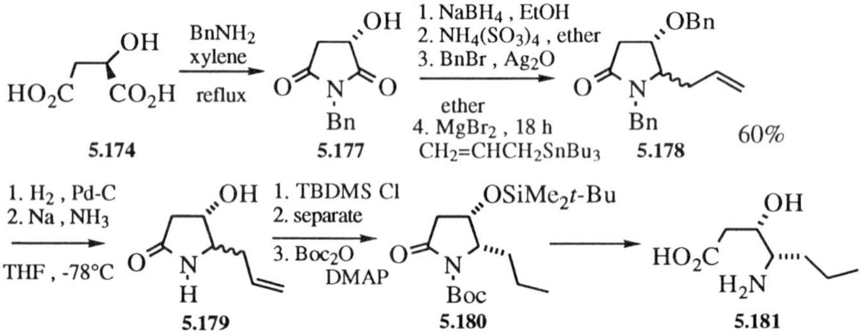

A closely related chiral template is *L*-β-malamidic acid (***5.182***), which was converted to ***5.183*** using a hypervalent iodine species.[81] Acid hydrolysis gave S-isoserine (3-amino-2-hydroxypropanoic acid, ***5.35***).

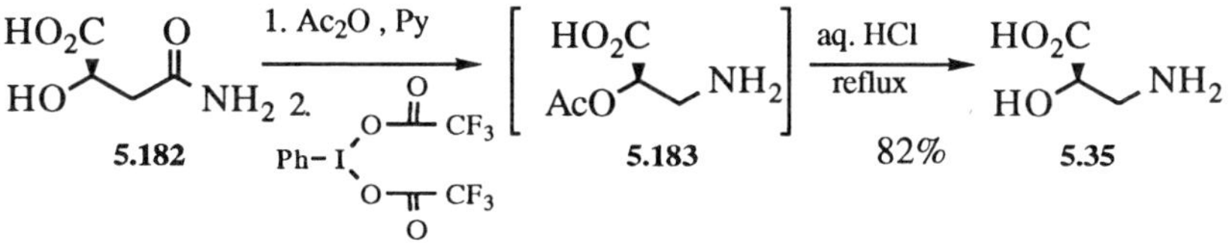

5.2.F. Transformations Using Carbohydrates as Chiral Templates

In Hanessian's work,[1] carbohydrates were important chiral templates for a variety of total syntheses. The chirality and functionality inherent to carbohydrates make them ideal for this purpose. One asymmetric synthesis of an amino acid began with a chiral, non-racemic tartrate derivative ***5.184***, which was converted to ***5.185***.[82] Reduction of the bromo-methyl moiety to a methyl group and manipulation of the protecting groups gave ***5.186***. Treatment of this mono-tosylate with methoxide generated an alkoxide that displaced the tosyl group to give an epoxide, and this was opened by azide and the alcohol protected to give ***5.187***. Hydrogenation of the azide gave methyl 3-amino-2-hydroxybutanoate (***5.188***). Other diastereomeric butanoic acid derivatives were prepared from the analogous tartrate derivative with high

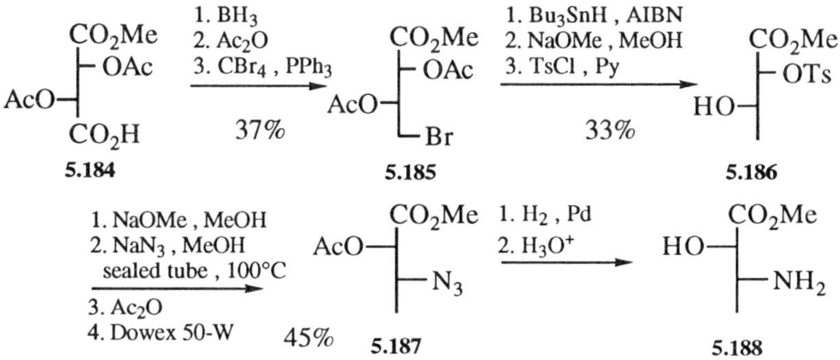

selectivity by this strategy. In another synthesis, *L*-(+)-tartaric acid (***5.187***) was converted to an amide-ester, and then acetylated to give ***5.190***.[83] Note that ***5.190*** is closely related to ***5.184***. Dehydration of the amide gave nitrile ***5.191*** and subsequent acid hydrolysis to give a mixture of 14% of ***5.192*** (4-amino-2,3-dihyroxybutanoic

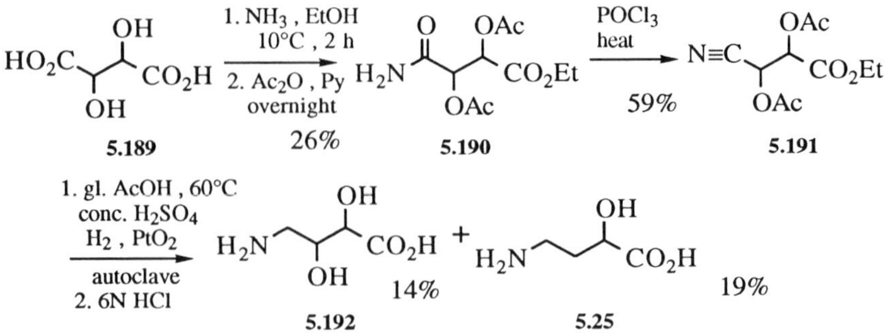

acid) and 19% of ***5.25*** (4-amino-2-hydroxybutanoic acid). The poor yields in the final steps, and the mixture of hydroxy amino acids make this approach less useful.

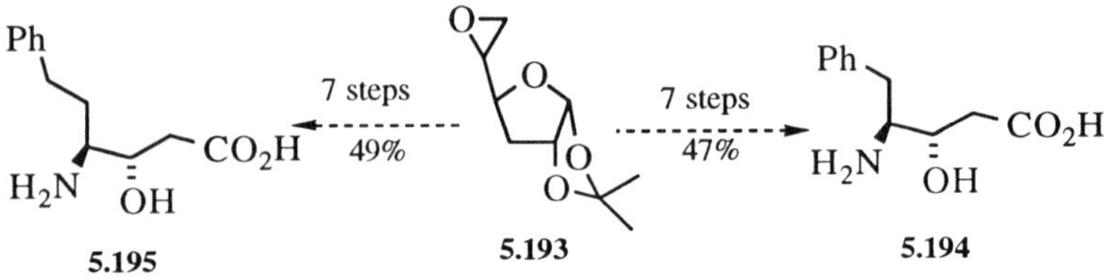

More complex carbohydrates can also be used. In another example, furanose ***5.193*** was converted to ***5.194*** in seven steps, and ***5.193*** was also converted to ***5.195***

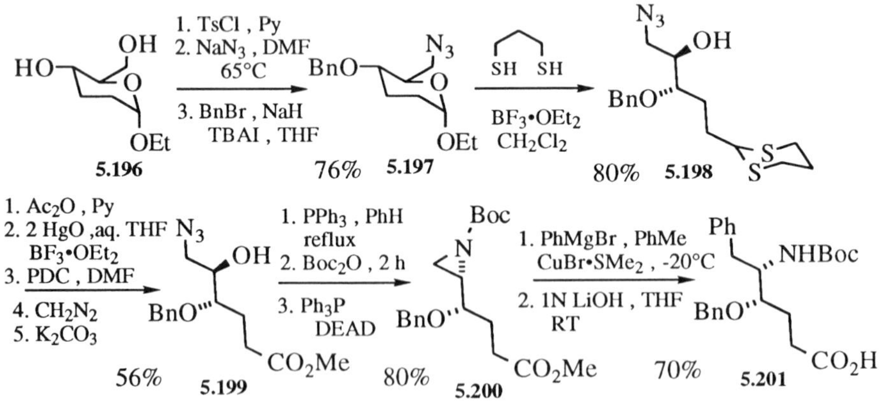

in seven steps by an identical process. Carbohydrate ***5.196*** was converted to azide ***5.197*** and the pyran ring was opened to give ***5.198***.[84] Unmasking the aldehyde and oxidation to an acid allowed its conversion to ***5.199***. Protection and an internal Mitsunobu reaction[85] led to aziridine ***5.200***. This was converted to 5-(N-Boc amino)-4-benzyloxy-6-phenylhexanoic acid (***5.201***) by opening the aziridine ring and protecting the functional groups.[84]

5.2.G. Miscellaneous Transformations Using Chiral Auxiliaries

This section contains synthetic approaches that do not easily fit into the other categories. They use chiral auxiliaries in the synthesis and are presented to show the variety that is possible in this general strategy. An example is the lengthy synthetic route that converted α-bromoacetic acid to isoserine. Chiral, nonracemic alcohol ***5.202*** (which is a menthol-like molecule), reacted with bromoacetic acid to give the

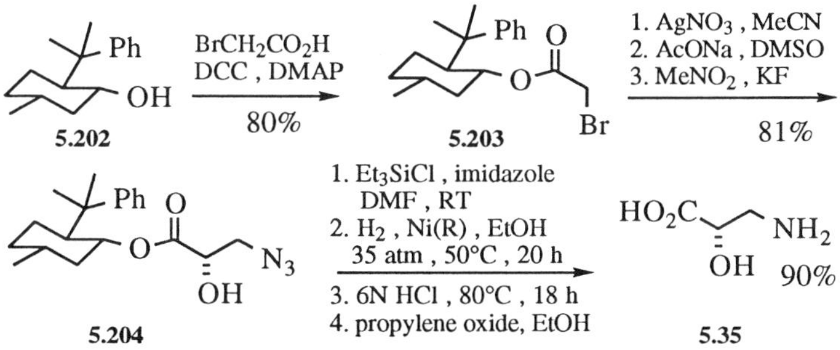

ester, ***5.203***. The "alcohol" becomes the chiral auxiliary in the chiral, non-racemic ester.[86] Reaction of the bromide with silver nitrate and then sodium acetate and nitromethane/KF led to azido-hydroxyester ***5.204***, with >95%de. Protection of the alcohol was followed by reduction of the azide. Hydrolysis and deprotection gave 3-amino-2S-hydroxypropanoic acid, *S*-(-)-isoserine (***5.35***).[86] This example demonstrates that even simple molecules can demand lengthy syntheses in order to produce a chiral, nonracemic product.

An achiral ketone (***5.205***) was used to prepare protected alcohol ***5.206*** via condensation with S-phenethylamine, and that protecting group was also an auxiliary.[87] Removal of the benzylic group and re-protection with Boc allowed Birch reduction[88] of the aromatic ring to give ***5.207***. Ozonolysis in methanol led to methyl 4-(N-Boc amino)-6-methyl-3-oxoheptanoate, ***5.208***.[87] In this synthesis, the phenyl ring functioned as the carboxyl surrogate and asymmetric induction in ***5.208*** arose from the chiral auxiliary introduced earlier in the sequence.

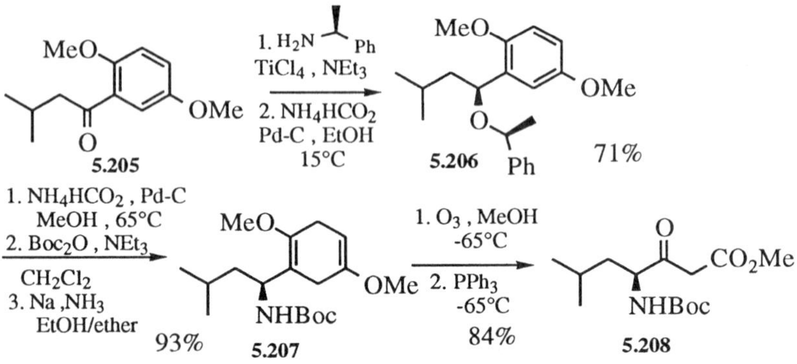

The reaction of S-phenethylamine and ethyl acetoacetate led to ***5.209*** (ethyl 3-(N-phenethyl)but-2-enoate)[89,90] as seen in chapter one, section 1.1.C.ii. Reduction of the double bond would give the ethyl 3-aminobutanoate derivative, and the presence of the chiral auxiliary would lead to asymmetric induction during that reduction.

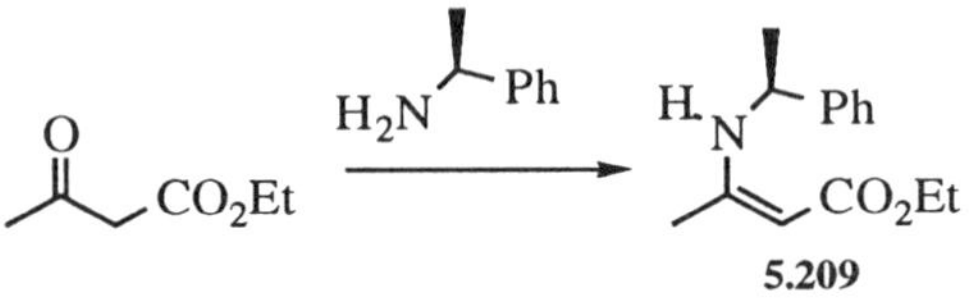

This was observed in the reduction of ***5.210*** with sodium cyanoboro-hydride, to give ***5.211***.[89] Catalytic hydrogenation cleaved the chiral N-benzylic auxiliary from nitrogen to give 3-amino-3-phenylpropanoic acid, ***5.161*** (2-89% ee in a variety of examples).

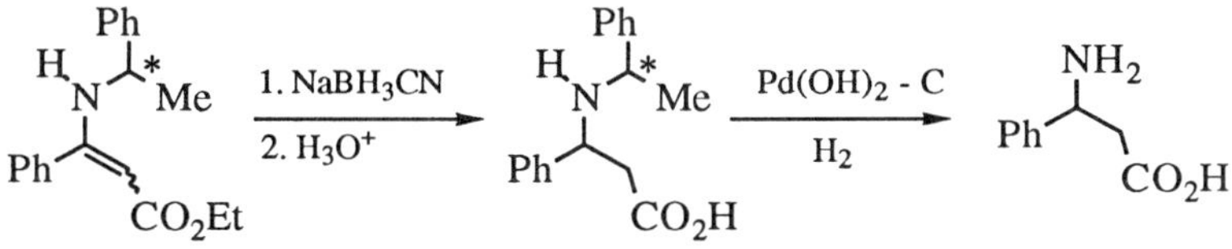

5.210 **5.211** **5.161**

Alkyne ***5.212*** was prepared from succinic anhydride in seven steps (19% overall yield). The auxiliary on the alcohol moiety allowed the preparation of chiral, non-racemic ***5.212***. Removal of the auxiliary gave the alcohol, ***5.213***.[91] Mitsunobu reaction[85] with phthalimide gave the phthalimidoyl group, with inversion of configuration (see ***5.214***). Subsequent steps deprotected the alcohol, allowing oxidation to the acid, and removed the protecting group from the alkyne and converted the phthalimidoyl group to an amine, giving 4-aminohex-5-ynoic acid, ***5.215***.[91]

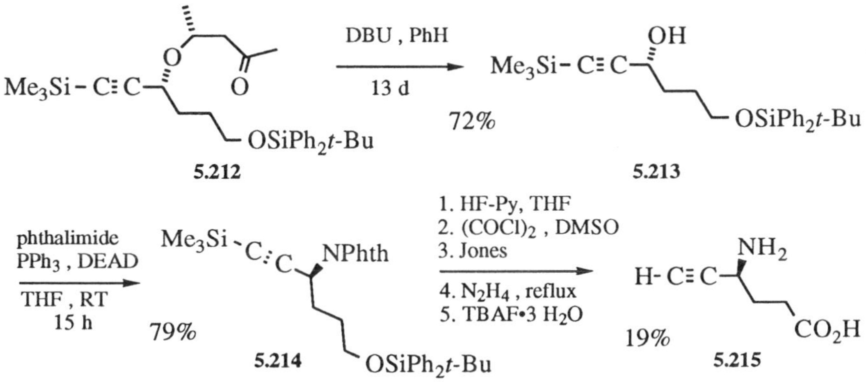

5.212 **5.213** **5.214** **5.215**

Seebach reported an asymmetric synthesis of a 3-aminobutanoic acid derivative using ***5.216*** as a chiral non-racemic precursor.[92] Reaction of ***5.216*** with benzyl-amine gave 3S-(N-benzylamino)butanoic acid, ***5.217a***, and the presence of the auxiliary that bridged the OH and ester moieties in the cyclic ***5.216*** led to good asymmetric induction. The stereochemistry of the final amino acid could be "reversed" by initial conversion of ***5.216*** to the β-lactone (***5.218***) followed by treatment with *n*-butyllithium and benzylamine, giving 3R-(N-benzylamino)butanoic acid, ***5.217b***.[92]

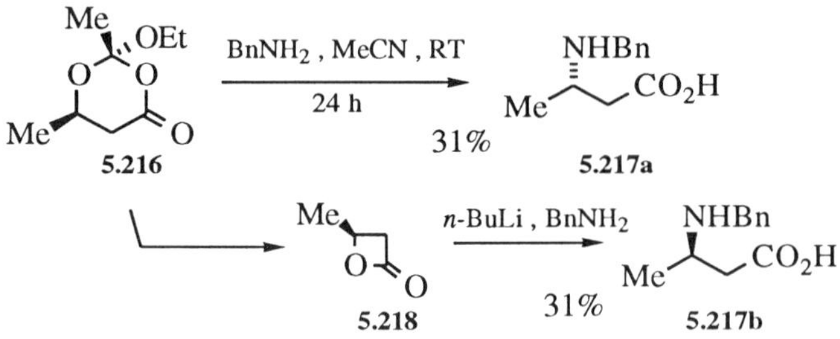

5.2.H. Miscellaneous Transformations Using Chiral Templates

As seen previously, a benzene ring can be oxidized to a carboxylic acid moiety, making aryl substituted alkyl amines putative sources of amino acids. A benzene ring can be converted to an acid moiety by less direct methods. In one interesting

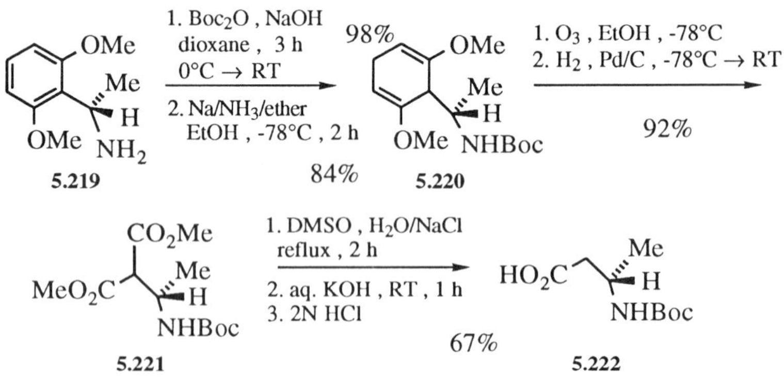

example chiral, nonracemic amine ***5.219*** was the chiral template, and the amine group was protected as its N-Boc derivative. This was followed by Birch reduction[88] to give ***5.220***.[93] Ozonolysis cleaved the vinyl ether moieties in the ring to esters, and catalytic hydrogenation gave ***5.221***. R-3-aminobutanoic acid (***5.222***) was the final product via deprotection, decarboxylation, and hydrolysis.[93]

Oxidative cleavage of alkenes offers another route to a carboxylic acid. An interesting example used citronellic acid (*5.223*, derived from pulegone)[94] as a template.

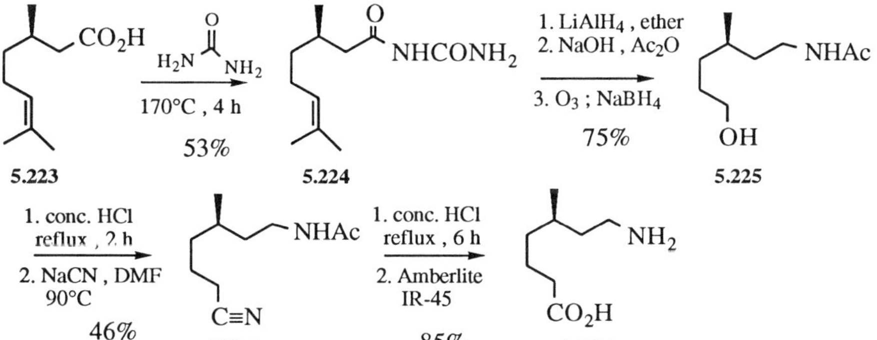

5.223 **5.224** **5.225** **5.226** **5.227**

The acid moiety was converted to an amine and the alkene moiety in *5.223* was converted to a nitrile and, thereby, to an acid.[95] This sequence for accomplishing this reacted *5.223* with urea to give *5.224*. Reduction followed by acylation and ozonolysis (with a reductive workup) gave *5.225*.[95] Conversion to the chloride and displacement with cyanide gave *5.226* and acid hydrolysis led to 5-methyl-7-aminoheptanoic acid, *5.227*.[95] In this lengthy sequence, the acid moiety in *5.223* functioned as the amine precursor and the alkenyl group functioned as the eventual acid moiety.

Another alkene cleavage method is illustrated by the conversion of the alcohol moiety in *5.228* (the template) to a mesylate and then to azide *5.229*.[96] Oxidative cleavage with periodate and ruthenium trichloride was followed by hydrogenation of the azide to give 3-amino-2-methylpentanoic acid, *5.230*.

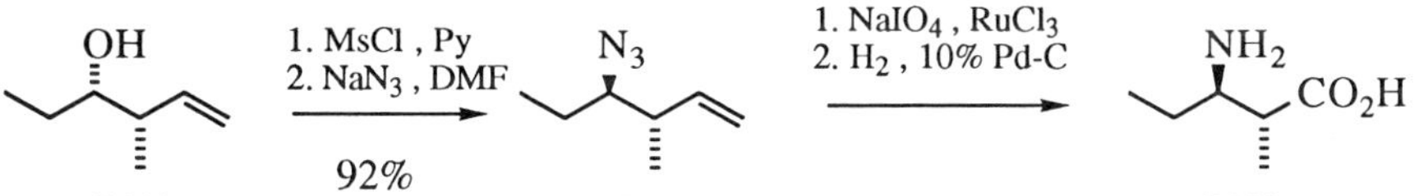

5.228 **5.229** **5.230**

The use of hydroxy-ester ***5.231*** as a template required its conversion to phenylthio amino derivative ***5.232***.[97] This carbamate was converted to oxazolidinone ***5.233***, and replacement of the phenylthio group with allyl (see ***5.234***) was followed by ring opening to give ***5.235***. As in the two previous examples, the alkenyl moiety was converted to the acid via oxidative cleavage, giving 4R-(N-Boc-N-methylamino)-5S-methyl-3R-methoxyheptanoic acid, ***5.236***.[97]

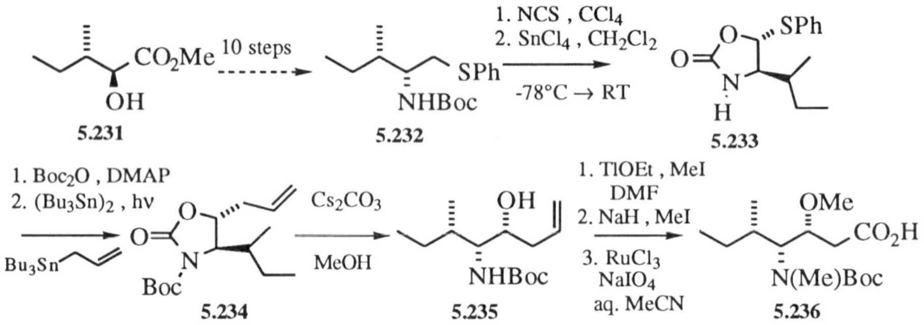

Displacement reactions in allylic systems are well known. In some cases, a nucleophile attacks the C=C moiety rather than the C-X moiety in an allylic system in what is known as an S_N2' reaction.[98] The use of allylic acetates with malonate

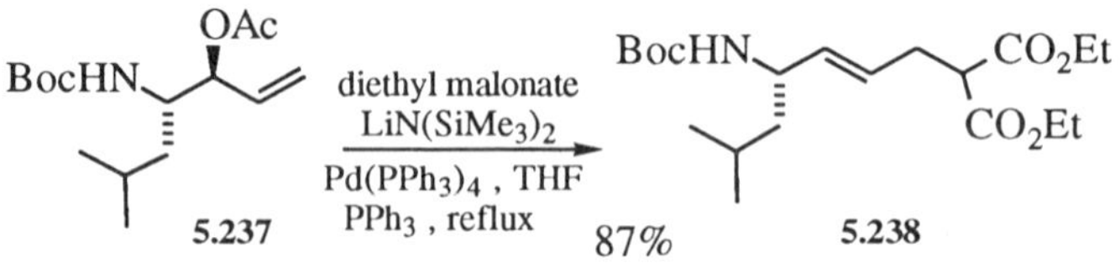

derivatives, catalyzed by palladium (0) is one example. Reaction of ***5.237*** with diethyl malonate, catalyzed by palladium, led to ethyl 6-(N-Boc amino)-2-carboethoxy-8-methylnon-4E-enoate, ***5.238***.[99] The preparation of ***5.237*** in enantiopure form allowed its use as a template in this reaction. The 6-amino-6-(4-benzyloxyphenyl) derivative was prepared in a similar manner.

Conjugated esters are also quite useful. Reaction of enantiopure ***5.239*** with the higher order cuprate derived from isopropyllithium gave methyl 5-(N-Boc amino)-

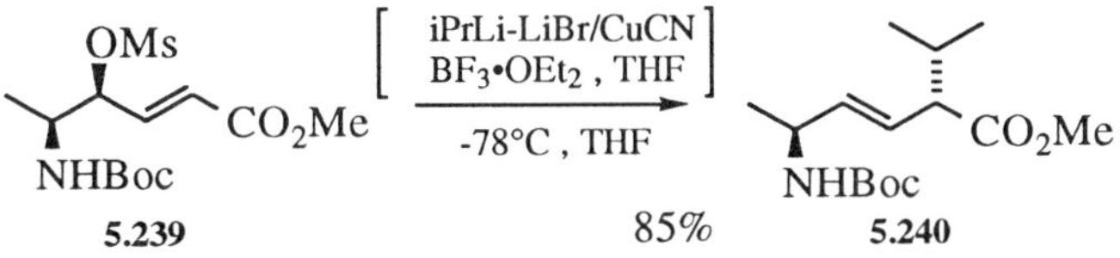

2-isopropylhex-3-enoate, ***5.240***.[100] The 2-methyl derivative was also prepared (in 93% yield), as was the 2-benzyl derivative (in 84% yield).

5.3. Catalytic Hydrogenation with Chiral, Nonracemic Catalysts

In several syntheses scattered throughout this book, an alkenyl amino acid was prepared. In all of those cases, reduction of the double bond leads to the saturated amino acid. If the double bond contains a substituent, the amino group or another

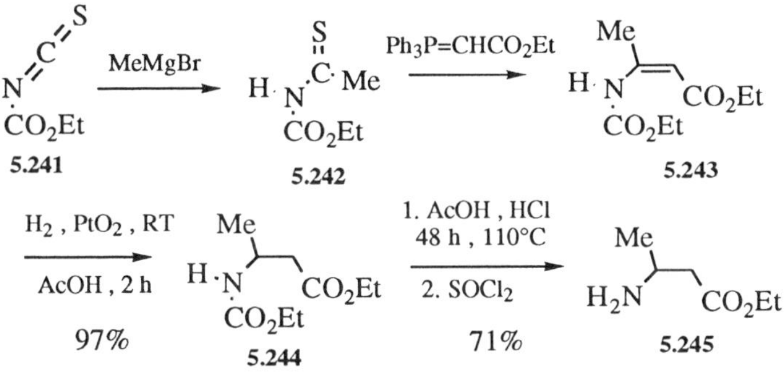

substituent, reduction leads to a new chiral center. This section will discuss catalytic hydrogenation of such alkene derivatives using chiral, nonracemic catalysts to produce chiral, nonracemic amino acids. In a highly specialized sequence, the requisite alkenyl amino-ester was prepared by reaction of Grignard reagents with isothiocyanate ***5.241*** to give ***5.242***. Wittig olefination[33] gave ***5.243***.[101a] The N-acyl derivative has also been used in this sequence.[101b] When N-carbamate ***5.243*** was hydrogenated over PtO_2 (Adams catalyst), ***5.244*** was obtained and hydrolysis gave ethyl 3-aminobutanoic acid (***5.245***), after esterification.[101] In this work, 3-amino-pentanoic acid, 3-aminoheptanoic acid and other substituted amino acids were prepared in good yield. Although ***5.245*** contains a chiral center, this synthetic sequence generated chiral, racemic material. Without the use of a chiral auxiliary, a chiral template, or a chiral catalyst no asymmetric induction was observed.[102,101]

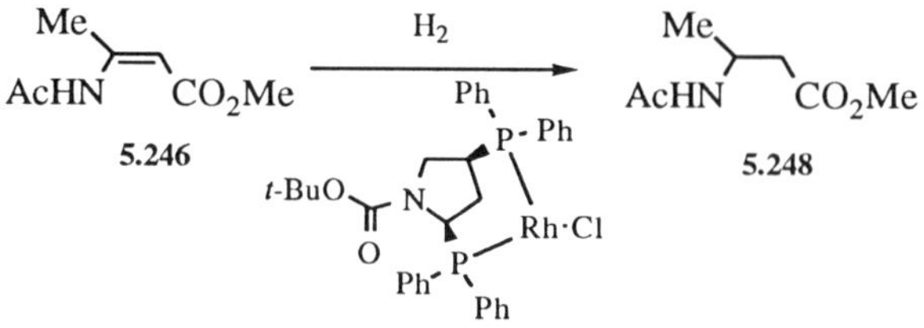

5.247

When the amino group is attached to the alkene as in ***5.246***, hydrogenation of the double bond can generate a chiral center if a chiral catalyst is used for hydrogenation. In this example, a chiral catalyst developed by Achiwa (***5.247***) allowed reduction of methyl 3-(N-acetylamino)but-2Z-enoate (***5.24)*** to give methyl 3-(N-acetylamino)butanoate (***5.248***), quantitatively (53%ee, R).[103] Similarly, methyl 3-(N-acetylamino)-3-phenyl-propanoate was prepared in 90% yield (47%ee, R).

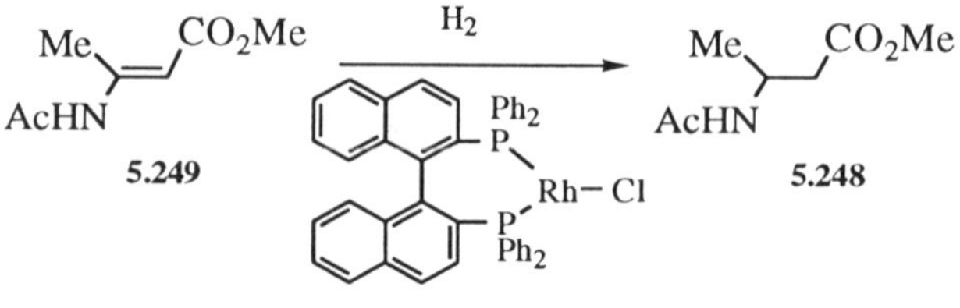

5.250

Noyori used the BINAP-Ru(II) catalyst [***5.250***, where BINAP is 2,2'-*bis*-(diphenylphosphino)-1,1'-binaphthyl] to reduce β-aminopropenoic acid as well as several derivatives.[104] Reduction of ***5.246*** with R-BINAP (***5.250***) gave a good yield of ***5.248***, but with only 5%ee, R. In several examples, the *E*-isomer showed greater selectivity during hydrogenation. Reduction of the *E*-isomer of ***5.249*** with ***5.250***, for example, gave a quantitative yield of ***5.248*** (96%ee, S). Similar hydrogenation of *E-**5.251*** led to a quantitative yield of **5.252** (in 90%ee, R).[104] Similar reduction of *Z-**5.251*** gave a quantitative yield of **5.252** but with only 9%ee (R). Another reduction of *E-**5.251*** with hydrogen in the presence of ***5.247*** was effective in terms of yield of product (a 91% yield of **5.252** was obtained from *E-**5.251***). The extent of asymmetric induction was only 47%ee (R), however, making it less effective than BINAP in this system.[103]

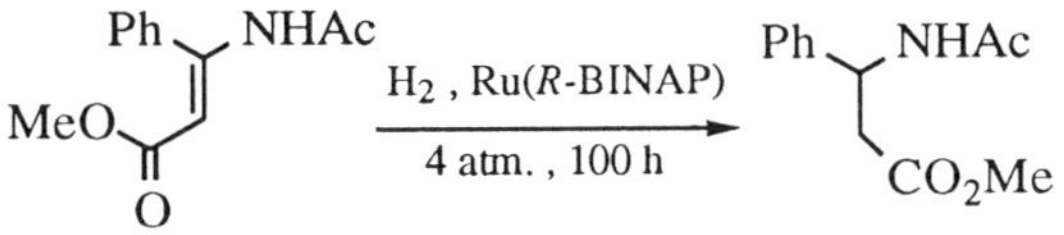

5.251 **5.252**

REFERENCES

1 (a) Hanessian, S., *Total Synthesis of Natural Products: The "Chiron" Approach*; Pergamon, ***1983***; (b) Hanessian, S. *Accts. Chem. Res.*, ***1979***, *12*, 159; (c) Hanessian, S.; Franco, J.; Larouche, B. *Pure Appl. Chem.*, ***1990***, *62*, 1887.

2 Juaristi, E.; Quintana, D.; Lamatsch, B.; Seebach, D. *J. Org. Chem.*, ***1991***, *56*, 2553.

3 (a) Juaristi, E.; Quintana, D. *Tetrahedron Asymmetry*, ***1992***, *3*, 723; (b) Juaristi, E.; Escalante, J. *J. Org. Chem.*, ***1993***, *58*, 2282.

4 Iizuka, K.; Kamijo, T.; Harada, H.; Akahane, K.; Kubota, T.; Etoh, Y.; Shimaoka, I.; Tsubaki, A.; Murakami, M.; Yamaguchi, T.; Iyobe, A.; Umeyama, H.; Kiso, Y. *Chem. Pharm. Bull*, ***1990***, *38*, 2487.

5 (a) Horner, L.; Hoffmann, H.; Wippel, J.H.; Klahre, G. *Ber.*, ***1959***, *92*, 2499; (b) Wadsworth,W.S. Jr.; Emmons, W.D. *J. Am. Chem. Soc.*, ***1961***, *83*, 1733; (c) Boutagy, J.; Thomas, R. *Chem. Rev.*, ***1974***, *74*, 87.

6 Kogen, H.; Nishi, T. *J. Chem. Soc., Chem. Commun.*, ***1987***, 311.

7 (a) Honoré, T.; Hjeds, H.; Krogsgaard-Larsen, P.; Christiansen, T.R. *Eur. J. Med. Chem. Chim.Ther.*, ***1978***, *13*, 429; (b) Grzonka, Z.; Lammek, B. *Synthesis*, ***1974***, 661; (c) also see Hashiguchi, S.; Kawada, A.; Natsugari, H.

Ibid, ***1992***, 403.

8 Wolf, J.-P.; Pfander, H. *Helv. Chim. Acta,* ***1986****, 69*, 918.

9 Wolf, J.-P.; Pfander, H. *Helv. Chim. Acta,* ***1987****, 70*, 116.

10 Jefford, C.W.; Wang, J. *Tetrahedron Lett.,* ***1993****, 34*, 1111.

11 (a) Takahashi, Y.; Hasegawa, S.; Izawa, T.;Kobayashi, S.; Ohno, M. *Chem. Pharm. Bull.,* ***1986****, 34*, 3020; (b) Takahashi, Y.; Yamashita, H.; Kobayashi, S.; Ohno, M. *Ibid,* ***1986****, 34*, 2732.

12 Ohshiro, S.; Kuroda, K.; Fujita, T. *Yakugaku Zasshi,* ***1967****, 87*, 1184 [*Chem. Abstr.,* ***1968****, 68*: 40031w].

13 Yoneta, T.; Shibahara, S.; Fukatsu, S.; Seki, S. *Bull. Chem. Soc., Jpn.,* ***1978****, 51*, 3296.

14 Sham, H.L.; Rempel, C.A.; Stein, H.; Cohen, J. *J. Chem. Soc., Chem. Commun.,* ***1987***, 683.

15 Gmeiner, P.; Lerche, H. *Heterocycles,* ***1990****, 31*, 9.

16 Gmeiner, P. *Tetrahedron Lett.,* ***1990****, 31*, 5717.

17 Gmeiner, P. *Liebigs Ann. Chem.,* ***1991***, 501.

18 (a) Mohan, R.; Chou, Y.-L.; Bihovsky, R.; Lumma, W.C. Jr.; Erhardt, P.W.; Shaw, K.J. *J. Med. Chem.,* ***1991****, 34*, 2402; (b) Nishizawa, R.; Saino, T.; Takita, T.; Suda, H.; Aoyagi, T.; Umezawa, H. *Ibid,* ***1977****, 20*, 510.

19 Miyazawa, T.; Akita, E.; Ito, T. *Agr. Biol. Chem.,* ***1976****, 40*, 1651.

20 Yang, W.; Pi, X.; Huang, Q.; Liu, Y. *Huaxue Shijie,* ***1991****, 32*, 516 [*Chem. Abstr.,* ***1992****, 117*: 29163a].

21 (a) Herdeis, C. *Synthesis,* ***1986***, 232; (b) Herdeis, C.; Dimmerling, A. *Arch. Pharm. (Weinheim),* ***1984****, 317*, 304.

22 Coppola, G.M.; Shuster, H.F. *Asymmetric Synthesis,* Wiley, New York, ***1987***, pp. 237-242.

23 Silverman, R.B.; Levy, M.A. *J. Org. Chem.,* ***1980****, 45*, 815.

24 Barlos, K.; Mamos, P.; Papaioannou, D.; Patrianakou, S. *J. Chem. Soc., Chem. Commun.,* ***1987***, 1583.

25 Rajashekar, B.; Kaiser, E.T. *J. Biol. Chem.,* ***1986****, 261*, 13617.

26 Andruszkiewicz, R.; Silverman, R.B. *J. Biol. Chem.,* ***1990****, 265*, 22288.

27 Smith, M.B.; Dembofsky, B.T.; Son, Y.C. *J. Org. Chem.,* ***1994****, 59*, 1719.

28 (a) Smith, A.L.; Williams, S.F.; Holmes, A.B.; Hughes, L.R.; Lidert, Z.; Swithenbank, C. *J. Am. Chem. Soc.,* ***1988****, 110*, 8696; (b) Holmes, A.B.;

Smith, A.L.; Williams, S.F.; Hughes, L.R.; Lidert, Z.; Swithenbank, C. *J. Org. Chem.*, ***1991***, *56*, 1393.

29 Hamada, Y.; Kawai, A.; Kohno, Y.; Hara, O.; Shiori, T. *J. Am. Chem. Soc.*, ***1989***, *111*, 1524.

30 (a) Rich, D.H.; Moon, B.J.; Boparai, A.S. *J. Org. Chem.*, ***1980***, *45*, 2288; (b) Peet, N.P.; Burkhart, J.P.; Angelastro, M.R.; Giroux, E.L.; Mehdi, S.; Bey, P.; Kolb, M.; Neises, B.; Schirlin, D. *J. Med. Chem.*, ***1990***,*33*, 394.

31 Johnson, R.L. *J. Med. Chem.*, ***1982***, *25*, 605.

32 Kogen, H.; Nishi, T. *J. Chem. Soc., Chem. Commun.*, ***1987***, 311.

33 (a) Wittig, G.; Rieber, M. *Ann.*, ***1949***, *562*, 187; (b) Wittig, G.; Geissler, G. *Ibid*, ***1953***, *580*, 44; (c) Wittig, G.; Schöllkopf, U. *Chem. Ber.*, ***1954***, *87*, 1318; (d) Gensler, W.J. *Chem. Rev.*, ***1957***, *57*, 191 (see p. 218).

34 Reetz, M.T.; Kanand, J.; Griebenow, N.; Harms, K. *Angew. Chem. Int. Ed., Engl.*, ***1992***, *31*, 1626.

35 Nishi, T.; Saito, F.; Nagahori, H.; Kataoka, M.; Morisawa, Y.; Yabe, Y.; Sakurai, M.; Higashida, S.; Shoji, M.; Matsushita, Y.; Iijima, Y.; Ohizumi, K.; Koike, H. *Chem. Pharm. Bull.*, ***1990***, *38*, 103.

36 Kamijo, T.; Harada, H.; Tsubaki, A.; Yamaguchi, T.; Iyobe, A.; Iizuka, K.; Kiso, Y. *Japan Kokai Tokkyo Koho*, JP 1-172365, ***1989*** [*Chem. Abstr*, ***1990***, *112:* 76461c].

37 Matsuda, F.; Matsumoto, T.; Ohsaki, M.; Ito, Y.; Terashima, S. *Chem. Lett.*, ***1990***, 723.

38 Tseng, C.C.; Terashima, S.; Yamada, S. *Chem. Pharm. Bull.*, ***1977***, *25*, 29.

39 Karrer, P.; Portmann, P.; Suter, M. *Helv. Chim. Acta* ***1948***, *31*, 1617.

40 (a) Metcalf, B.W.; Casara, P. *Tetrahedron Lett.*, ***1975***, 3337; (b) Danzin, C.; Claverie, N.; Jung, M.J. *Biochem. Pharmacol.*, ***1984***, *33*, 1741.

41 DeCamp, A.E.; Kawaguchi, A.T.; Volante, R.P.; Shinkai, I. *Tetrehedron Lett.*, ***1991***, *32*, 1867.

42 Evans, B.E.; Rittle, K.E.; Homnick, C.F.; Springer, J.P.; Hirshfield, J.; Veber, D.F. *J. Org. Chem.*, ***1985***, *50*, 4615.

43 Iizuka, K.; Kamijo, T.; Harada, H.; Akahane, K.; Kubota, T.; Umeyama, H.; Kiso,Y. *J. Chem. Soc., Chem. Commun.*, ***1989***, 1678.

44 Iizuka, K.; Kamijo, T.; Harada, H.; Akahane, K.; Kubota, T.; Umeyama, H.; Ishida, T.; Kiso, Y. *J. Med. Chem.*, ***1990***, *33*, 2707.

45 Plata, D.J.; Leanna, M.R.; Morton, H.E. *Tetrahedron Lett.*, ***1991***, *32*, 3623.

46 Shimohigashi, Y.; Waki, M.; Izumiya, N. *Mem. Fac. Sci., Kyushu Univ., Ser. C*, ***1979***, *11*, 217 [*Chem. Abstr.*, ***1979***, *91*: 39804u].

47 Shimohigashi, Y.; Waki, M.; Izumiya, N. *Bull. Chem. Soc., Jpn.*, ***1979***, *52*, 949.

48 Pons, D.; Savignac, M.; Genet, J.-P. *Tetrahedron Lett.*, ***1990***, *31*, 5023.

49 Garner, P. *Tetrahedron Lett.*, ***1984***, *25*, 5855.

50 (a) Danishefsky, S. *Accts. Chem. Res.*, ***1981***, *14*, 400; (b) Idem, *Aldrichimica Acta*, ***1986***, *19*, 59.

51 Estermann, H.; Seebach, D. *Helv. Chim. Acta*, ***1988***, *71*, 1824.

52 (a) Rich, D.H.; Sun, E.T.; Boparai, A.S. *J. Org. Chem.*, ***1978***, *43*, 3624; (b) Stanfield, C.F.; Parker, J.E.; Kanellis, P. *J. Org. Chem.*, ***1981***, *46*, 4797.

53 (a) Wolff, L. *Ann.*, ***1912***, *394*, 23 (see p. 25); (b) Kirmse,W. *Carbene Chemistry, 2nd Ed.*, Academic Press, New York, ***1971***, pp. 475-492; (c) Mundy, B.P.; Ellerd, M.G. *Name Reactions and Reagents in Organic Synthesis*, Wiley, New York, ***1988***, pp. 232-233

54 Lucente, G.; Piccinnia, G.; Romeo, A. *Gazz. Chim. Ital.*, ***1966***, *96*, 1380.

55 Papaioannou, D.; Stavropoulos, G.; Sivas, M.; Barlos, K.; Francis, G.W.; Aksnes, D.W.; Maartmann-Moe, K. *Acta Chem. Scand.*, ***1991***, *45*, 99.

56 Thaisrivongs, S.; Tomasselli, A.G.; Moon, J.B.; Hui, J.; McQuade, T.J.; Turner, S.R.; Strohbach, J.W.; Howe, W.J.; Tarpley, W.G.; Heinrikson, R.L. *J. Med. Chem.*, ***1991***, *34*, 2344.

57 Shue, Y.-K.; Carrera, G.M. Jr.; Nadzan, A.M. *Tetrahedron Lett.*, ***1987***, *28*, 3225.

58 Seebach, D.; Estermann, H. *Tetrahedron Lett.*, ***1987***,*28*, 3103.

59 Devant, R.M.; Radunz, H.-E. *Tetrahedron Lett.*, ***1988***, *29*, 2307.

60 (a) Reformatsky, S. *Ber.*, ***1887***, *20*, 1210; (b) Rathke, M.W. *Org. React.*, ***1975***, *22*, 423; (c) Diaper, D.G.M.; Kuksis, A. *Chem. Rev.*, ***1959***, *59*, 89.

61 Andrés, C.; González, A.; Pedrosa, R.; Pérez-Encabo, A. *Tetrahedron Lett.*, ***1992***, *33*, 2895.

62 Correa, A.; Denis, J.N.; Greene, A.E. *PCT Int. Appl.* WO 91 17,977 [*Chem. Abstr.*, ***1992***,*116*: P129621w].

63 Davies, S.G.; Ichihara, O. *Tetrahedron Asymmetry*, ***1991***, *2*, 183.

64 Bellasio, E.; Vigevani, A.; Crostiani, G. *Farmaco Ed. Sci.*, ***1970***, *25*, 409.

65 (a) Matsumoto, K.; Morris, A.R. *Organic Synthesis at High Pressure*, Wiley,

New York, ***1991***; (b) Matsumoto, K.: Sera, A.; Uchida, T. *Synthesis, **1985***, 1; (c) Matsumoto, K.; Sera, A. *Ibid, **1985***, 999.

66 d'Angelo, J.; Maddaluno, J. *J. Amer. Chem. Soc., **1986**, 108*, 8112.

67 Chong, J.M.; Sharpless, K.B. *J. Org. Chem., **1985**, 50*, 1560.

68 Kitazume, T.; Yamamoto,T.; Ohnogi, T.; Yamazaki, T. *Chemistry Express, **1989**, 4*, 657 [*Chem. Abstr., **1990**, 112:* 199011p].

69 Cardillo, G.; Orena, M.; Penna, M.; Sandri, S.; Tomasini, C. *Syn. Lett, **1990***, 543.

70 Amoroso, R.; Cardillo, G.; Tomasini, C.; Tortoreto, P. *J. Org. Chem., **1992**, 57*, 1082.

71 Harris, K.J.; Sih, C.J. *Biocatalysis, **1992**, 5*, 195.

72 Srairi, D.; Maury, G. *Bull. Soc. Chim., Fr., **1987***, 297.

73 Andruszkiewicz, R.; Barrett, A.G.M.; Silverman, R.B. *Synth. Commun., **1990**, 20*, 159.

74 Chiba, T.; Ishizawa, T.; Sakaki, J.; Kaneko, C. *Chem. Pharm. Bull., **1987**, 35*, 4672.

75 Milewska, M.J.; Poloński, T. *Synthesis, **1988***, 475.

76 (a) Curtius, T. *J. Prakt. Chem., **1894**, 50*, 275; (b) Smith, P.A.S. *Org. React., **1946**, 3*, 337; (c) Saunders, J.H.; Slocombe, R.J. *Chem. Rev., **1948**, 43*, 203.

77 Liebman, K.C.; Fellner, S.K. *J. Org. Chem., **1962**, 27*, 438.

78 Roncari, G.; Kuryło-Borowska, Z.; Craig, L.C. *Biochemistry, **1963**, 5*, 2153.

79 Bernardi, A.; Micheli, F.; Potenza, D.; Scolastico, C.; Villa, R. *Tetrahedron Lett., **1990**, 31*, 4949.

80 (a) Katsuki, T.; Yamaguchi, M. *Bull. Chem. Soc., Jpn., **1976**, 49*, 3287; (b) Jouin, P.; Castro, B. *J. Chem. Soc., Perkin Trans. I, **1987***, 1177; (c) Klutchko, S.; O'Brien, P.; Hodges, J.C. *Synth. Commun., **1989**, 19*, 2573; (d) Andrew, R.G.; Conrow, R.E.; Elliott, J.D.; Johnson, W.S.; Ramezani, S. *Tetrahedron Lett., **1987***, 6535.

81 Andruszkiewicz, R.; Czeruinski, A.; Grzybowska, J. *Synthesis, **1983***, 31.

82 Umemura, E.; Tsuchiya, T.; Umezawa,S. *J. Antibiot., **1988**, 41*, 530.

83 Yokoo, A.; Akutagawa, S. *Bull. Chem. Soc., Jpn., **1962**, 35*, 644.

84 Chakraborty, T.K.; Gangakhedkar, K.K. *Tetrahedron Lett., **1991**, 32*, 1897.

85 Mitsunobu, O. *Synthesis, **1981***, 1.

86 Solladie-Cavallo, A.; Khiar, N. *Tetrahedron Lett., **1988**, 29*, 2189.

87 Bringmann, G.; Künkel, G.; Geuder, T. *Syn. Lett.*, ***1990***, 253.

88 (a) Birch, A.J. *Quart. Rev. (London),* ***1950***, *4*, 69; (b) Birch, A.J.; Smith, H. *Ibid,* ***1958***, *12*, 17; (c) Watt, G.W. *Chem. Rev.*, ***1950***, *46*, 317.

89 Furukawa, M.; Okawara,T.; Noguchi, Y.; Terawaki, Y. *Chem. Pharm. Bull.*, ***1979***, *27*, 2223.

90 (a) Glickman, S.A.; Cope, A.C. *J. Am.Chem.Soc.*, ***1945***, *67*, 1017; (b) Michaelis, A. *Ann.*, ***1909***, *366*, 324 (see p. 337); (c) Brühl, J.W. *Z. Physik. Chem.*, ***1895***, *16*, 216.

91 (a) Tabor, A.B.; Holmes, A.B.; Baker, R. *J. Chem. Soc., Chem. Commun.*, ***1989***, 1025; (b) Holmes, A.B.; Tabor, A.B.; Baker, R. *J. Chem. Soc., Perkin Trans. I,* ***1991***, 3301.

92 Griesbeck, A.; Seebach, D. *Helv. Chim. Acta,* ***1987***, *70*, 1326.

93 Bringmann, G.; Geuder, T. *Synthesis,* ***1991***, 829.

94 Lukeš, R.; Zabáčová, A.; Plešek, J. *Croat. Chem. Acta,* ***1957***, *29*, 201 [*Chem. Abstr.*, ***1959*** *53*: 17898c].

95 Overberger, C.G.; Takekoshi, T. *Macromolecules,* ***1968***, *1*, 1.

96 Bates, R.B.; Gangwar, S. *Tetrahedron Asymmetry,* ***1993***, *4*, 69.

97 Kano, S.; Yuasa, Y.; Shibuya, S. *Heterocycles,* ***1990***, *31*,1597.

98 (a) Magid, R.M. *Tetrahedron,* ***1980***, *36*, 1901; (b) Bordwell, F.G.; Pagani, G.A. *J. Am. Chem. Soc.*, ***1975***, *97*, 118; (c) Bordwell, F.G.; Mecca, T.G. *Ibid,* ***1975***, *97*, 123, 127; (d) Bordwell, F.G.; Wiley, P.F.; Mecca, T.G. *Ibid,* ***1975***, *97*, 132.

99 Thompson, W.J.; Tucker, T.J.; Schwering, J.E.; Barnes, J.L. *Tetrahedron Lett.*, ***1990***, *31*, 6819.

100 Ibuka, T.; Habashita, H.; Otaka, A.; Fujii, N. *J. Org. Chem.*, ***1991***, *56*, 4370.

101 (a) Slopianka, M.; Gossauer, A. *Synth. Commun.*, ***1981***, *11*, 95; (b) Idem, *Liebigs Ann. Chem.*, ***1981***, 2258.

102 Aberhart, D.J.; Lin, H.-J. *J. Org. Chem.*, ***1981***, *46*, 3749.

103 Achiwa, K.; Soga, T. *Tetrahedron Lett.*, ***1978***, 1119.

104 Lubell, W.D.; Kitamura, M.; Noyori, R. *Tetrahedron Asymmetry,* ***1991***, *2*, 543.

6

Biologically Important Amino Acids

There are many examples of non-α-amino acids that possess important biological properties. A handful of these were mentioned in previous chapters. This chapter will discuss the synthesis of several types of these amino acids in order to illustrate both the impetus for synthesizing amino acids and also the various structural types that are to be found. The synthetic approaches used in this chapter span the range of methods presented in the first five chapters.

6.1. GABA-ergic Compounds

4-Aminobutanoic acid (γ-aminobutyric acid, known as GABA) is an important mammalian neurotransmitter.[1] There are several amino acids that have similar biological activities (GABA-ergic compounds: possessing GABA-like activity) and they are collected into this section. This section also illustrates the utility of structurally simple amino acids.

As mentioned, GABA is an important neural transmitter and deficiencies in GABA are associated with diseases that exhibit neuromuscular dysfunction such as epilepsy, Huntington's disease and Parkinson's disease.[1] 5-Aminopentanoic acid (δ-aminovaleric acid, DAVA) is also a neurotransmitter and used for treatment of neuromuscular disease. In at least one study, 3-alkyl-4-aminobutanoic acid derivatives were shown to be *in vitro* activators of *L*-glutamic acid decarboxylase and they showed anti-convulsant activity.[2]

Baclofen [4-amino-3-(*p*-chlorophenyl)butanoic acid, ***6.1***] is an important GABA-ergic compound used in the treatment of epilepsy.[3] One synthesis began with conjugate addition of the anion of nitromethane to the double bond in ***6.1***.[4] Catalytic hydrogenation of the nitro group in the Michael product (***6.2***) gave an amino-ester that spontaneously cyclized to a lactam. This required acid hydrolysis to give baclofen, ***6.3***.[4] Similarly prepared were 3-(4-bromophenyl)-, 3-(4-tolyl)-, and the 3-(4-hydroxyphenyl)-derivatives, as well as 3-pyridyl and 2-pyridyl-4-aminobutanoic acids.[4]

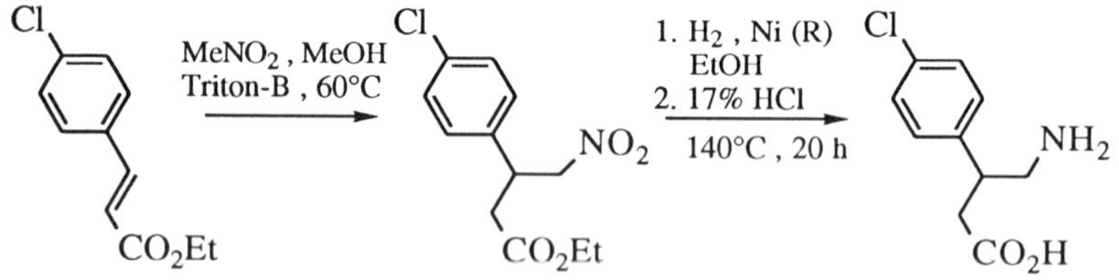

6.1 **6.2** **6.3**

When ***6.4*** reacted with phenyl isocyanate to give a nitrile N-oxide, addition to 1-hexene gave dihydroisoxazole ***6.5***.[5] Hydrolysis of the ester and catalytic hydrogenation opened the ring to give 4-amino-6-hydroxydecanoic acid, ***6.6***. When compared to GABA, ***6.6*** was shown to be a more lipophilic GABA-ergic compound[1] that inhibited GABA binding to synaptic membranes. Similarly prepared were 8-methyl-4-amino-6-hydroxynonanoic acid, 4-amino-6-hydroxydecanoic acid, 4-amino-6-

hydroxydodecanoic acid, 4-amino-6-hydroxytetradecanoic acid, 4-amino-6-hydroxypentadecanoic acid, and 4-amino-6-hydroxyhexadecanoic acid.[5] In general, all of these amino acids were more lipophilic than GABA and retained GABA-ergic activity.[5]

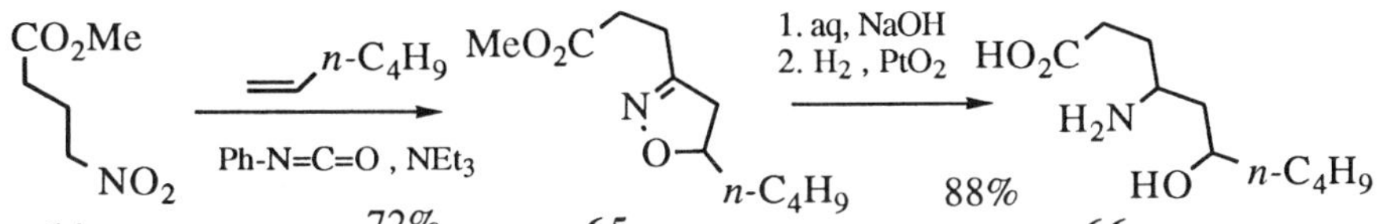

6.4 **6.5** **6.6**

A fluorinated derivative was prepared by conversion of keto-ester ***6.7*** to enamino ester ***6.8*** and isolated as a 1:1 mixture of *cis:trans* isomers.[6] Reduction of the enamine and removal of the protecting groups gave ***6.9*** , which was shown to be an

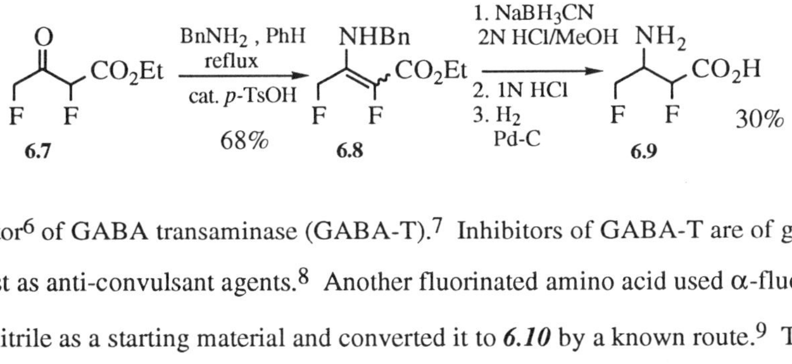

6.7 **6.8** **6.9**

inhibitor[6] of GABA transaminase (GABA-T).[7] Inhibitors of GABA-T are of general interest as anti-convulsant agents.[8] Another fluorinated amino acid used α-fluoroacetonitrile as a starting material and converted it to ***6.10*** by a known route.[9] The bromide moiety in ***6.10*** was converted to an alcohol moiety in ***6.11***.[10] Oxidation of the alcohol to an acid and deprotection gave ***6.12***, which was also shown to be an inhibitor of GABA-T.

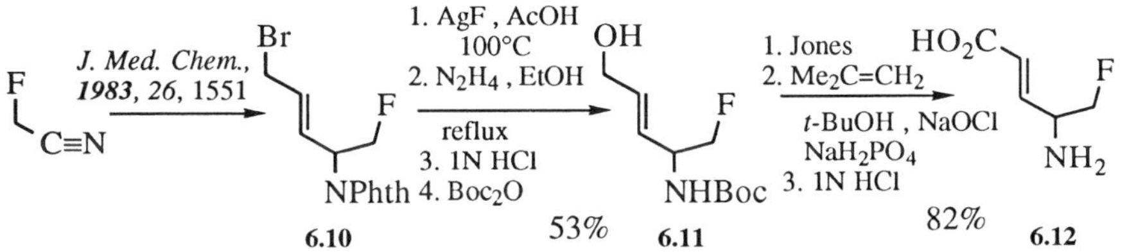

6.10 **6.11** **6.12**

2-Pyrrolidinone derivatives can be useful synthons. Ethyl pyroglutamate (*13*), derived from *L*-glutamic acid (see chapter five, section 5.1.C.ii), was converted to cyanomethyl-2-pyrrolidinone ***6.14***.[11] Reduction to the amine[11b] and Cope elimination[12] gave ***6.15***, which was hydrolyzed to 4-aminohex-5-enoic acid (***6.16***), also an

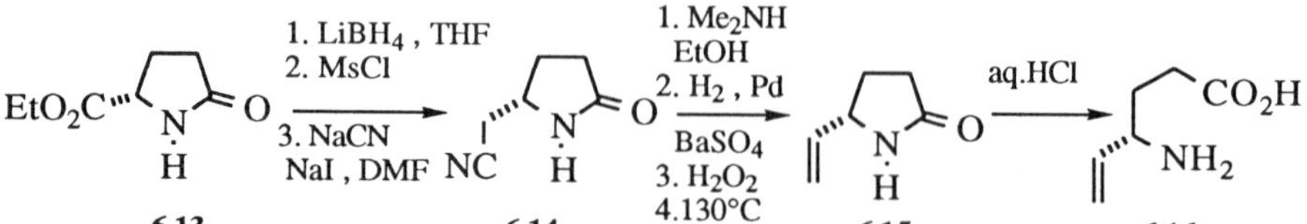

6.13 **6.14** **6.15** **6.16**

inhibitor of GABA transaminase (GABA-T).[7,8] An alternative synthesis converted ***6.13*** to aldehyde ***6.73***, using an interesting N-butenyl protecting group that inhibited racemization at C_6.[13] Formation of this aldehyde allowed Wittig olefination[14] to

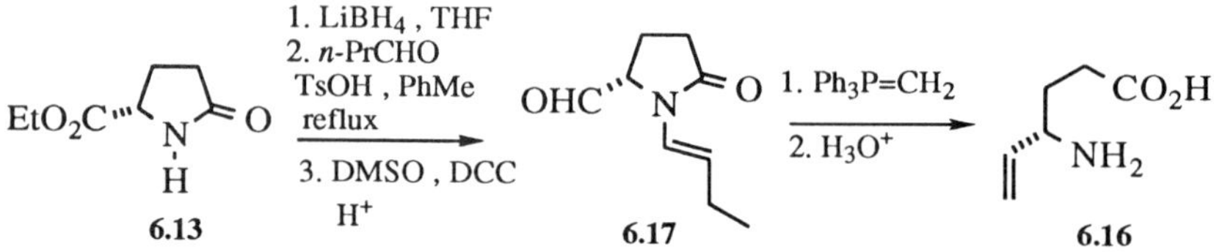

6.13 **6.17** **6.16**

give the alkene (N-butenyl ***6.15***) and acid hydrolysis gave ***6.16***.[15] It is noted that ***6.15*** has been used as a key synthon for the asymmetric synthesis of several chiral, nonracemic pyrrolizidine alkaloids.[16] A quite different approach to ***6.16*** converted cyclopropane dicarboxylic acid ***6.18*** to ***6.19***.[17] Reaction with ammonia gave lactam ***6.20***, which was hydrolyzed to ***6.16***.

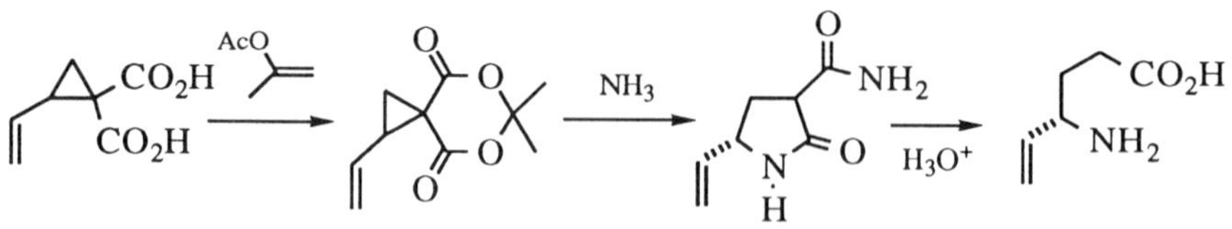

6.18 **6.19** **6.20** **6.16**

Alkynyl amino acids also show GABA-ergic activity. An example is the reaction of 3-amino-1-propyne (***6.21***) with benzaldehyde to give Schiff base ***6.22***.[18] Treatment with ethylmagnesium bromide gave the alkyne anion, which reacted with chlorotrimethyl-silane to give the C-$SiMe_3$ derivative, ***6.23***. Treatment with *n*-butyllithium gave the imine carbanion and subsequent conjugate addition to methyl acrylate gave ***6.24***. Acid hydrolysis liberated 4-aminohex-5-ynoic acid, ***6.25***.[18,19] This amino acid showed properties as a sedative, an antidepressant, and a GABA-T

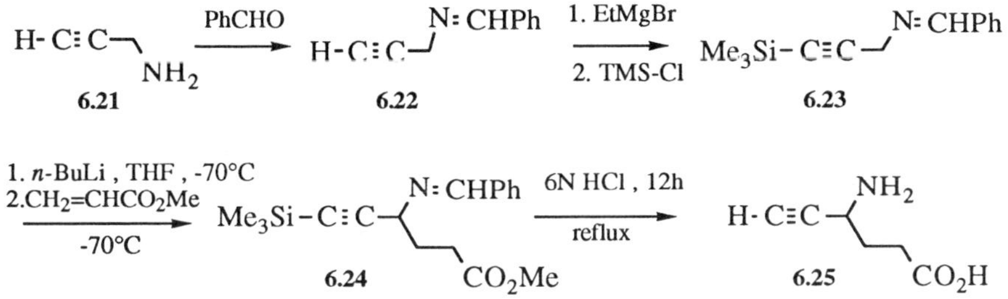

inhibitor.[18b] The ethyl ester of ***6.25*** (ethyl 4-aminohex-5-ynoate) was prepared by essentially the same route and shown to be a plant growth inhibitor.[20] This route was also used to prepare ***6.25*** and cyclize it to 5-ethynyl-2-pyrrolidinone.[21] Smith also prepared this particular alkyne by a different route (from ***6.15***) and used it in the synthesis of chiral, nonracemic pyrrolizidine alkaloids.[16c] A common use for alkynes in the synthesis of alkenyl amino acids is catalytic hydrogenation of the triple bond, generating either an alkenyl amino acid or a saturated amino acid. Palladium catalyzed hydrogenation[22] of N-acetyl ***6.25*** [4-(N-acetylamino)hex-5-ynoic acid, ***6.26***], for example, gave ***6.16***.[23]

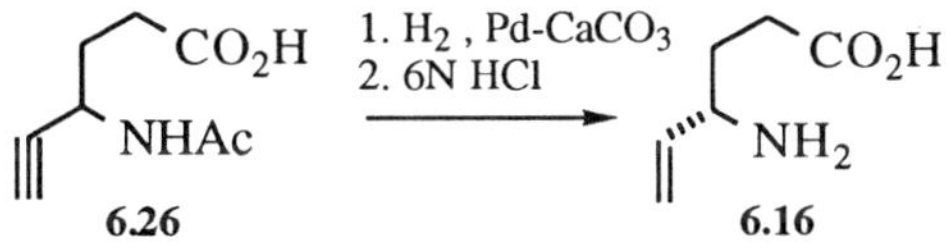

6.2. GABOB and Carnitine

This section will focus on 4-amino-3-hydroxybutanoic acid (GABOB, ***6.27***) and its trimethylammonium derivative (carnitine, ***6.28***). The R and S enantiomers of each amino acid are designated ***6.27a*** and ***6.27b*** or ***6.28a*** and ***6.28b***, respectively. These amino acids are generally prepared by the methods described in previous chapters, but their biological importance has led to a variety of synthetic approaches.

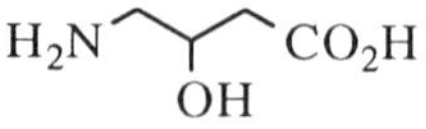

6.27

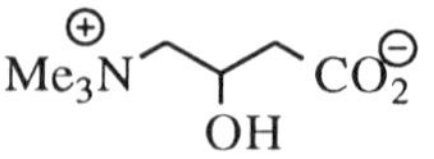

6.28

R-(-)-GABOB is an inhibitory neurotransmitter[24] that has been used in the treatment of human epilepsy.[25] It also has hypertensive properties.[26] Interestingly, GABOB was first prepared by Tomita in 1923.[27] Carnitine is used to treat myocardial ischemia[28] and carnitine deficiency.[28a] It is also important for the transport of long chain fatty acids through the mitochondrial membrane[29] and is used for the treatment of myopathic deficiency.[30]

6.2.A. Synthetic Strategies Based on Functional Group Interchange Reactions

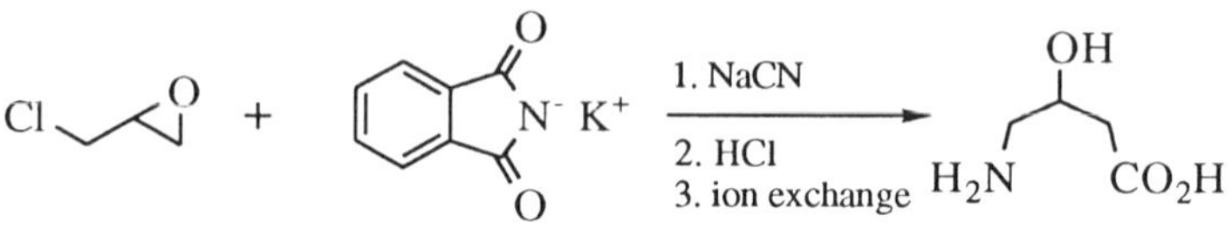

6.29 **6.27**

A common synthetic approach to GABOB uses epichlorohydrin (***6.29***) as a starting material. In a simple example, ***6.29*** was treated with potassium phthalimide and sodium cyanide to give GABOB (***6.27***), after hydrolysis and removal of the phthalimidoyl group.[31] In one variation of this reaction, the intermediate phthalimidoyl product was isolated prior to treatment with cyanide.[32] The reaction can be modified in several other ways. A variation of this sequence used the reaction of ***6.29*** with potassium cyanide and then ammonia. This was followed by hydrolysis of the nitrile to give GABOB.[33] Another variation converted the nitrile to the corresponding amide prior to reaction with ammonia. The amide was then hydrolyzed to give GABOB.[34]

A strikingly different strategy reacted ***6.29*** with methanolic carbon monoxide to give ***6.30*** via reaction at the epoxy carbon. When the chlorine moiety in ***6.30***

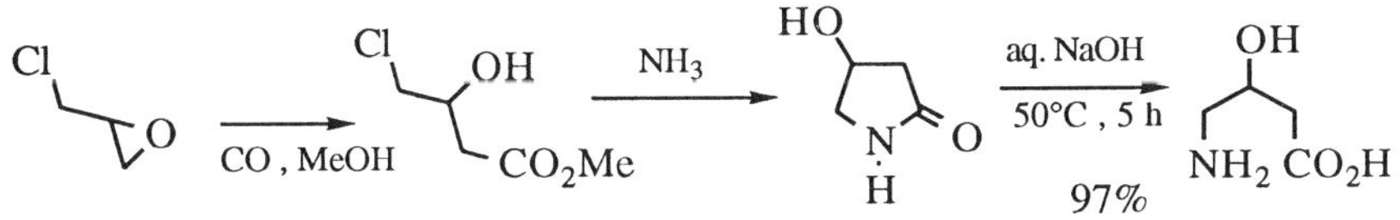

6.29 **6.30** **6.31** **6.27**

was displaced by ammonia, spontaneous cyclization occurred to give lactam ***6.31*** [35] When ***6.31*** was opened with aqueous sodium hydroxide, GABOB (***6.27***) was formed. The cyano analog of ***6.30*** (replace CO_2Me with CN) has also been prepared, from allyl bromide, and converted to carnitine, ***6.28***.[36]

Halo-acid derivatives are useful precursors to GABOB. Chlorination of 4-aminobutanoic acid (GABA - see section 6.1) gave 3-chloro-4-aminobutanoic acid (***6.32***)

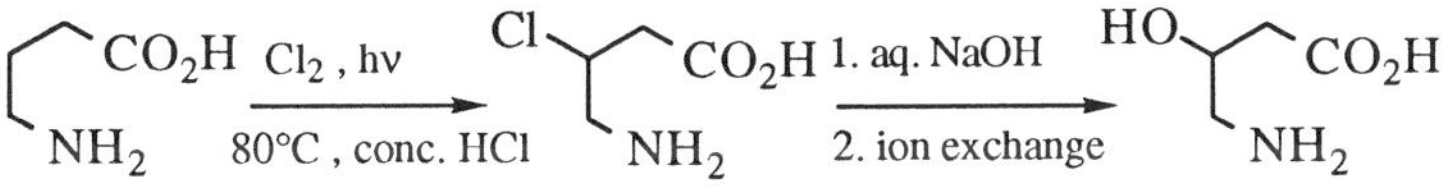

6.32 **6.27**

in good yield.[37] Subsequent treatment with aqueous hydroxide and passage through an ion exchange resin interchanged the chlorine with a hydroxyl group to give GABOB. In a closely related sequence, the chlorine moiety in methyl 4-chloro-3-hydroxybutanoate was displaced by ammonia, and subsequent hydrolysis of the ester GABOB in 68% yield.[38]

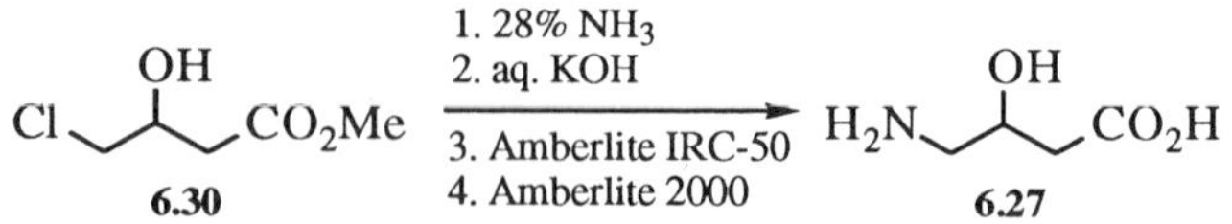

6.30 6.27

A related study used azide as an amine surrogate. Azide ***6.33*** is derived from a species analogous to ***6.30.*** Hydrolysis of the ester was followed by reduction of the azide, and exhaustive methylation to give carnitine (***6.28***).[39]

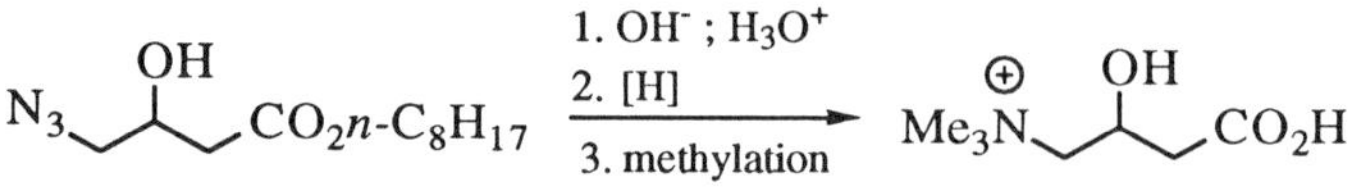

6.33 6.28

A somewhat related strategy used an α-chloro ketone as a starting material. The ketone moiety in keto-amide ***6.34*** was reduced quantitatively with sodium borohydride and then reacted with ammonia to give ***6.35***.[40] Aqueous acid hydrolysis led to GABOB, ***6.27***. An α-bromo ketone is a closely related precursor to ***6.30***. It is

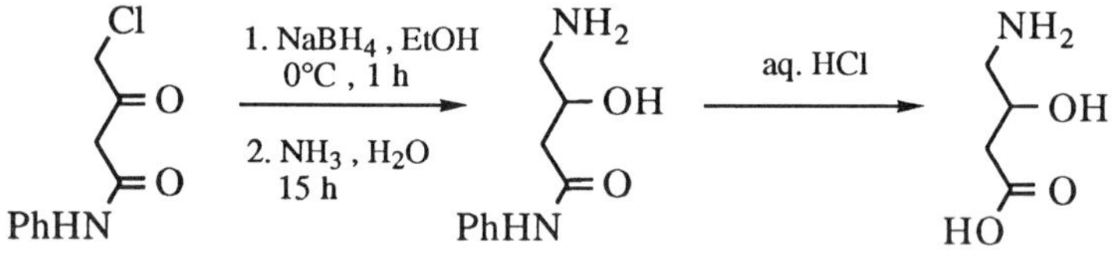

6.34 6.35 6.27

possible to react such a species with amines as well as ammonia. Ethyl acetate was α-brominated and the ketone carbonyl in that product was reduced with sodium borohydride (a mild reducing agent to prevent reduction of the C-Br bond) to give alcohol ***6.158***.[41] Reaction with trimethylamine displaced the bromine and hydrolysis of the ester gave carnitine, ***6.28***.

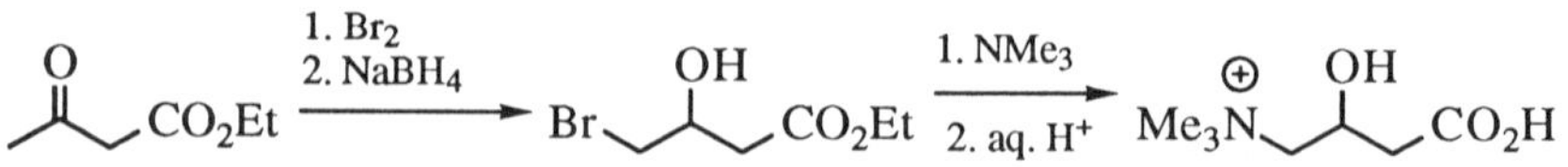

6.36 6.28

Conjugated acids such as crotonic acid can be used as precursors to halides that can be converted to GABOB via reaction with N-bromosuccinimide to give an allylic bromide, as in ***6.37***.[42] Reaction with ammonium hydroxide converted the bromide to an amine (***6.38***) and this was followed by reaction with an acidic resin in water to convert the alkenyl moiety to an alcohol, giving GABOB.

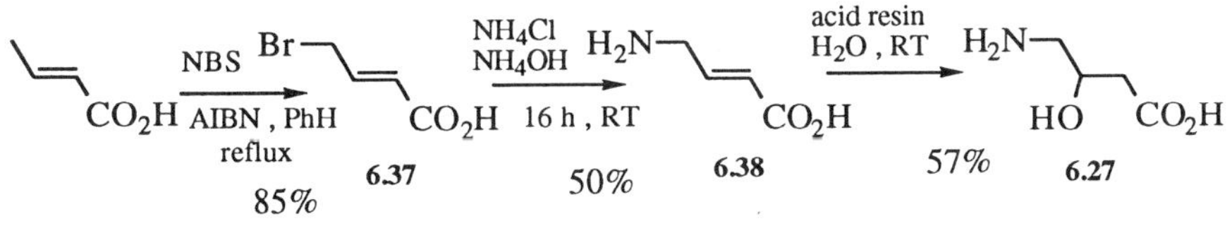

GABOB can be prepared by using a "heterocyclic" strategy. A [3+2]-cycloaddition reaction between ***6.39*** (prepared from the amine of glyoxylic acid)[43] and N-acetyl allylamine generated dihydroisoxazole ***6.40***.[44] Treatment with methanolic potassium carbonate gave lactam ***6.41***, and catalytic hydrogenation over Raney nickel opened the ring to give the N-acetyl methyl ester of GABOB, ***6.42***. Acid hydrolysis liberated GABOB.[44]

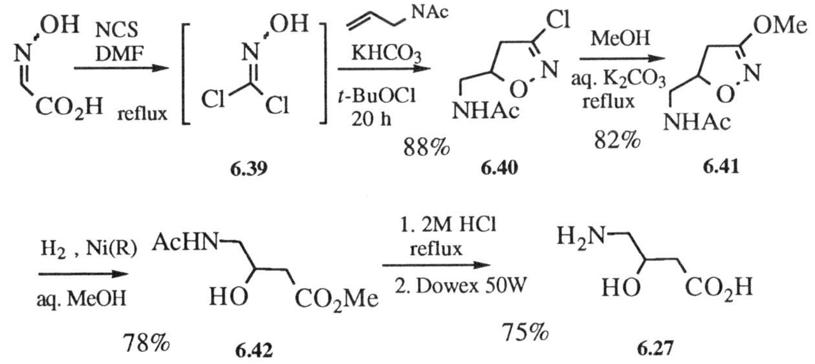

Aldehydes and ketones are useful sources of GABOB when condensed with acid derivatives. Aldehyde ***6.43*** was prepared by initial reaction of phthalic anhydride and glycine, via conversion of the resulting phthalimido acid to an acid chloride and catalytic hydrogenation. Subsequent condensation with malonic acid and pyridine (the Doebner condensation)[45] gave ***6.44*** as a mixture of *E*- and *Z*-isomers.[46] Aqueous acid hydrolysis led to GABOB.

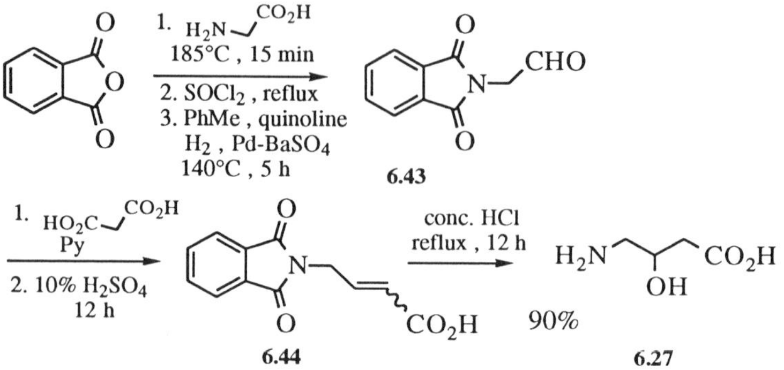

6.43 **6.44** **6.27**

6.2.B. Enzymatic Transformations

There are several synthetic approaches that rely on enzymatic transformations for the preparation of chiral, nonracemic GABOB or carnitine. Several variations of this strategy are presented in this section, usually varying by the organism used as a "reagent." In one example, *threo* β-hydroxyl glutamic acid (***6.45***) was synthesized since it was not commercially available. Once in hand, ***6.45*** was selectively decarboxylated by *E. coli* to give GABOB (***6.27***).[47] Exhaustive methylation gave carnitine (***6.28***).

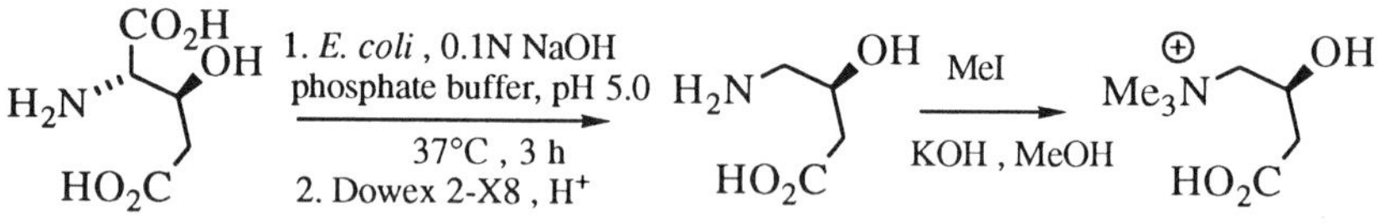

6.45 **6.27** **6.28**

Selective reduction of ketones to give chiral, nonracemic alcohols is another important application of enzymatic reactions. Reduction of ***6.46*** with baker's yeast provided ***6.47*** with high selectivity.[48] Reaction with trimethylamine followed by acid hydrolysis gave carnitine (***6.28a***).

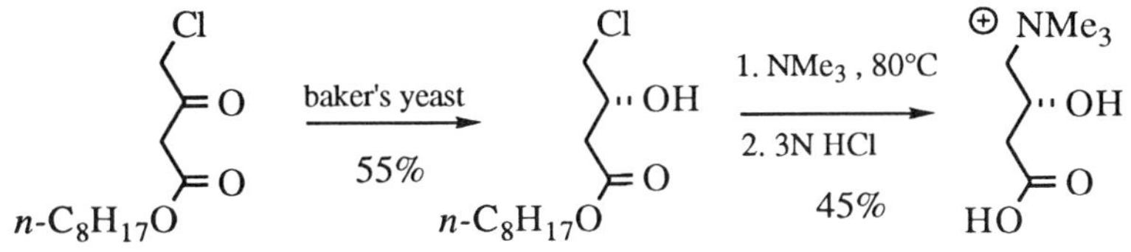

6.46 **6.47** **6.28a**

The kinetic resolution of key intermediates is probably the most common enzymatic transformation applied to the synthesis of GABOB. But-2-enoic acid was epoxidized to give racemic ***6.48***.[49] Kinetic resolution with pig liver esterase led to a 30% isolated yield of ***6.49*** after its separation from unreacted ester. Reaction of ***6.49*** with ammonia opened the epoxide to give S-GABOB, ***6.27b***.

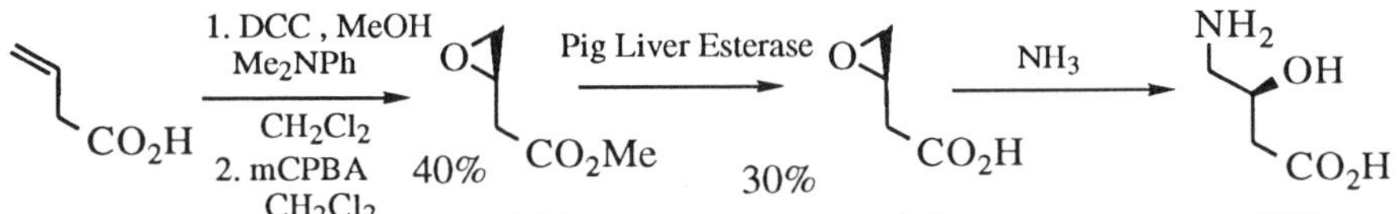

6.48 **6.49** **6.27b**

The diethyl ester of citric acid (***6.50***) was similarly converted to the half-ester (***6.51***) with *Arthobacter sp* (ATCC 19140).[50] Acetylation of the alcohol and conversion to the acid chloride was followed by reaction with sodium azide to give acyl

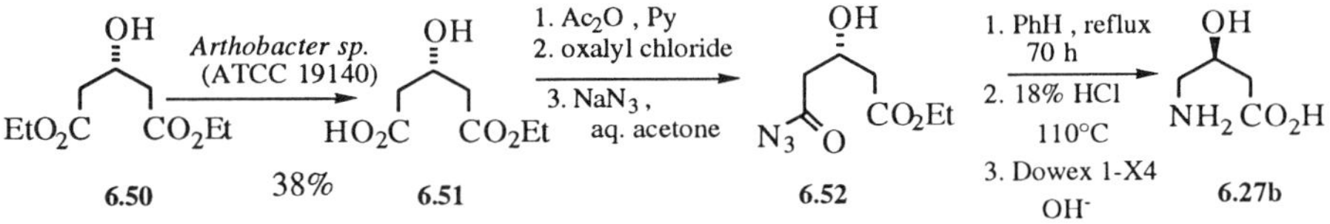

6.50 **6.51** **6.52** **6.27b**

azide ***6.52***. Heating induced a Curtius rearrangement[51] and acid hydrolysis of the resulting product led to S-GABOB (***6.27b***, 36% overall yield from ***6.51***). Carnitine was prepared, as above, by exhaustive methylation. Similar results were obtained from citric acid when ***6.51*** was formed in 70% yield using *Corynebacterium equi* (IFO-3730).[50]

Diethyl acetal derivative ***6.53*** also served as a precursor to GABOB, via initial conversion to cyanohydrin ***6.54***. Acetylation and selective partial hydrolysis of the *R*-acetate by *Candida cylindacea* lipase (CCL) gave a mixture of ***6.56*** and ***6.55b***.[52] After separation, reduction of the nitrile moiety in ***6.55a*** was followed by hydrolysis to give the ethyl ester of *R*-GABOB (***6.57a***). In this particular study, that ester was hydrolyzed and converted to carnitine.[52] The opposite antipode of GABOB was obtained by treatment of ***6.56*** with porcine pancreatic lipase (PPL), which gave a 48% yield (60% conversion) of ***6.55b***. This nitrile was quantitatively reduced to give the ethyl ester of *S*-GABOB, ***6.57b***.

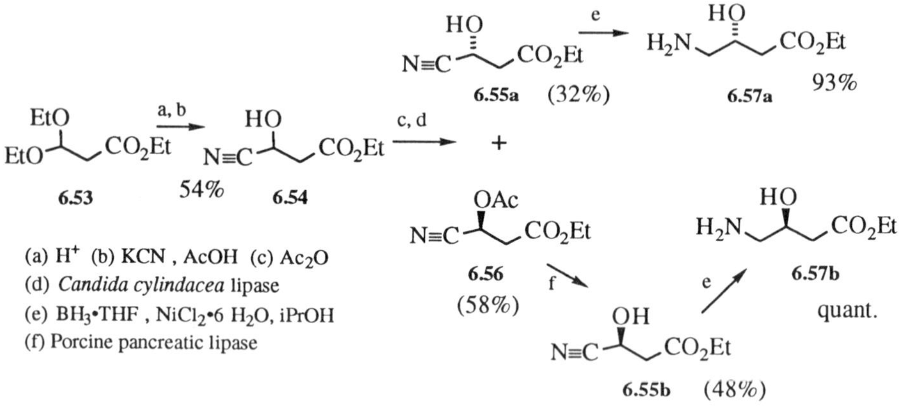

(a) H^+ (b) KCN , AcOH (c) Ac_2O
(d) *Candida cylindacea* lipase
(e) BH_3•THF , $NiCl_2$•6 H_2O, iPrOH
(f) Porcine pancreatic lipase

6.2.C. Transformations Using Chiral Auxiliaries

There are several reaction sequences that use a chiral auxiliary to produce chiral, nonracemic GABOB, but are rather disparate in their approaches. In one, cyclobutanone ***6.58***[53] reacted with R-phenethylamine to give an imine, and this was oxidized with *m*-chloroperoxybenzoic acid to give spiro-oxaziridine derivative ***6.59***.[54] Photolysis of ***6.59*** led to a mixture of lactams ***6.60*** and ***6.61***.[54] Separation of these diastereomers was followed by treatment of ***6.61*** with ammonia to cleave the auxiliary protecting group, giving ***R-6.31.*** Subsequent hydrolysis with aqueous HCl opened the lactam ring to give R-GABOB (***6.27a***).

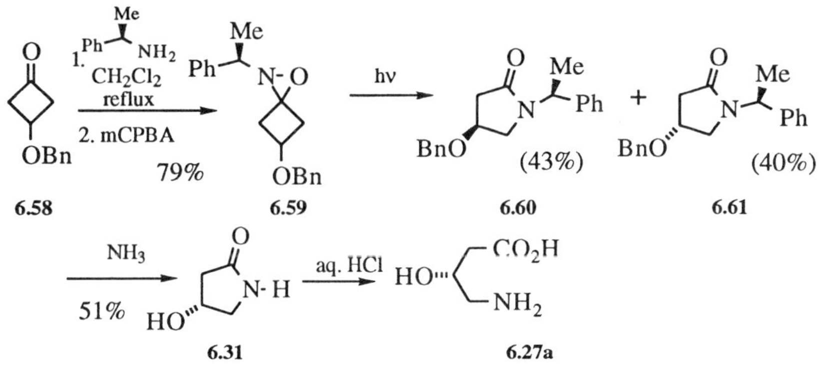

Enantiomer ***6.62*** also contains a phenethylamine auxiliary, and reaction with iodine gave a 1:1 mixture of ***6.63*** and ***6.64***, which were chromatographically separated.[55] The iodide moiety in ***6.63*** was replaced with hydrogen by treatment with tri-*n*-butyltin hydride. Removal of the auxiliary-protecting group from nitrogen, hydrolysis to open the ring, and conversion of the ester to an acid gave R-GABOB, ***6.27a***.

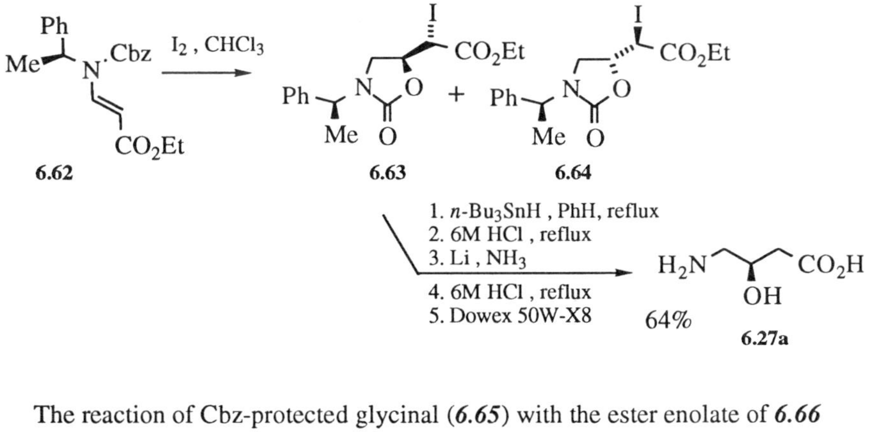

The reaction of Cbz-protected glycinal (***6.65***) with the ester enolate of ***6.66***

(bearing the chiral auxiliary) gave ***6.67***.[56] Hydrolysis of the ester removed the auxiliary and removal of the Cbz group by hydrogenation over palladium gave R-GABOB (***6.27a***) with 82%ee.

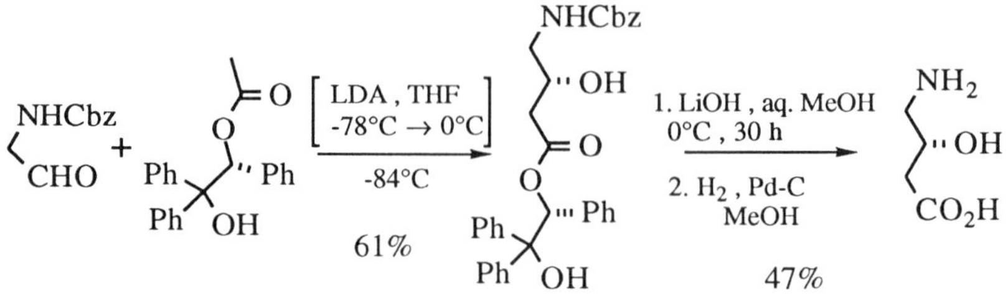

6.65 **6.66** **6.67** **6.68**

6.2.D. Transformations Using Chiral Templates

There are several reaction sequences that produce chiral, nonracemic GABOB or carnitine that begin with an enantiopure molecule. A variety of approaches using such chiral templates are collected into this section.

6.2.D.i. Carbohydrate Templates

The potassium salt of *D*-erythronate (***6.68***) is an example of a carbohydrate template, and it was converted to ***6.69***.[57] Subsequent reaction with sodium azide, catalytic hydrogenation to reduce the azide to the amine, and acid hydrolysis led to *R*-GABOB (***6.27a***). Using potassium *L*-erythronate as a precursor likewise led to *S*-GABOB (***6.27b***) in 55% overall yield.[57]

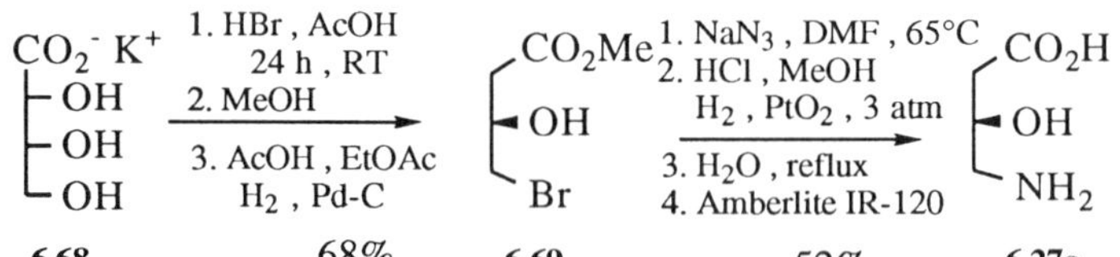

6.68 **6.69** **6.27a**

Another carbohydrate template is ascorbic acid (***6.70***), which was converted to ***6.71*** in eight steps. Conversion to the mesylate and reaction with potassium azide gave ***6.72***. Hydrogenation of the azide moiety and hydrolysis gave GABOB.[58]

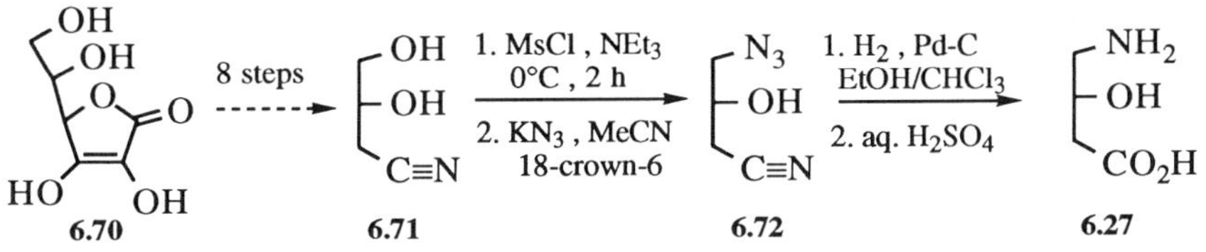

6.2.D.ii. Amino Acid Based Templates

The methyl ester of 4-hydroxyproline (***6.73***) was prepared[59] and used as a chiral template. Deprotonation and treatment with *t*-butyl hypochlorite gave an N-chloramine, which was dehydrohalogenated to imine ***6.74*** upon treatment with triethylamine.[60] The ester was hydrolyzed and the resulting imino acid was oxidized with hydrogen peroxide to "add" an oxygen and effect decarboxylation to generate lactam ***6.116.*** Subsequent ring opening with aqueous acid gave R-GABOB and the overall yield of ***6.27a*** from ***6.73*** was 29%.

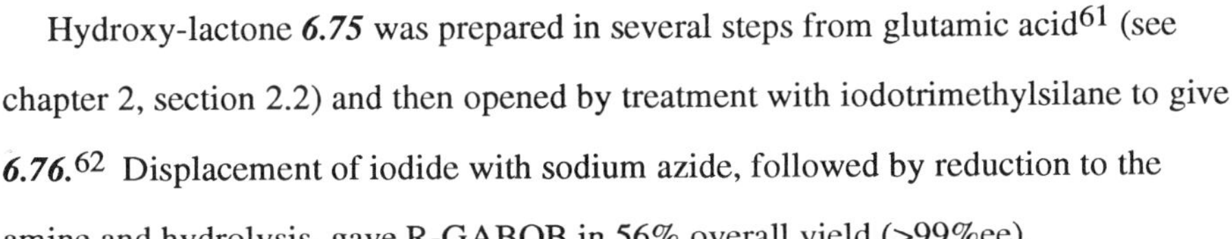

Hydroxy-lactone ***6.75*** was prepared in several steps from glutamic acid[61] (see chapter 2, section 2.2) and then opened by treatment with iodotrimethylsilane to give ***6.76.***[62] Displacement of iodide with sodium azide, followed by reduction to the amine and hydrolysis, gave R-GABOB in 56% overall yield (>99%ee).

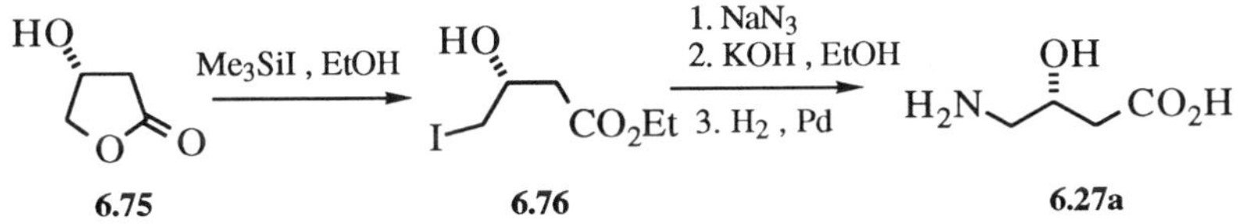

A chiral, non-racemic oxazolidine derivative (***6.77***) was obtained by resolution of the racemic material prepared from serine.[63] Electrolytic conversion of the acid moiety to an O-acetate (***6.78***) followed by thermal elimination of acetic acid gave ***6.79***. Reaction with HCl gave ***6.80***, which was condensed with the enolate anion of dibenzyl malonate to give ***6.81*** after deprotection and decarboxylation (25% yield for both steps, with the displacement step proceeding in only 27%). Opening the ring with aqueous acid gave S-GABOB, ***6.27b***.

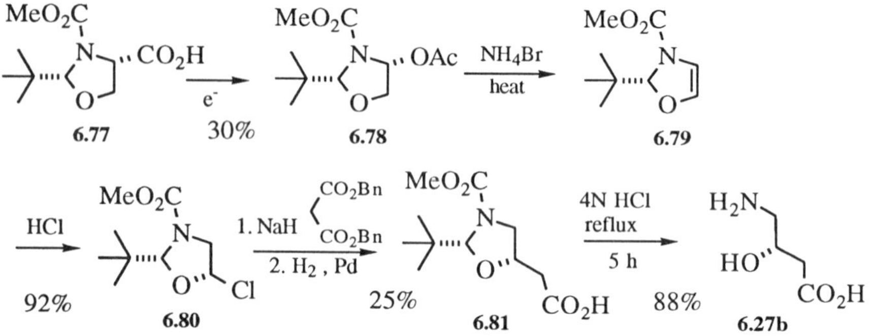

6.2.D.iii. Miscellaneous Chiral Templates

A rather long, asymmetric synthesis of *R*-(-)-GABOB was reported in which β-pinene (***6.82***) was converted to ***6.83*** in eight steps.[64] Oxidative cleavage of the

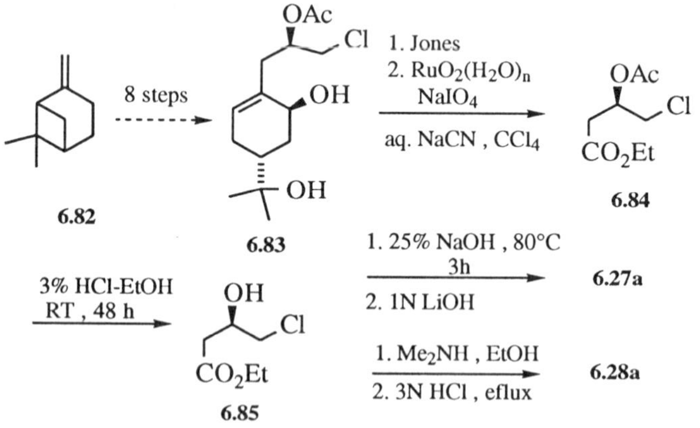

cyclohexene ring led to ***6.84***, which was hydrolyzed to ***6.85*** with good enantiomeric purity. Ester ***6.85*** was converted into R-GABOB (***6.27a***) and into R-carnitine, ***6.28a***. This preparation of GABOB from ***6.85*** involved initial conversion to hydroxy-lactam followed by hydrolytic ring opening of that lactam intermediate.[64]

Epichlorohydrin (***6.29***) was used in section 6.2.A to prepare GABOB. Enantiopure R or S-epichlorohydrin is also available for use as a chiral template. *R*-Epichlorohydrin (***6.29a***), for example, was treated with potassium cyanide and then with sodium azide to give ***R-6.86***.[65] The product of the first step (reaction with cyanide) produced 4-chloro-3-hydroxybutanenitrile, which reacted with sodium azide and is a useful synthetic intermediate. Reduction of the azide moiety and hydrolysis of the nitrile led to ***R-6.27a*** , making the azide the amine surrogate and the nitrile the acid

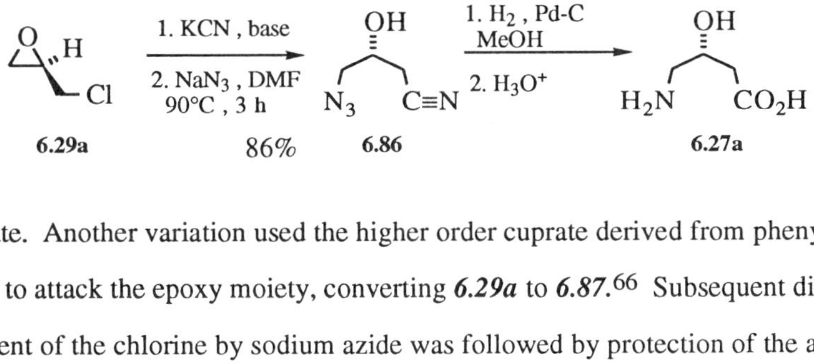

6.29a **6.86** **6.27a**

surrogate. Another variation used the higher order cuprate derived from phenyllithium to attack the epoxy moiety, converting ***6.29a*** to ***6.87***.[66] Subsequent displacement of the chlorine by sodium azide was followed by protection of the alcohol as the carbonate (see ***6.88***). In this sequence, the phenyl ring is a carboxyl surrogate,

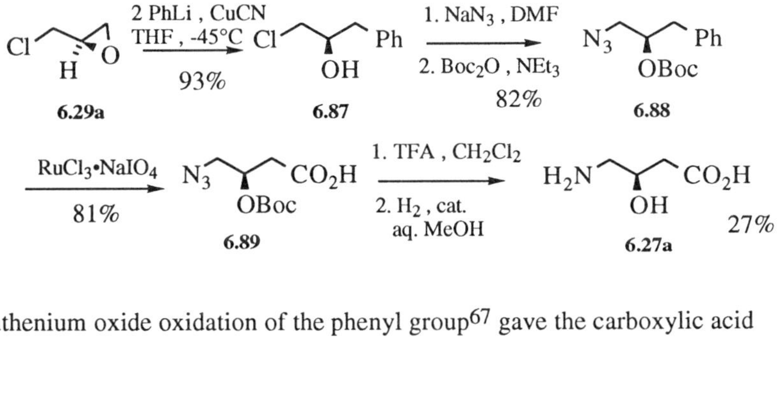

and ruthenium oxide oxidation of the phenyl group[67] gave the carboxylic acid moiety in ***6.89***.[66] Removal of the Boc group proceeded in very poor yield (30%) but hydrogenation of the resulting hydroxy-azide gave R-GABOB, ***6.27a***.

A related synthesis used chiral, nonracemic epoxy-alcohol ***6.90***, and involved oxidation of the alcohol moiety to an acid moiety (to give ***6.91***).[68] Opening the epoxide with ammonia led to a 66% overall yield of R-GABOB (49%ee). It is noted that allyl oxirane was recently converted to R-GABOB in two steps.[69]

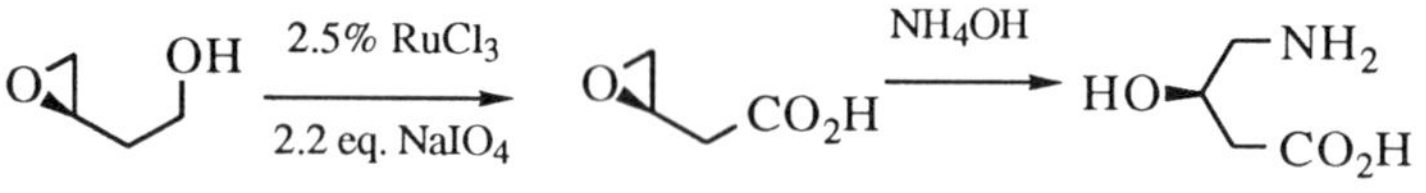

6.90 **6.91** **6.27a**

Allylic alcohol ***6.92*** was prepared in about seven steps from tartaric acid and once in hand, treatment with sodium hydride and benzyl bromide gave ***6.93***.[70] Hydroboration of the alkene moiety and Jones oxidation of the resultant alcohol gave a carboxylic acid (***6.94***). Deprotection and ring opening gave R-GABOB.[70]

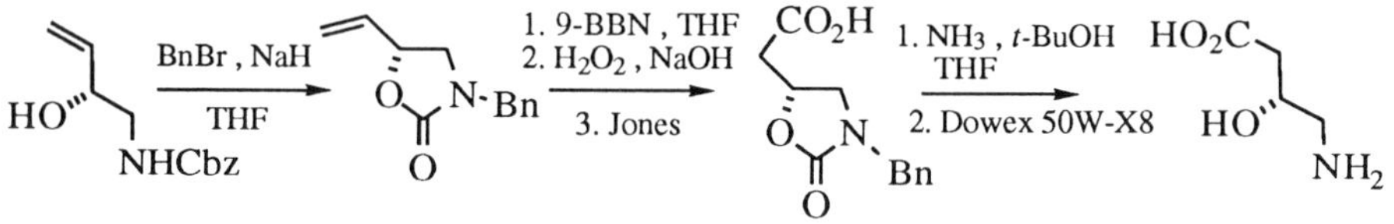

6.92 **6.93** **6.94** **6.27a**

6.3. Statine and Related Compounds

Statine (4-amino-3-hydroxy-6-methylheptanoic acid, ***6.95***) is an important hydroxy amino acid with biological activities that have attracted considerable

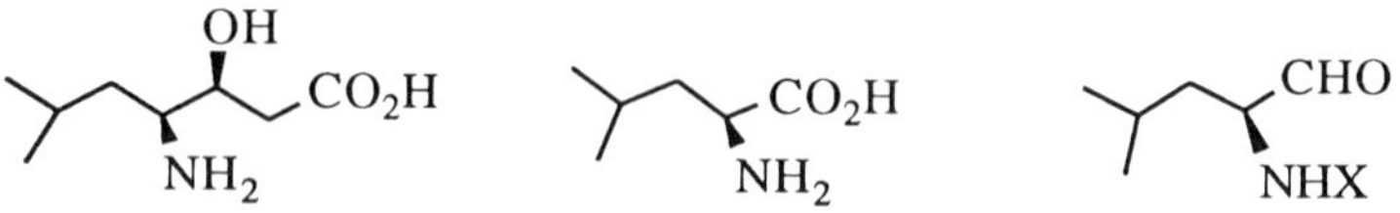

6.95 **6.96** **6.97**

attention. Statine is a key constituent of pepstatin and related inhibitors and also of the blood pressure regulatory enzyme renin.[71] It is a potential anti-hypertensive medicinal agent since it is a component of renin inhibitors.[71] It is also a constituent of cathepsin D.[72]

6.3.A. Leucine Based Syntheses

The most important method for preparing statine involves the conversion of leucine (***6.96***) to leucinal derivatives (***6.97***). This aldehyde could then be condensed with a variety of nucleophilic reagents or with an ylid. In both cases, the carbon chain is extended and an acid moiety is incorporated.

An early and typical synthesis shows the conversion of N-Boc-*L*-leucine methyl ester (***6.98***) to aldehyde ***6.99***.[73] This aldehyde was condensed with the lithium enolate of ethyl acetate to give ***6.100***. Alcohol ***6.100*** was produced as a mixture of diastereomers and one of the important challenges of this strategy is to control the diastereoselectivity of the condensation. In this example, the 3S,4S diastereomer of ***6.100*** was separated and isolated (in 40% yield based on ***6.99***) and acid hydrolysis gave N-Boc statine (***6.101***).[73]

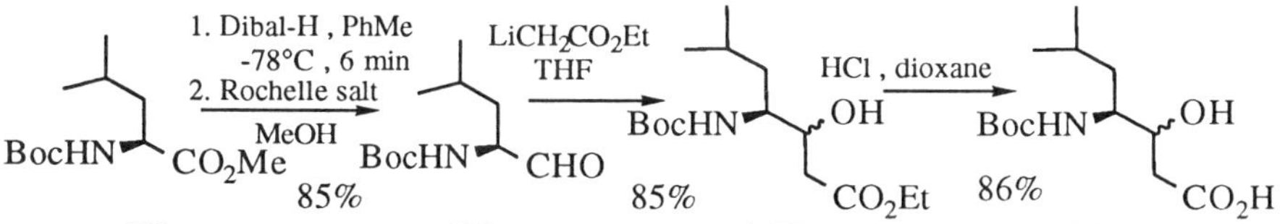

6.98 **6.99** **6.100** **6.101**

Other methods for the preparation of ***6.99*** from leucine involved reduction of the acid moiety with borane and then oxidation to an aldehyde with chromium trioxide and pyridine.[74] Another variation added zinc bromide to the enolate condensation reaction,[75] which led to greater selectivity for the *anti* diastereomer.

Wuts improved the selectivity of the condensation using ***6.66*** (which bears a chiral auxiliary; see section 6.2.C) in a condensation reaction with ***6.99***.[76] This reaction gave an 8:1 mixture of ***6.102*** and ***6.103***.[76] Saponification gave a 9.2:1 mixture of diastereomeric N-Boc statine derivatives, ***6.101a*** and ***6.101b***.

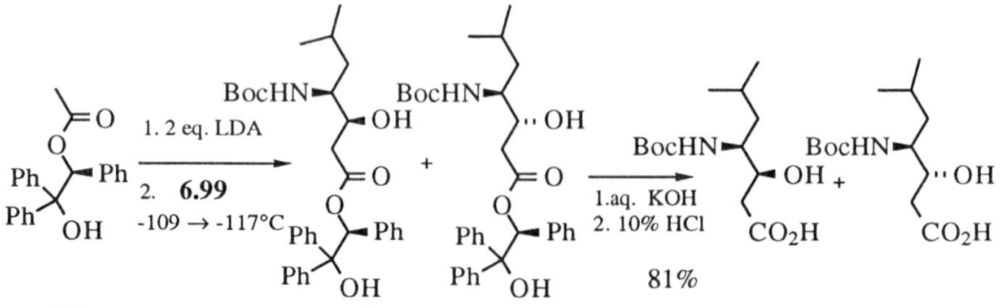

6.66 6.102 6.103 6.101a 6.101b

The Mukaiyama reaction[77] provides a good synthetic route to statine. *L*-Leucine methyl ester (***6.104***) was converted to the N-isopropylcarbamate aldehyde (***6.105***).[78] Mukaiyama reaction with the trimethylsilyl enolate of methyl acetate, catalyzed by

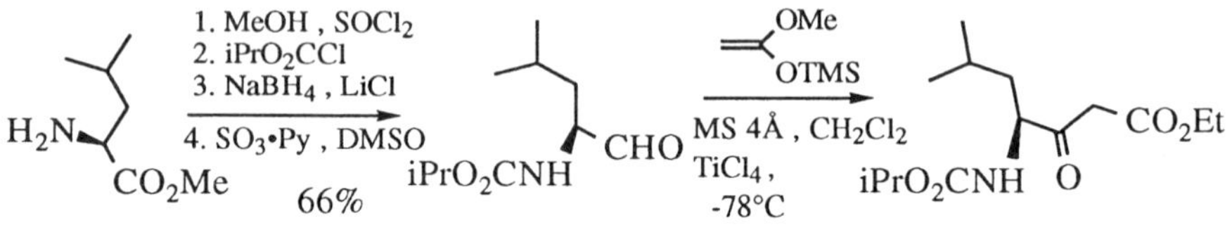

6.104 6.105 6.106

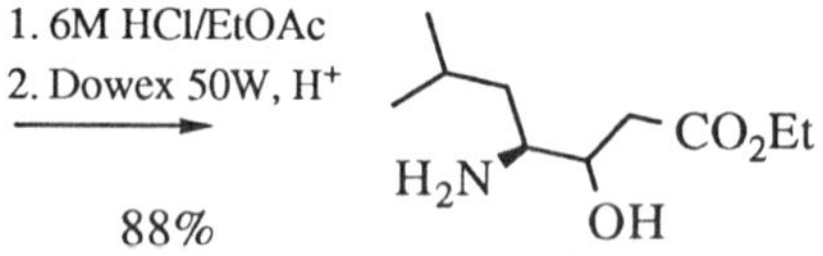

6.107

titanium tetrachloride, gave ***6.106*** with good diastereoselectivity (94:6 *anti:syn*).[78] The *anti* diastereomer was converted to the ethyl ester of statine, ***6.107***.

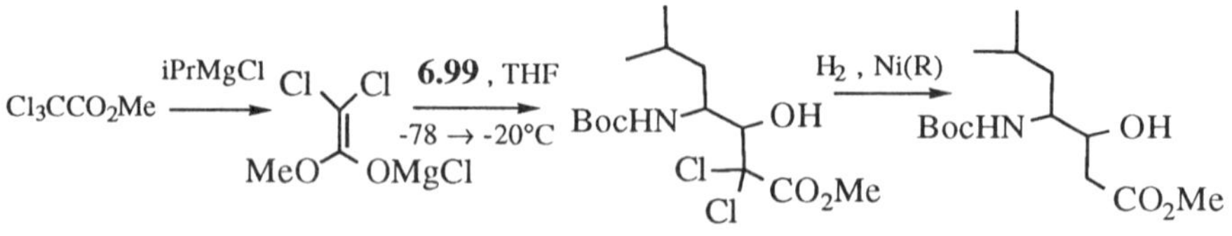

6.108 6.109 6.110

An alternative approach condensed ***6.99*** with the magnesium enolate of methyl trichloroacetate, ***6.108***. This reaction gave a 70:30 mixture of 3R,4S:3S,4S ***6.109***.[79] Separation of the diastereomers and catalytic hydrogenation led to hydrogenolysis of the α-chloro moiety to give 3S,4S-***6.110*** in 90% yield.

Another type of condensation reaction used N-phthaloyl leucinal (***6.111***) in a Reformatsky reaction[80] with zinc and bromo *t*-butyl acetate. The product was ***6.112***, obtained as a mixture of diastereomers.[81] Separation of the requisite diastereomer was followed by conversion to statine (***6.95***) by deprotection of nitrogen and the

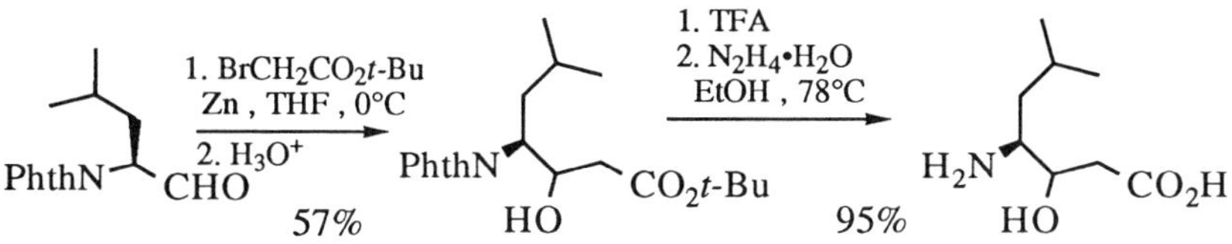

6.111 **6.112** **6.95**

carboxyl group. A modification of this sequence converted N-phthaloyl leucine to the acid chloride (with thionyl chloride) and condensed it with the magnesium salt of mono-ethyl malonate.[82] The resultant ketone was reduced and decarboxylated to give statine, but in only 3.2% overall yield.[82] This strategy was reported earlier in the patent literature.[83]

Other leucine derivatives can be used to prepare statine. The acid moiety of N-Boc leucine reacted with imidazole dicarbonyl to form a highly reactive acyl

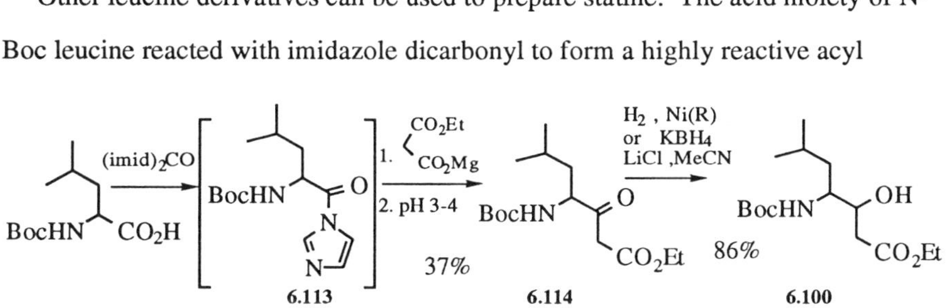

6.113 **6.114** **6.100**

imidazoyl derivative, which was then condensed with ester enolates to form oxo derivatives. N-Boc-Leucine was converted to ***6.113***,[84] *in situ*, and then condensed with the magnesium enolate of mono-ethyl malonate to give ***6.114***.[85] Raney nickel hydrogenation (or alternatively, reduction with $LiBH_4$ generated from KBH_4-LiCl) converted the ketone moiety in ***6.114*** to an alcohol in ***6.100***. Deprotection gave statine. It is noted that in this work Klostermeyer also reacted ***6.114*** with $LiCH_2CO_2Et$ to give ***6.100*** in 80% yield.[85] This strategy has been used by others to prepare statine.[86] The ketone moiety in ***6.114*** can be hydrogenated with high diastereoselectivity by using a $RuBr_2$[R-BINAP] catalyst, giving the *syn*-amino alcohol with >99:1 selectivity and in 97% yield.[87] This approach was also used to prepare cyclohexyl derivative ***6.188***[88] (see section 6.3.D). Schuda used this basic approach to prepare a cyclohexyl derivative from phenylalanine and used sodium cyanoborohydride to reduce the ketone intermediate.[89,86e]

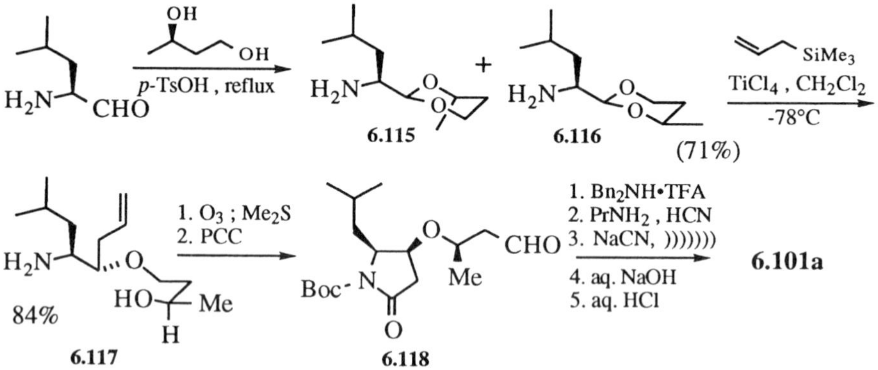

There are several strategies that involve the use of transition metal catalysts in addition to the Mukaiyama reaction. Reaction of leucinal with 1,3-butanediol gave a mixture of ***6.115*** (in 9% yield) and ***6.116*** (in 71% yield).[90] After the chromatographic separation of ***6.116***, the titanium tetrachloride catalyzed reaction with 3-trimethylsilyl-1-propene gave ***6.117***. Ozonolysis and oxidation gave lactam ***6.118***, which was converted to ***6.101a*** in five steps (overall yield was 32% from ***6.117***).

Note that in this case, the coupling reaction with sodium cyanide required the use of ultrasound [)))))))].

Another transition metal mediated route involved formation of an oxazolidinone via coupling an allyl palladium reagent with amide ***6.119***. This reaction gave a 15:1 mixture of ***6.120:6.121***.[91] Hydroboration of the vinyl group was followed by hydrolysis to give ***6.122***. The primary alcohol moiety was oxidized to the acid moiety in ***6.101a*** using platinum oxide and oxygen.

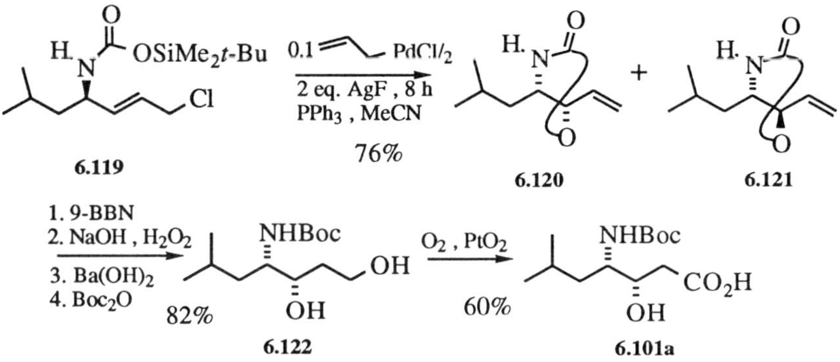

It should be noted that statine derivatives can be prepared from statine itself. Statine derivative ***6.123***, for example, was transformed into the 2-ethyl derivative

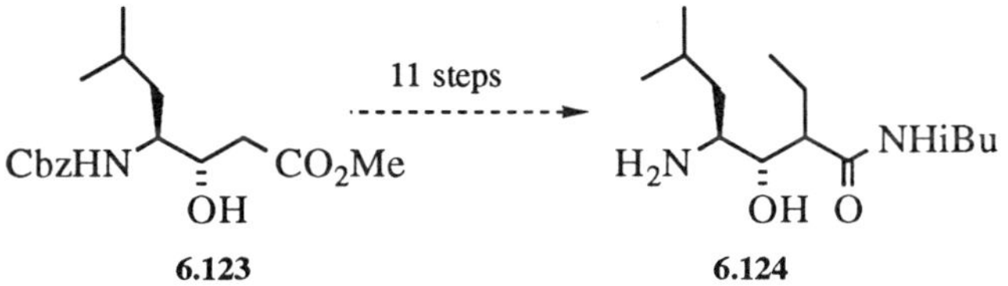

6.124 in eleven steps (22% overall yield).[92]

There are several interesting transformations of leucine that ultimately lead to statine. One method converted *L*-leucine ethyl ester to amide ***6.125***.[93] A Dieckmann cyclization[94] followed by decarboxylation gave keto-lactam ***6.126*** from

6.125. Reduction of the ketone moiety and ring opening with aqueous HBr gave statine[93] This fundamental approach was later used to prepare both statine and an analogous cyclohexyl derivative[95]

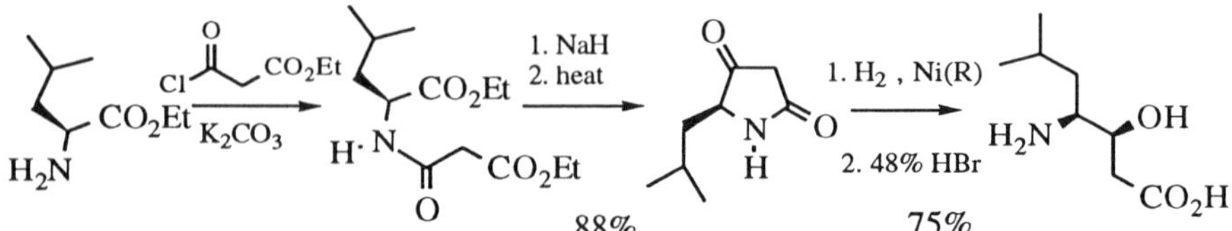

6.125 **6.126** **6.95**

A "heterocyclic" strategy used an oxazolidone ring as a protected carboxyl (an amide) which also served as a chiral auxiliary. Treatment of ***6.127*** with a boron triflate, for example, gave boron enolate reagent ***6.128 in situ***. and it reacted with N-Boc leucinal to form ***6.129***.[96] Boron enolates have been used in many syntheses as an alternative to lithium or sodium enolate in Aldol reactions.[97] Hydrogenolysis of the methylthio moiety gave a methylene moiety and treatment of the amide auxiliary with base gave ***6.95*** (24% overall yield).

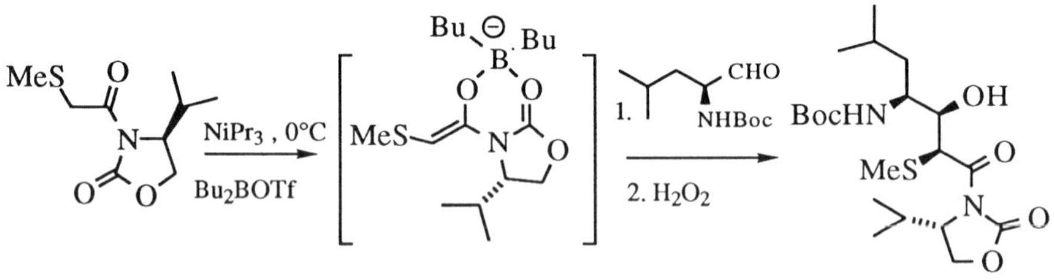

6.127 **6.128** **6.129**

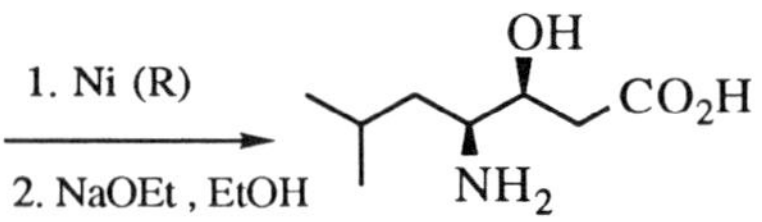

6.130

Leucinal derivatives can react with carbonyl-bearing ylids to form an alkenyl amino acid. The alkene moiety can be converted to an alcohol, allowing a synthesis of statine. Conjugated ester ***6.130*** was prepared from N-Cbz-leucinal with good selectivity (>30:1 *cis:trans*).[98] The ester moiety was reduced to an aldehyde and the double bond epoxidized, allowing reduction with Red-Al to give ***6.131***.[98] The primary alcohol moiety was oxidized with oxygen and platinum oxide to give

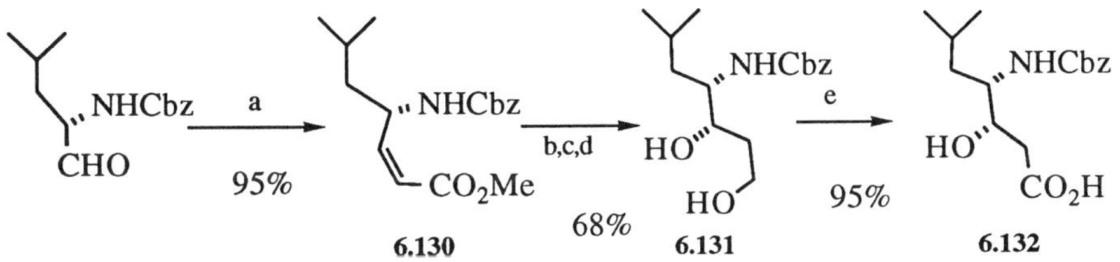

(a) $[CF_3CH_2O]_2P(O)CH_2CO_2Me$, TMS_2NK , THF , -78°C (b) Dibal-H , CH_2Cl_2 , -78°C (c) mCPBA (d) Red-Al , THF (e) PtO_2 , O_2 , $NaHCO_3$

N-Cbz statine (***6.132***). A number of other derivatives were prepared by this method, including 4-amino-3-hydroxypentanoic acid, 4-amino-3-hydroxy-7-methyloctanoic acid, and 4-amino-3-hydroxy-5-phenylpentanoic acid.[98]

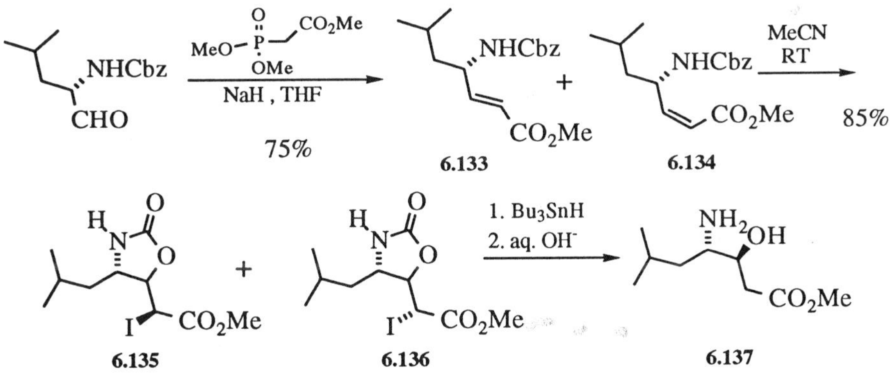

In a similar manner, N-Cbz leucinal reacted with the ylid of the phosphonate ester shown to give a 15:1 mixture of ***6.133:6.134.*** Stirring this mixture in acetonitrile gave a 98:2 mixture of ***6.135:6.136.***[99] Reduction of the iodo moiety with tri-*n*-butyltin hydride, followed by basic hydrolysis, gave statine methyl ester in 45%

overall yield. A larger percentage of the *trans*-alkene (***6.133***) could be produced (25:1 *trans:cis*) by changing the ylid to a trifluoroethyl ethoxy phosphonate ester.[99]

6.3.B. Carbohydrate Based Syntheses

Carbohydrates are important chiral, non-racemic precursors (chiral templates) to statine and its derivatives. Aldehyde ***6.138*** [100] was converted to alcohol ***6.139*** by reaction with isobutylmagnesium bromide.[101] After conversion to a benzoate (***6.140***), displacement with sodium azide gave ***6.141***. The tetrahydrofuran ring was opened and oxidation gave ***6.142***.[101] Oxidation of the aldehyde and then reduction of the azide moiety led to (+)-statine (3R,4S-***6.95***). Reaction of ***6.139*** with mesyl chloride and then sodium benzoate gave the benzoate (***6.140***) of opposite absolute configuration, which was converted to (-)-statine (3S,4R-***6.95***) by the same sequence.

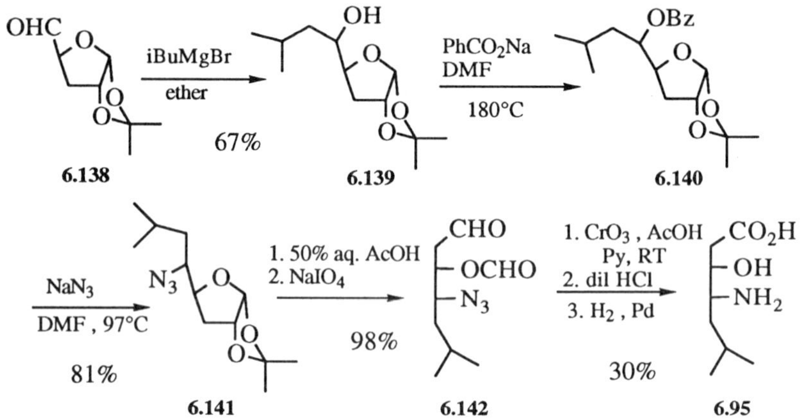

Another approach used ***6.143*** (rather than the aldehyde moiety in ***6.138***). Opening the epoxide with isopropylmagnesium cuprate gave the substituted alcohol, which was converted to the mesylate, ***6.144***.[102] Reaction with azide displaced the mesylate and hydrolysis deprotected the acetonide and opened the ring, allowing oxidation to acid ***6.145***. Catalytic hydrogenation of the azide gave 3S,4S-***6.95***. Using cyclopentylmagnesium bromide, cyclohexylmagnesium bromide, cycloheptylmagnesium bromide, benzylmagnesium bromide, or 2-phenylethylmagnesium bromide rather than isopropylmagnesium bromide in this sequence led to a variety of

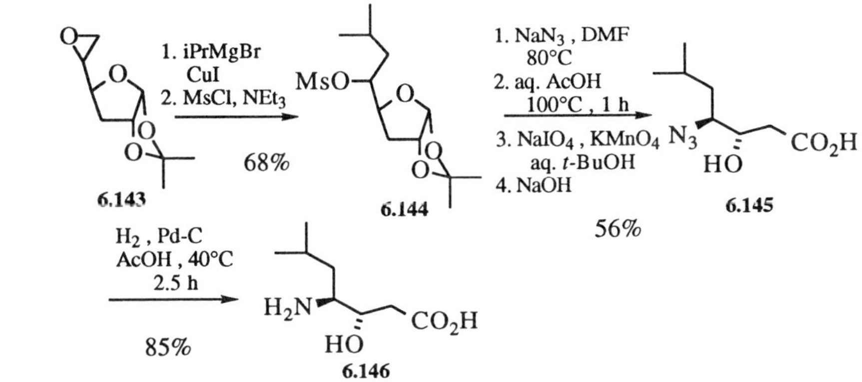

hydroxy amino acid derivatives [4-amino-5-cyclopentyl-3-hydroxypentanoic acid; 4-amino-5-cyclohexyl-3-hydroxypentanoic acid; 4-amino-5-cycloheptyl-3-hydroxypentanoic acid; 4-amino-3-hydroxy-6-phenylhexanoic acid; and, 4-amino-3-hydroxy-6-phenylheptanoic acid, respectively].

6.3.C. Other Synthetic Strategies

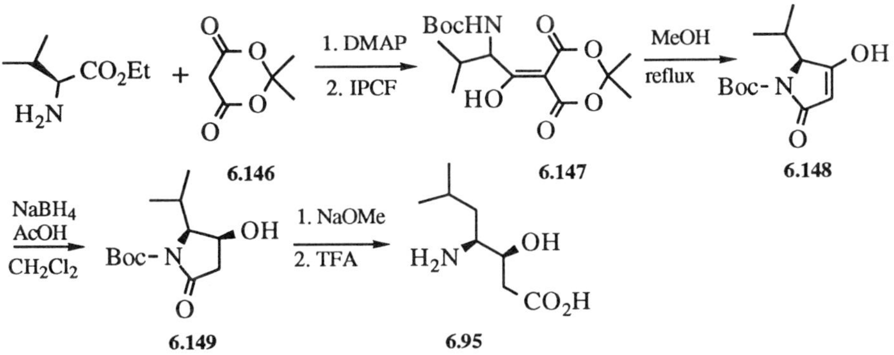

The α-amino acid valine was converted to statine via initial condensation with ***6.146*** to give ***6.147***.[103] Heating this adduct in methanol removed the protecting group and led to lactam ***6.148***, which was reduced to ***6.149***. Hydrolysis gave 3S,4S-statine (***6.95***), in 63% overall yield from valine.

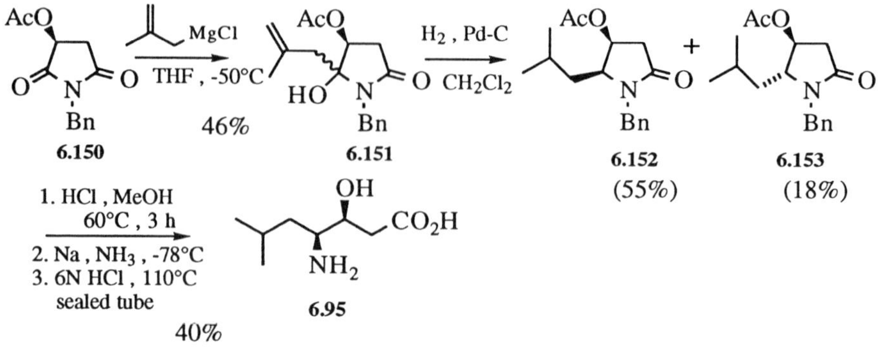

6.150 **6.151** **6.152** **6.153** **6.95**

The valine synthesis just presented utilized lactam intermediates. A quite different 'lactam driven' synthesis reacted imide ***6.150*** with isobutenylmagnesium chloride to give ***6.151***.[78a] Catalytic hydrogenation gave ***6.152*** along with 18% of

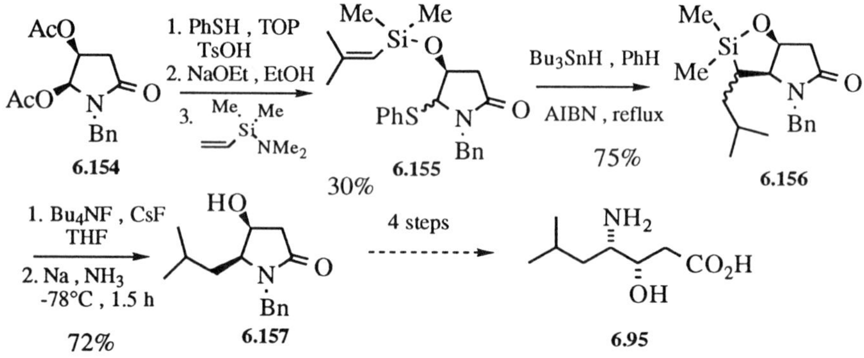

6.154 **6.155** **6.156** **6.157** **6.95**

6.153. Separation of these products allowed the conversion of ***6.152*** into 3S,4S-statine, ***6.95***. This fundamental approach was used by Speckamp except that a vinyl silane was used in the condensation reaction with the imide, to form an isobutylene moiety.[104] Speckamp introduced a modification that allowed stereocontrol of the amine and alcohol moieties. Lactam ***6.154*** was converted to vinyl silane ***6.155*** and subsequent treatment with tri-*n*-butyltin hydride gave ***6.156***.[104] Treatment with fluoride and de-benzylation with sodium in ammonia gave ***6.157***, which was converted to 3S,4S-***6.95*** in four steps.

An allyl tin driven strategy converted ***6.158***[105] to allyl derivative ***6.159***.[106] The allyl moiety was oxidatively cleaved and then esterified with diazomethane to give ***6.160***. The methoxy moiety was then converted to an isobutenyl group (see ***6.161***, in 33% overall yield) via coupling with the allyl-tin reagent shown. Catalytic hydrogenation followed by acid hydrolysis gave statine, ***6.95***. A key feature of this synthesis is the use of an oxazolidone intermediate as a chiral template. It is noted that an analog of ***6.159*** was prepared from *D*-glucosamine, and converted into 3R,4R-***6.95***.[107]

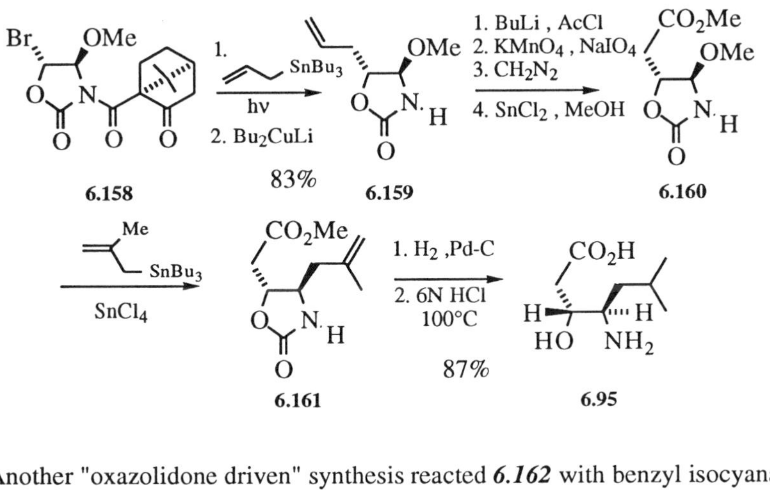

Another "oxazolidone driven" synthesis reacted ***6.162*** with benzyl isocyanate to give ***6.163***.[108] Reduction of the aldehyde to an alcohol and adjustment of pH led to formation of ***6.164***, but with poor diastereoselectivity (isolated as a 1:1 *cis:trans* mixture). Incorporation of the isobutenyl group and ring opening gave ***6.165***. Hydrogenation of the alkenyl moiety was followed by formation of ***6.166***. A $RuCl_2/NaIO_4$ oxidation of the phenyl group gave ***6.167***, and hydrolysis gave 3R,4R-***6.95***.

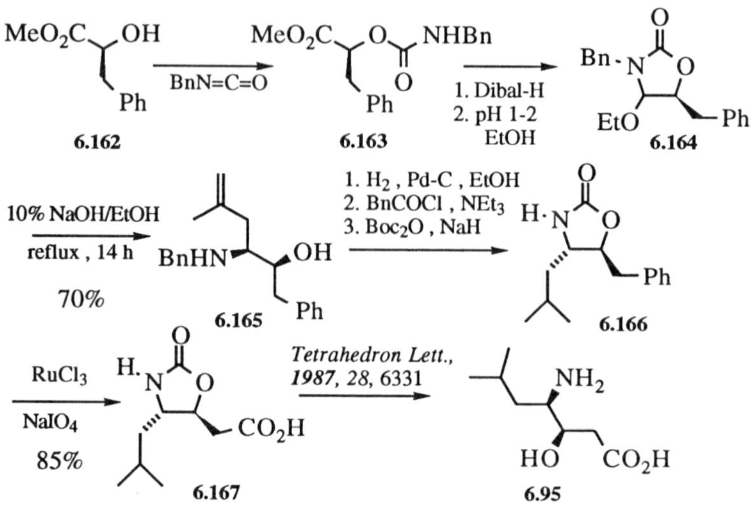

A different heterocyclic strategy used an isoxazoline intermediate. When 5-methyl-2E-hexen-1-ol (***6.168***) was treated with trichloroacetonitrile and heated,

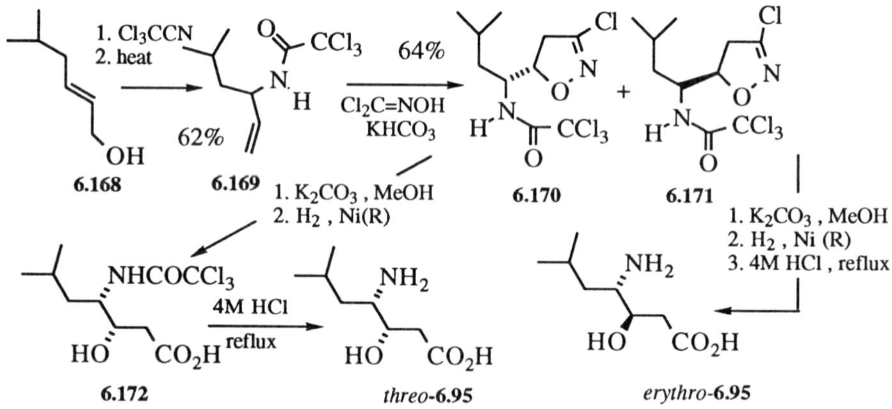

amide ***6.169*** was formed.[109] Reaction with the oxime of phosgene led to a 1.4:1 mixture of ***6.170:6.171***. Conversion to the methoxy derivative allowed ring opening by catalytic hydrogenation to give ***6.172***. Acid hydrolysis gave the *threo* derivative, *threo-**6.95*** (84% yield from ***6.170***). Likewise, after separation from ***6.170, 6.171*** was converted to the *erythro* diastereomer, *erythro-**6.95***.

Epoxides, derived from alkenes, are important synthetic precursors to statine. Sharpless asymmetric epoxidation[110] of 5-methyl-1-hexen-3-ol (***6.173***) gave ***6.174***

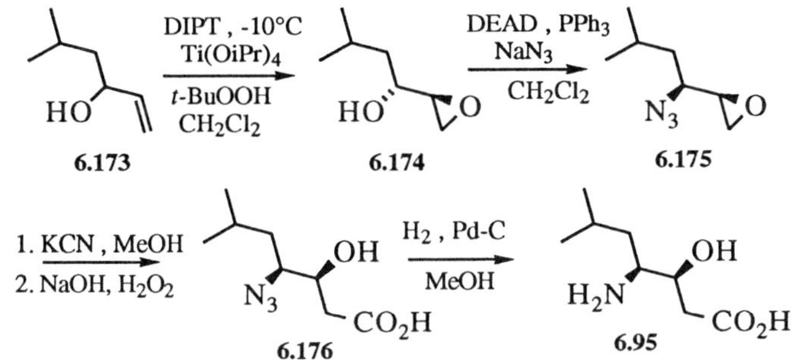

and Mitsunobu inversion[111] using azide as a nucleophile gave ***6.175***.[112] Opening the epoxide with potassium cyanide followed by hydrolysis of the resulting nitrile gave ***6.176***. This was converted to 3S,4S-***6.95*** by catalytic hydrogenation of the

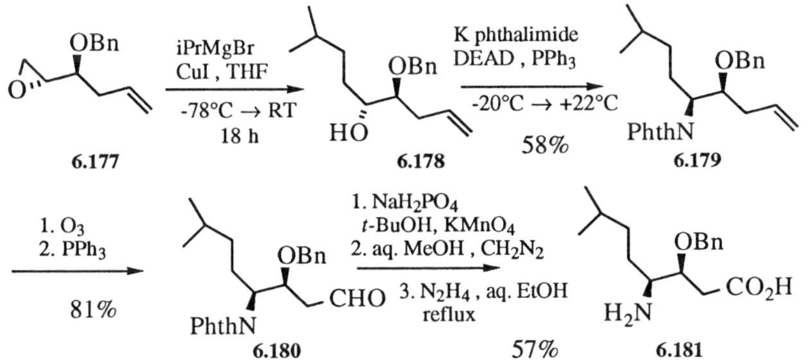

azido group.[112] A related study converted epoxide ***6.177*** [113] to ***6.178***.[114] Mitsunobu inversion[111] using phthalimide gave ***6.179***, and subsequent ozonolysis of the alkene gave aldehyde ***6.180***. Selective oxidation of the aldehyde to the carboxylic acid was followed by removal of the phthaloyl group to give O-benzyl statine (3S,4S-***6.181***).

6.3.D. Synthesis of Statine Analogs and Isosteres

A chemical isostere relates to two or more molecules with similarities in physiochemical properties of atoms, groups, or radicals with similar electronic

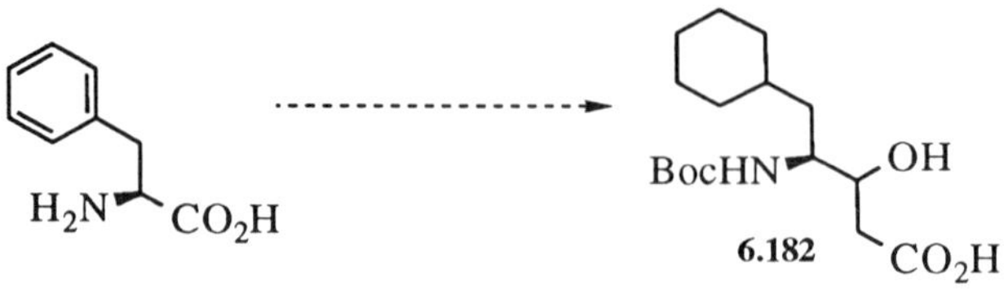

structures.[115] Other properties that contribute to being an isostere include basicity, bond angles, acidity, electronegativity, polarizability, bonds angles, size, shape of molecular orbitals, electron density, and partition coefficients. It essentially refers to replacing one group in a molecule with another that will give similar chemical or

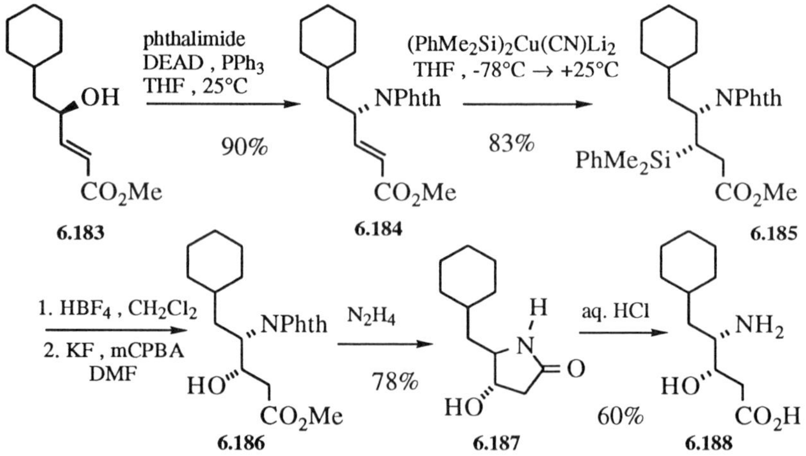

physiological properties.[115] An example of an isostere is cyclohexyl derivative ***6.182***, which can be prepared from phenylalanine via reduction of the phenyl ring and enolate condensation (33% overall yield).[116,88]

Another route to this cyclohexyl derivative involves the preparation of ***6.183*** via an olefination reaction with a substituted phenylalinal derivative followed by reduction of the phenyl ring. The alcohol moiety in ***6.183*** was converted to the N-phthaloyl derivative (***6.184***) by a Mitsunobu inversion.[111] Conjugate addition with a higher order silyl cuprate gave ***6.185***.[117] The silyl moiety was converted to an alcohol (***6.186***) by treatment with tetrafluoroboric acid and then KF/*m*-chloroperoxybenzoic acid. Removal of the phthalimidoyl group with hydrazine led to an aminoester, which cyclized to lactam ***6.187***.[117] Hydrolysis gave 4-amino-5-cyclohexyl-3-hydroxypentanoic acid (***6.188***) in 60% yield.

Other isosteres of statine are those amino acids derived from diastereomers of *L*-leucine. The general reactions used for *L*-leucine can be applied to its isomers. One

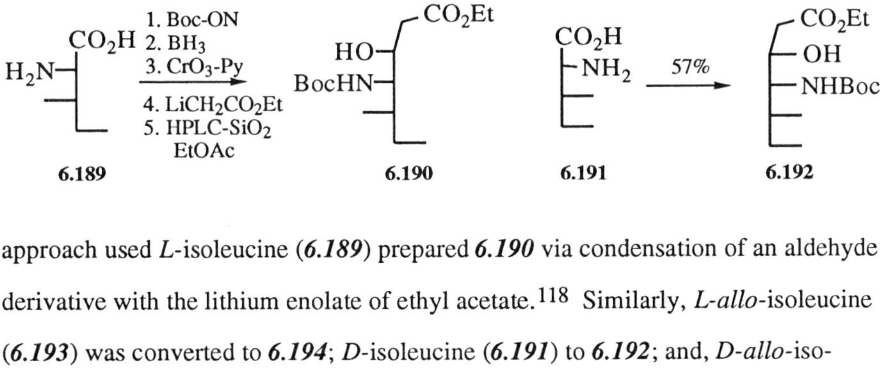

6.189 **6.190** **6.191** **6.192**

approach used *L*-isoleucine (***6.189***) prepared ***6.190*** via condensation of an aldehyde derivative with the lithium enolate of ethyl acetate.[118] Similarly, *L-allo*-isoleucine (***6.193***) was converted to ***6.194***; *D*-isoleucine (***6.191***) to ***6.192***; and, *D-allo*-isoleucine (***6.195***) to ***6.196***.[118a]

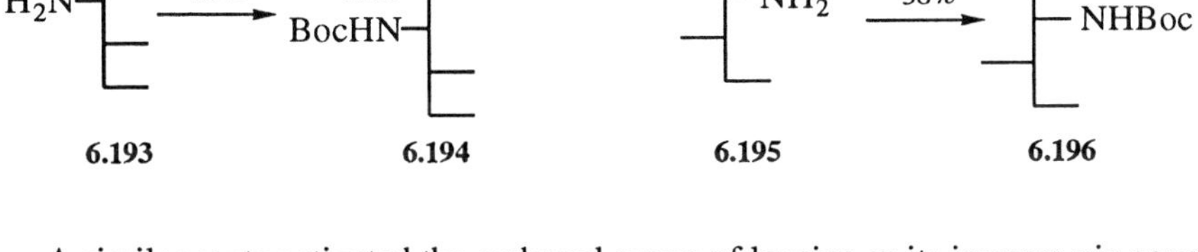

6.193 **6.194** **6.195** **6.196**

A similar route activated the carboxyl group of leucine or its isomers via conversion to an acyl imidazole, which was then condensed with ester enolates. The reaction of N-Boc isoleucine with 2,2'-carbonyl-*bis*-(3,5-dioxo)-4-methyl-1,2,4-

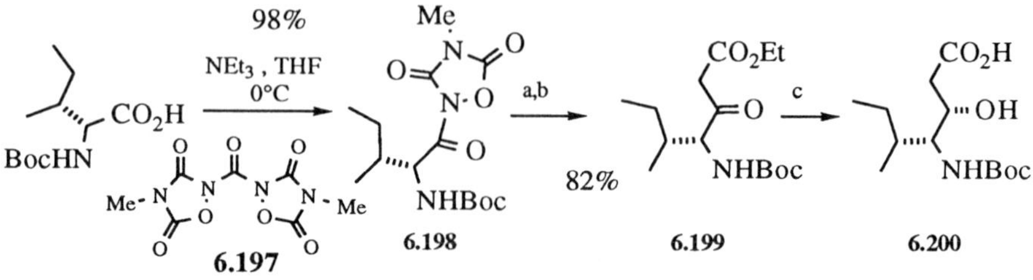

(a) LDA, EtOAc, THF, -78°C (b) H_3O^+ (c) $NaBH_4$

oxadiazolidine (***6.197***) gave ***6.198***.[119] Condensation of ***6.198*** with the lithium enolate of ethyl acetate gave ketone ***6.199***. Reduction of the ketone moiety in ***6.199***

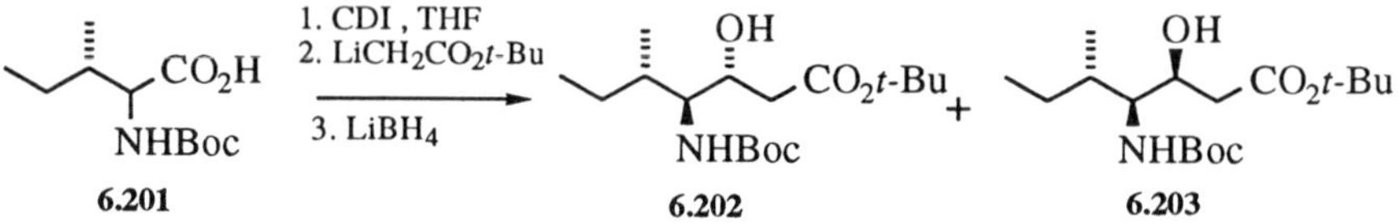

with sodium borohydride gave the alcohol ***6.200*** [4-(N-Boc amino)-3-hydroxy-5-methylheptanoic acid], as in previous examples of this method. In a similar manner,

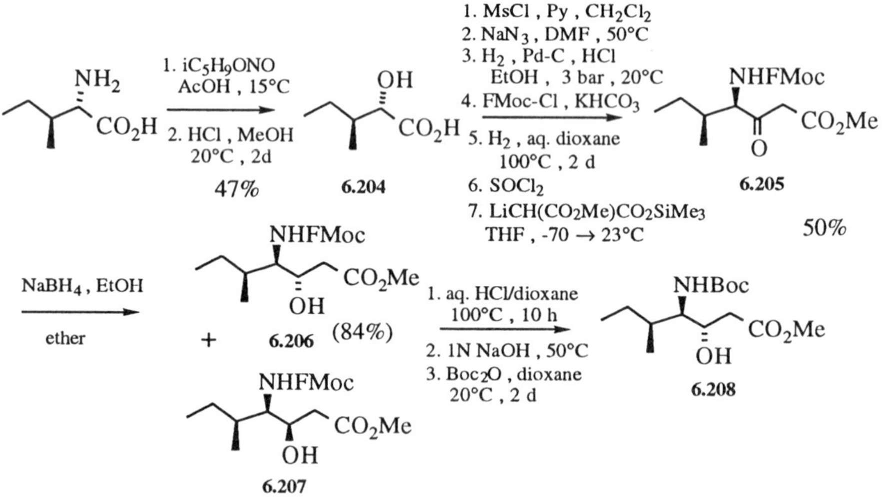

6.201 to a 1:1.5 mixture of ***6.202:6.203***, which are isosteres of statine.[86d]

A quite different method converted the amino moiety of isoleucine to an hydroxyl group (see ***6.204***).[120] Conversion to the azide, reduction to the amine and condensation of the ester with lithio methyl (2-trimethylsilyloxycarboxy) acetate gave ***6.205.*** Sodium borohydride reduction gave ***6.206***, along with 9% of ***6.207.*** Hydrolysis and re-protection of ***6.206*** gave N-Boc isostatine, ***6.208***. In this work, it was noted that amino acid ***6.208*** is a component of didemnin cyclodipeptide.[121]

6.4. Syntheses of Natural Products and Important Non-Natural Products

There are many other biologically important molecules that have a non-α-amino acid as an important structural component, or use the amino acid as an important synthetic precursor. This section collects several different types of these targets. They are loosely categorized by their biological activity.

6.4.A. Synthesis of Antibiotics

There are a wide variety of antibiotics, often with very different structural features. Several syntheses of antibiotics that contain a non-α-amino acid moiety. In some cases, the molecule is not used clinically as an antibiotic since other biological properties (such as being an anti-viral or an anti-cancer agent) are more important. Nonetheless, they are included in this section.

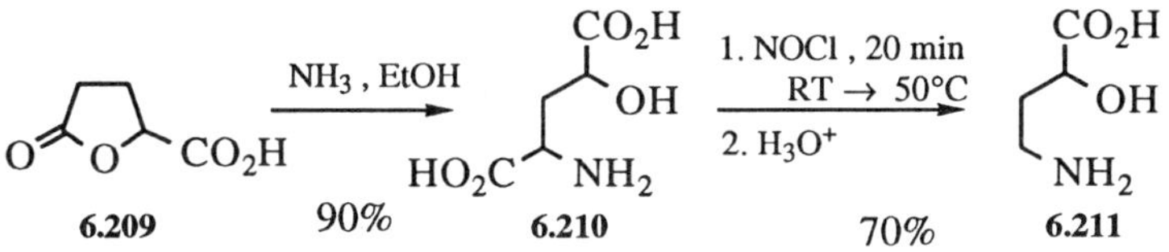

Lactone ***6.209*** (α-carboxy-γ-butyrolactone) is an interesting and useful synthetic intermediate.[122] Lactone ***6.209*** is readily converted to 2-hydroxy glutamic acid (***6.210***) and then treated with NOCl to give 2-hydroxy-4-aminobutanoic acid

(***6.211***).[123] Interestingly, even this simple amino acid is biologically important since it is a structural component of the antibiotic butirosin B[124] and also of amikacin BBK-8.[125]. It is noted that the carboxyl moiety in ***6.209*** can be reduced to give a hydroxymethyl group at that position.[122] That alcohol derivative was converted to 5-amino-3-hydroxy-2-methylpentanoic acid.[126] Amino acid ***6.211*** was also prepared from glutamic acid by treatment with basic cupric carbonate to give ***6.214*** after reaction with ***6.213***.[127] The amino group was converted to an hydroxyl group by reaction with HONO (giving ***6.215***) and removal of the phthalimidoyl group by reaction with hydrazine gave ***6.211***.

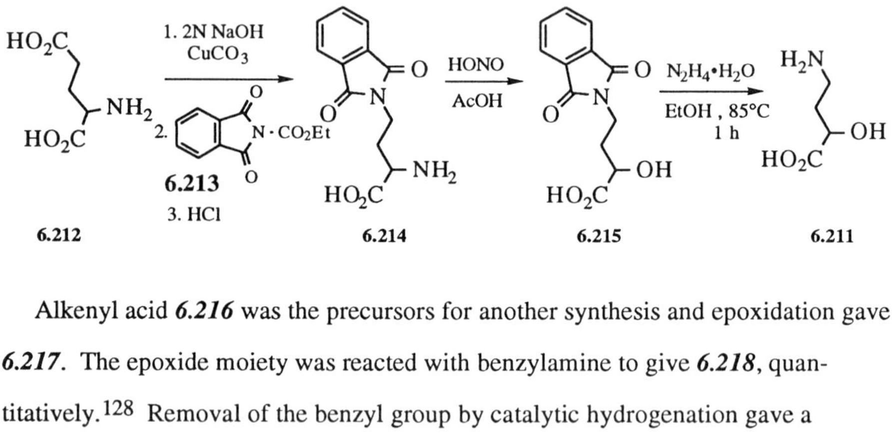

6.212 **6.214** **6.215** **6.211**

Alkenyl acid ***6.216*** was the precursors for another synthesis and epoxidation gave ***6.217***. The epoxide moiety was reacted with benzylamine to give ***6.218***, quantitatively.[128] Removal of the benzyl group by catalytic hydrogenation gave a quantitative yield of 3-amino-2-hydroxybutanoic acid (***6.219***), which is a component of the antibiotic kanamycin.[128b]

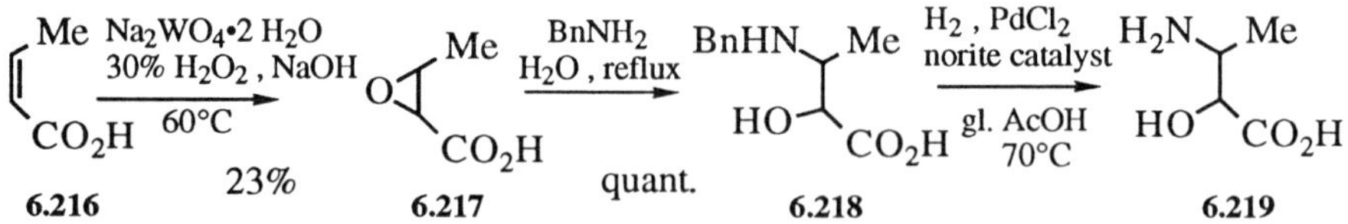

6.216 **6.217** **6.218** **6.219**

Relatively complex nitrones react with acrylate derivatives via a [3+2]-cycloaddition to give the corresponding isoxazolidine. A chiral non-racemic nitrone was used in this manner to give a chiral non-racemic amino acid that was a key intermediate for a synthesis of thienamycin.[129]

Oxazolidinones can be used as synthetic intermediates. An example is the conversion of ***6.220*** to ***6.221*** , which was followed by ring opening to give ***6.222.***[130] Oxidation to the aldehyde and Wittig olefination[14] gave ***6.223.*** This conjugated ester was "chain extended" in six steps (18% yield), which included condensation with

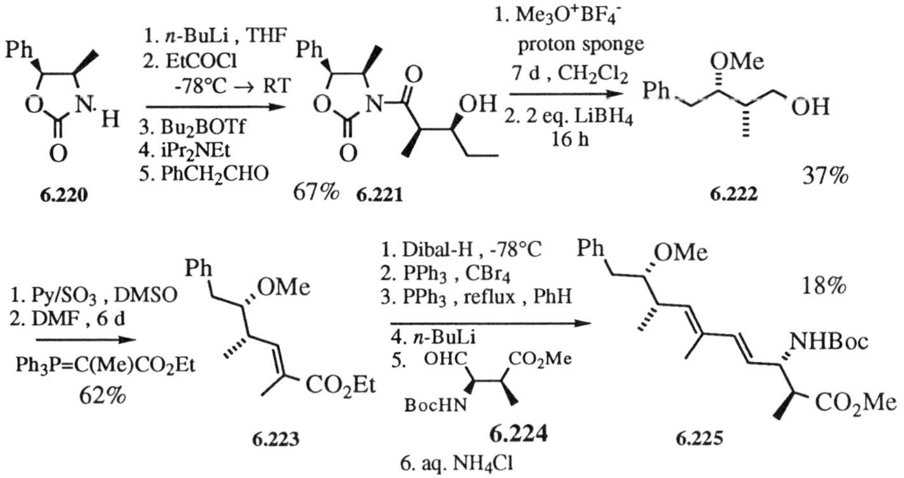

6.220 **6.221** **6.222** **6.223** **6.224** **6.225**

6.224 to give methyl 3-(N-Boc amino)-2,6,8-trimethyl-9-methoxy-10-phenylnonan-4,6-dienoate, ***6.225***. This particular alkenyl amino acid was shown to be hepatotoxic and a component of microcyctins.[131] It and other related amino acids produced in this study were also components of nodularin.[132]

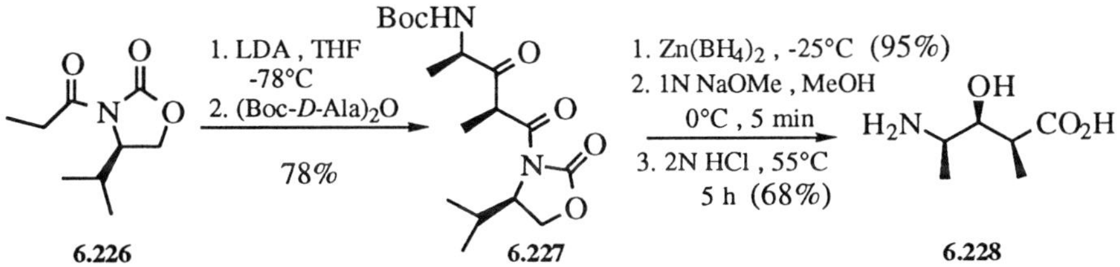

6.226 **6.227** **6.228**

Oxazolinone ***6.226***, which is also a chiral auxiliary for the propanoic acid moiety, was converted to its enolate anion and condensed with N-Boc alanine anhydride, to give ***6.227***.[133] Reduction of the ketone moiety and acid hydrolysis led to 2S,3S,4R-4-amino-3-hydroxy-2-methylpentanoic acid (***6.228***).[133] This amino acid is an acid hydrolysis product of bleomycin A_2.[134,135] Although bleomycin can be categorized as an antibiotic, it is best known for its anti-cancer activity.[134,135]

The same amino acid was prepared from N-phthaloyl alanine (***6.229***), which was converted to ***6.230***.[136] Reduction of the ketone moiety led to ***6.231***, which is the phthalimide derivative of ***6.228***.[134,135]

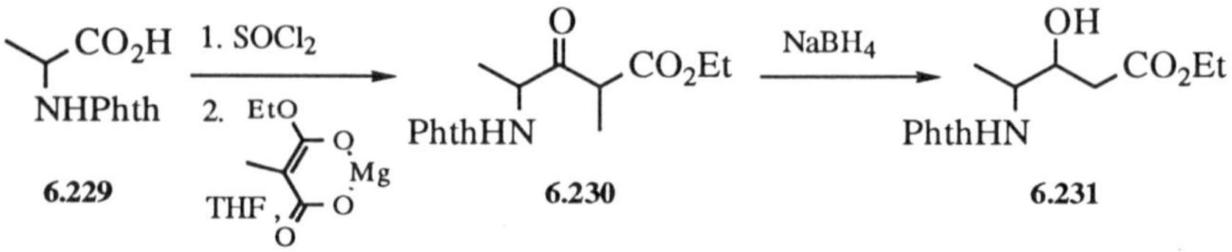

6.229 **6.230** **6.231**

As just shown, amino acids are key synthetic intermediates for the synthesis of non-α-amino acids. *S*-(+)-Phenylglycine (***6.232***) was converted to ***6.233***.[137] Reduction to the aldehyde, condensation with vinylmagnesium bromide, O-alkylation and oxidative cleavage of the alkenyl moiety gave ***6.233***, which was used in the synthesis of baccatin-III derivatives.[137]

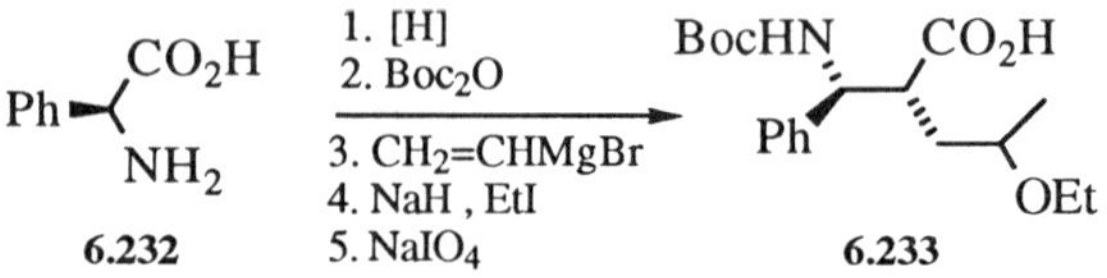

6.232 **6.233**

Carbohydrates can be used as chiral templates, as seen many times in previous sections. As part of a bleomycin A_2 synthesis,[134,135] Hecht converted 5-deoxy-*L*-arabano-γ-lactone (***6.234***) to ***6.235***.[135e] Reaction with azide, hydrogenation of the azide and an unusual N-benzylation reaction gave the acetonide of N-benzyl 4-(N-benzylamino)-2,3-dihydroxypentanamide, ***6.236***.

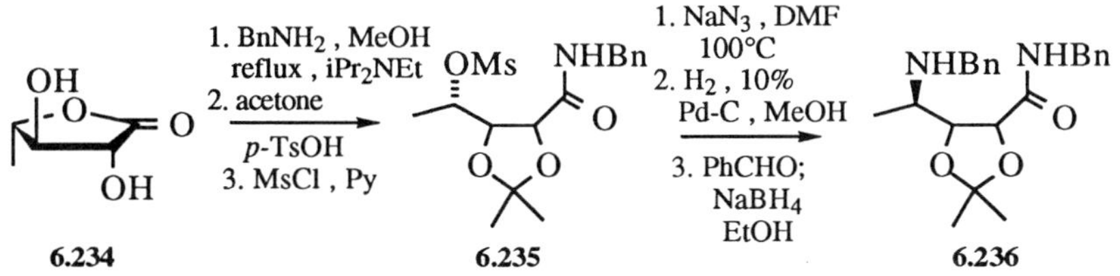

A quite different synthetic approach added, NOCl to ***6.237***,[138] giving ***6.238*** as a reactive intermediate.[139] Basic hydrolysis gave the keto-oxime (***6.239***) and catalytic hydrogenation gave a *erythro*-5-hydroxy-4-aminohexanoic acid (***6.240***) which was used in a synthesis of kasugamycin.

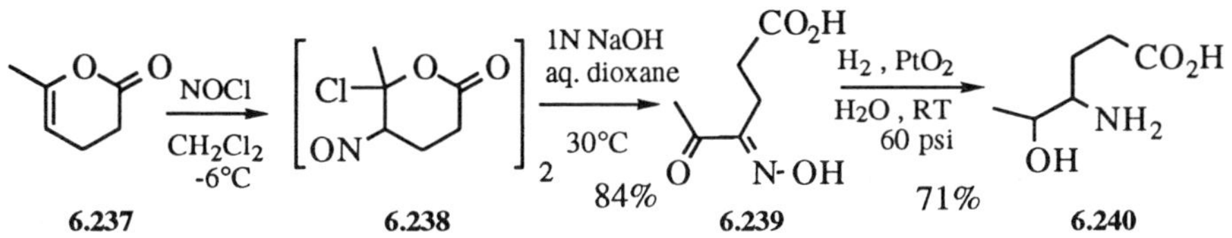

6.4.B. Synthesis of Peptide and Dipeptide Inhibitors

As seen for the synthesis of statine, highly substituted hydroxy amino acids may be peptidase inhibitors. Several analogs have been prepared with the goal of preparing new peptidase inhibitors. Some of these amino acid syntheses are collected into this section.

N-Boc leucine was converted to an acid chloride by treatment with isobutyl chloroformate. Subsequent treatment with diazomethane gave the diazoketone (***6.241***), which was treated with silver benzoate. This led to a Wolff rearrangement[140] and formation of ***6.242***.[141] Reduction of the ester moiety to an aldehyde with diisobutylaluminum hydride allowed condensation with the lithium enolate of ethyl acetate to give ***6.243***. Saponification led to N-Boc-5-amino-3-hydroxy-7-methyloctanoic acid (***6.244***, given the pseudonym AHMOA).[141] Amino acid

AHMOA is a key component of a renin inhibitory angiotensin peptide fragment.[141]

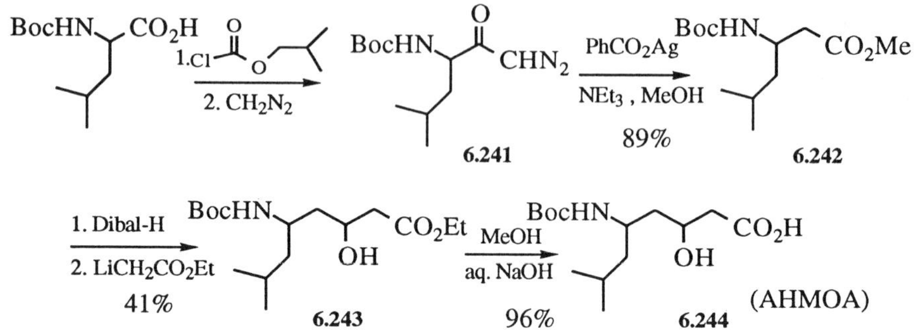

A lactone-based synthesis converted keto-alkyne ***6.245*** to carboxylic acid ***6.246*** in four steps.[142] Formation of lactone ***6.247*** was followed by conversion to azide ***6.248.*** Reaction with butylamine and hydrogenation gave N-butyl 5-amino-6-cyclohexyl-2-isopropylhexanamide, ***6.249.*** This amino acid product is considered to an hydroxy-ethylene dipeptide isostere.[115,143] In the case of ***6.249***, the term isostere essentially refers to replacing one group in a molecule with another that will give similar chemical or physiological properties.[115]

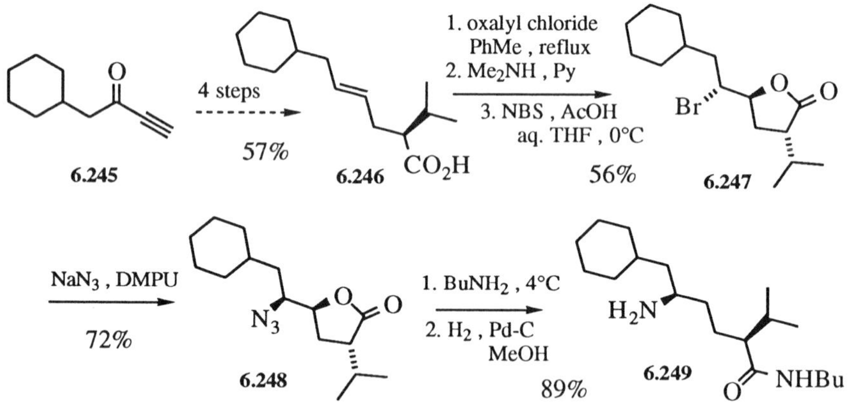

Another inhibitor was prepared from the N-Cbz derivative of phenylalanine, which was converted to ***6.250*** and then reduced to give alcohol ***6.251.***[144] Reaction

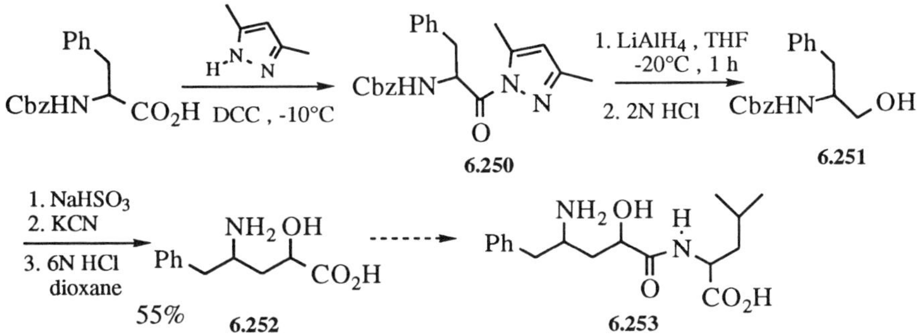

6.250 6.251 6.252 6.253

with sodium thiosulfate and then potassium cyanide gave the cyanohydrin, which was hydrolyzed to 4-amino-2-hydroxy-5-phenylpentanoic acid (***6.231***).[144] This was converted to bestatin (***6.253***), a competitive inhibitor of aminopeptidase B and of leucine aminopeptidase.[145]

Epoxide ***6.254***[146] was condensed with the enolate anion of ***6.255***. In this case, high diastereoselectivity was achieved when alcohol ***6.347*** was produced in >90% yield.[147] The high diastereoselectivity is due to using enantiopure ***6.254*** as a chiral template and "equipping" ***6.255*** with the auxiliary shown. Amino-amide ***6.256*** (a derivative 5-amino-2-benzyl-4-hydroxy-6-phenylhexanoic acid) is an intermediate in a synthetic route to hydroxy ethylene dipeptide isostere inhibitors of HIV-1 protease.

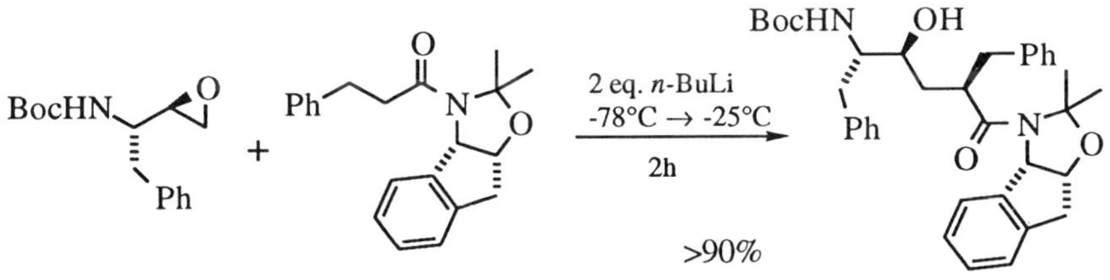

6.254 6.255 6.256

6.4.C. Other Natural Products

As the name of this section implies, the synthetic targets are difficult to categorize. This section will help show the variety of applications available for using non-α-amino acids.

When *D*-pantolactone (***6.257***)[148] was treated with potassium phthalimide,

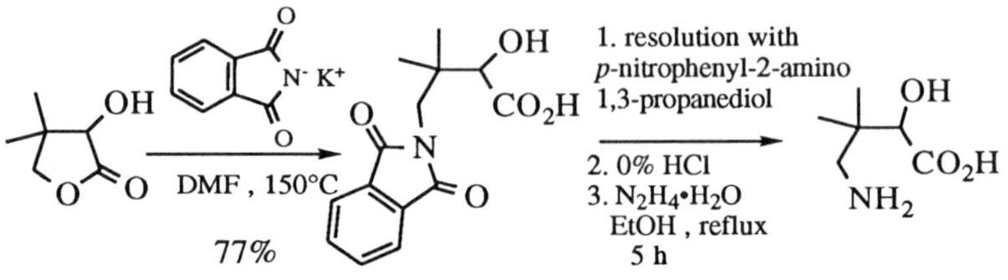

6.257 **6.258** **6.259**

racemic ***6.258*** was obtained.[149] Resolution and isolation of both antipodes of the ***6.258*** racemate was followed by hydrolysis and removal of the phthalimido group with hydrazine to give 4-amino-2-hydroxy-3,3,-dimethylbutanoic acid (***6.259*** is the 4-amino-4-deoxy analog of pantothenic acid).[149]

Dimethyl tartrate (***6.260***) is also an important chiral template. The first step was protection of the diol moiety to give acetonide ***6.261.*** A multi-step transformation followed which gave 4-amino-2,3-dihydroxybutanoic acid, ***6.262.***[150] This dihydroxyamino acid was converted to anthopleurine (4-amino-2,3-dihydroxybutanoic acid, or 4-[N,N,N-trimethylammonium]-2,3-dihydroxybutanoic acid, ***6.263***), an alarm pheromone of the sea anemone *Anthopleura elegantissima.*[150,151]

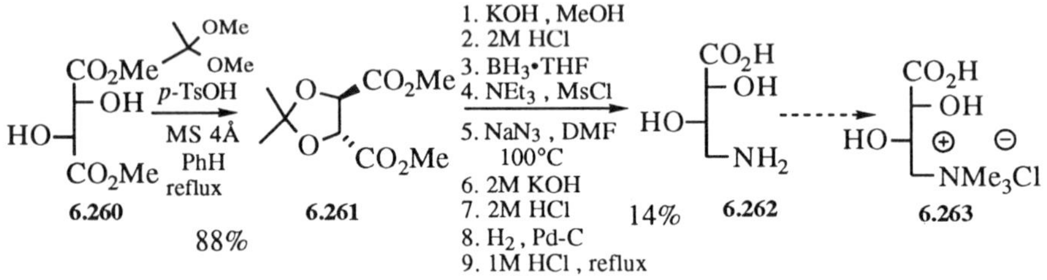

6.260 **6.261** **6.262** **6.263**

Amino acids can be converted to aldehydes, as amply demonstrated in previous sections. These precursors can also be used for the preparation of polyhydroxy amino acids. Horner-Wadsworth-Emmons olefination[152] of ***6.264***, for example, gave ***6.265***.[153] Epoxidation, oxidation and Wittig olefination[14] gave ***6.266***.

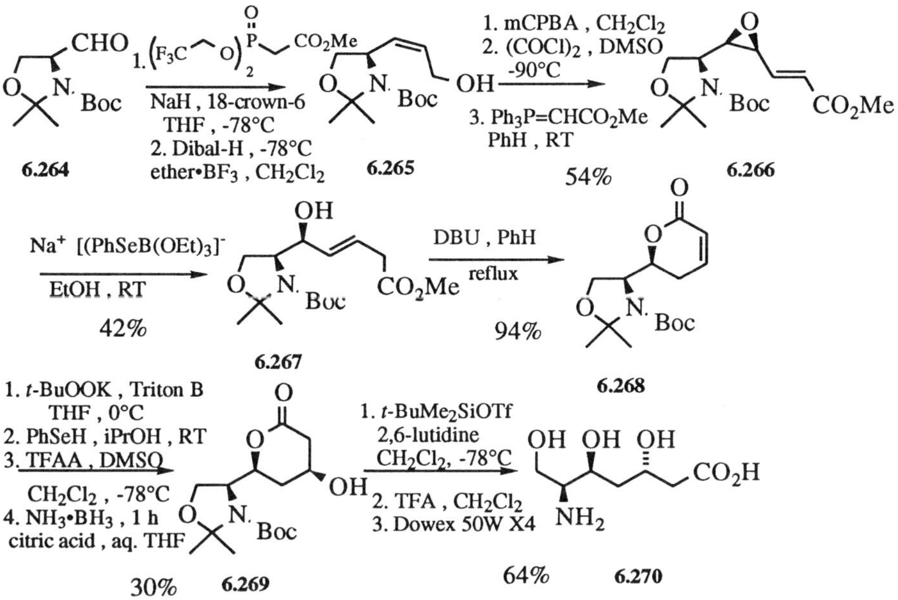

Isomerization of the double bond out of conjugation (to give ***6.267***) followed by treatment with base gave lactone ***6.268***. Introduction of the β-hydroxyl group in ***6.269*** led to the final product, (-)-galantinic acid (6-amino-3,5,7-trihydroxyheptanoic acid, ***6.270***).[153]

Conjugated ester ***6.271*** was treated with diisobutylaluminum hydride to reduce the ester moiety to an alcohol. This allylic alcohol was subjected to Sharpless asymmetric epoxidation.[110] Opening the epoxide (***6.272***) with azide was followed by protection of the diol moiety to give ***6.273***.[154] Reduction of the azide, protection of the amine, and oxidation gave lactam ***6.274***.[154] Acid hydrolysis gave 4-amino-2,3-dihydroxy-3-methylbutanoic acid, ***6.275***, a degradation product of carzinophilin.[155d]

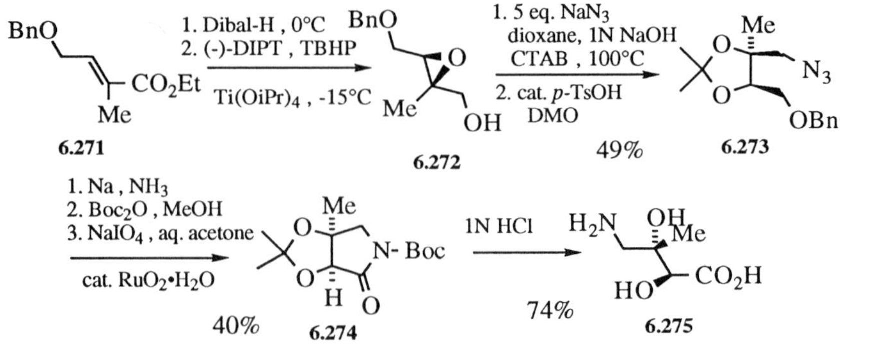

6.271 6.272 6.273 6.274 6.275

D-Glucose (*6.276*) was converted to *6.277* in six steps.[156] The aldehyde moiety in *6.277* was oxidized to an acid, esterified, and the diolide was hydrolyzed to give

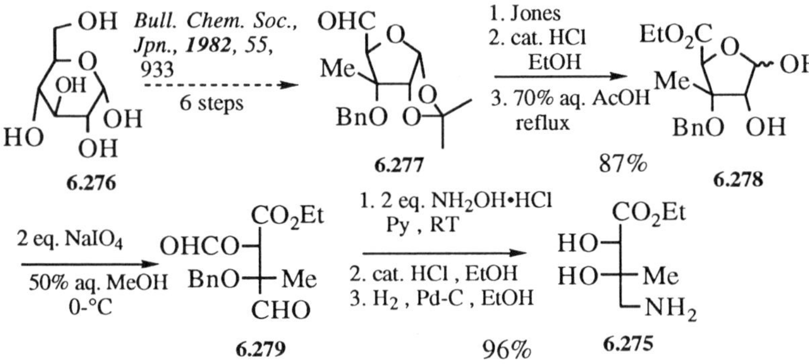

6.276 6.277 6.278 6.279 6.275

6.278.[157] Oxidation to *6.279* allowed conversion to *6.275*, which is an intermediate for the synthesis of the C_1-C_{18} segment of carzinophilin A (also see above).[155]

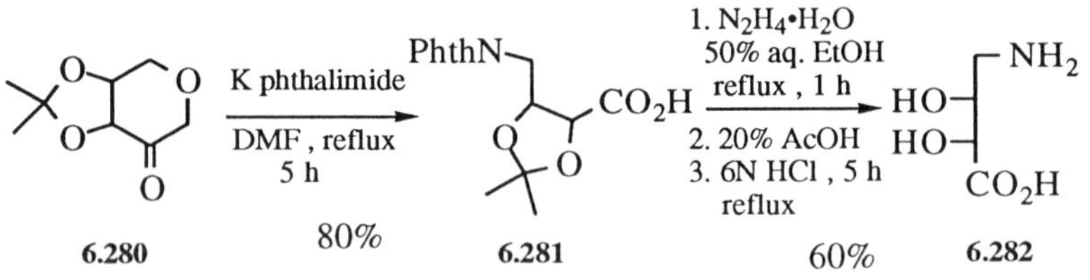

6.280 6.281 6.282

Another carbohydrate, 2,3-O-isopropylidene-*D*-erythronolactone (*6.280*), was prepared from isoascorbic acid[158] and then converted to phthaloyl derivative

6.281.[159] Deprotection provided 4-amino-4-deoxy-*D*-erythronic acid (4-amino-2,3-dihydroxybutanoic acid, ***6.282***), which was used in a synthesis of eritadenine.[159] This is a structural isomer of ***6.262*** from above.

The last example used a simple amino alcohol (3-aminopropanol, ***6.283***) as a starting material. Reaction of ethyl bromoacetate gave ***6.284***.[160] Swern oxidation[161] of the alcohol gave an aldehyde, and Wittig olefination[14] gave ethyl 5-(N-benzoyl-N-carboethoxymethyl)pent-2E-enoate (***6.285***), which was converted to α-kainic acid.

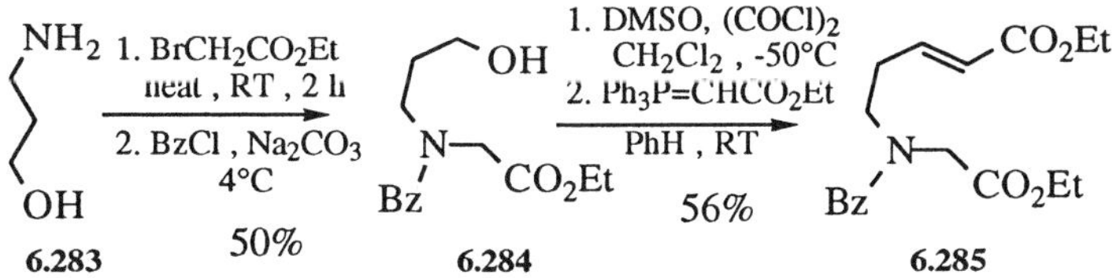

6.283 **6.284** **6.285**

REFERENCES

1 Enna, S.J. *The GABA Receptors*, The Humana Press, Clifton, New Jersey, ***1983***.

2 Taylor, C.P.; Vartanian, M.G.; Andruszkiewicz, R.; Silverman, R.B. *Epilepsy Res.*, ***1992***, *11*, 103 [*Chem. Abstr.*, ***1992***, *117*: 783s].

3 see Reference 1; pp. 5, 8-9, 14-16, 39, 63, 70, 75, 132-133, 152-153, 195, 198-200, 305.

4 Uchimaru, F.; Sato, M.; Kosasayama, E.; Shimizu, M.; Takashi, H. *Jpn.* 70 16,692 [*Chem. Abstr.*, ***1970***, *73*: P77617w].

5 Borea, P.A.; Bonora, A.; Baraldi, P.G.; Simoni, D. *Farmaco Ed. Sci.*, ***1983***, *38*, 411.

6 Bey, P.; Gerhart, F.; Jung, M.; Schirlin, D. *Eur. Pat. Appl.* EP 46,711 [*Chem. Abstr.*, ***1982***, *97*: P39394f].

7 Roberts, E. *Biochem. Pharm.*, ***1974***, *23*, 2637.

8 Hammond, E.J.; Wilder, B.J. *Gen. Pharmacol.*, ***1985***, *16*, 441.

9 Bey, P.; Gerhart, F.; Van Dorsselaer, V; Danzin, C. *J. Med. Chem.*, ***1983***, *26*, 1551.

10 Bey, P.; Gerhart, F.; Jung, M. *J. Org. Chem.*, ***1986***, *51*, 2835.

11 (a) Silverman, R.B.; Levy, M.A. *J. Org. Chem.*, ***1980***, *45*, 815; (b) Frieben,

W.; Gerhart, F. *Brit. UK Patent Appl.* GB 2,133,002 [*Chem. Abstr.*, ***1984***, *101*: P231027j].

12 (a) Cope, A.C.; Foster, T.T.; Towle, P.H. *J. Am. Chem. Soc.*, ***1949***, *71*, 3929; (b) Cope, A.C.; Pike, R.A.; Spencer, C.F. *Ibid*, ***1953***, *75*, 3212; (c) DePuy, C.H.; King, R.W. *Chem. Rev.*, ***1960***, *60*, 431.

13 Smith, M.B.; Wang, C.J.; Keusenkothen, P.F.; Dembofsky, B.; Fay, J.N.; Zezza, C.A.; Kwon, T.W.; Sheu, J.; Son, Y.C.; Menezes, R.F.; *Chem. Lett.*, ***1992***, 247.

14 (a) Wittig, G.; Rieber, M. *Ann.*, ***1949***, *562*, 187; (b) Wittig, G.; Geissler, G. *Ibid*, ***1953***, *580*, 44; (c) Wittig, G.; Schöllkopf, U. *Chem. Ber.*, ***1954***, *87*, 1318; (d) Gensler, W.J. *Chem. Rev.*, ***1957***, *57*, 191 (see p. 218).

15 Kwon, T.W.; Keusenkothen, P.; Smith, M.B. *J. Org. Chem.*, ***1992***, *57*, 6169.

16 (a) Keusenkothen, P.F.; Smith, M.B. *Tetrahedron*, ***1992***, *48*, 2977; (b) Idem, *Tetrahedron Lett.*, ***1989***, *30*, 3369; (c) Idem, *J. Chem. Soc., Perkin Trans. I*, ***1994***, 2485; (d) Idem *J. Mol. Structure*, ***1994***, *319*, 289.

17 Gittos, M.W.; Letertre, G.J. *Belg. Patent* 873,766 [*Chem. Abstr.*, ***1979***, *91*: P91499n].

18 Metcalf, B.W.; Casara, P. *Tetrahedron Lett.*, ***1975***, 3337.

19 (a) Metcalf, B.W.; Jung, M. *U.S. Patent* 3,959,356 [*Chem. Abstr.*, ***1976***,*85*: P142651s]; (b) Idem *U.S. Patent* 4,041,041 [*Chem. Abstr.*, ***1977***, *87*: P200807b].

20 Estreicher, H.; Flint, D.H.; Silverman, R.B. *U.S. Patent* 4,582,529 [*Chem. Abstr.*, ***1986***, *105*; P56355y].

21 (a) Lundkvist, J.R.M.; Hacksell, U. *Acta Pharm. Suec.*, ***1986***, *23*, 416; (b) Lundkvist, J.R.M.; Ringdahl, B.; Hacksell, U. *J. Med. Chem.*, ***1989***, *32*, 863.

22 (a) Lindlar, H. *Helv. Chim. Acta*, ***1952***, *35*, 446; (b) Fieser, L.F.; Fieser, M. *Reagents for Organic Synthesis, Vol. 1*, Wiley, New York, ***1967***, p. 566.

23 (a) Nishizawa, R.; Saino, A.; Setani, K. *Jpn. Kokai Tokkyo Koho* 80 00,355 [*Chem. Abstr.*, ***1980*** *92*: P163584g]; (b) Metcalf, B.W. *U.S.Patent* 3,960,927 [*Chem. Abstr.*, ***1976***, *85*: P143512j].

24 (a) Roberts, E.; Krause, D.N.; Wong, E.; Mori, A. *J. Neurosci.*, ***1981***, *1*, 132; (b) Otsuka, M.; Obata, K.; Miyata, Y.; Tanaka, Y. *Ibid*, ***1971***, *18*, 287; (c) Otsuka, M.; Miyata, Y. *Advances in Biochemical Psychopharmacology, Vol. 6*, Raven Press, New York, ***1972***, p 61.

25 (a) *Unlisted Drugs*, ***1964***, *16*, 6k; (b) *Unlisted Drugs*, *1973*, *25*, 141d; (c) Hayashi, T.; Nagai, K. in *Nervous Inhibition: Proceedings of the 2nd Faraday Harbor Symposium*, Pergamon Press, Forey (Ed), London, ***1961***, pp. 389-394; (d) Hayashi, T. *Ibid*, pp.378-388.

26 (a) DeMaio, D.; Madeddu, A.; Faggioli, L. *Acta Neurol.*, ***1961***, *16*, 366; (b) Buscaino, G.A.; Ferrari, E. *Ibid*, ***1961***, *16*, 748.

27 Tomita, M. *Z. Physiol Chem.*, ***1923***, *124*, 253.

28 (a) Bahl, J.J.; Bressler, R. *Ann. Rev. Pharmacol. Toxicol.*, ***1987***, *27*, 257; (b) DiPalma, J.R.; Ritchie, D.M.; McMichael, R.F. *Arch. Int. Pharmacodyn. Ther.*, ***1975***, *217*, 246; (c) Rebuzzi, A.G.; Schiavoni, G.; Amico, C.M.; Montenero, A.S.; Meo, F.; Manzoli, U. *Drugs Exptl. Clin. Res.*, ***1984***, *10*, 219; (d) Imai, S.; Matsui, K.; Nakazawa, M.; Takatsuka, N.; Takeda, K.; Tamatsu, H. *Brit. J. Pharmac.*, ***1984***, *82*, 533.

29 Bremer, J. *Physiol. Rev.*, ***1983***, *63*, 1420.

30 Borum, P.R. *Nutr. Rev.*, ***1981***, *39*, 385.

31 Hebron, S.A. *Spanish Patent* 391,718 [*Chem. Abstr.*, ***1974***, *80*: P47461t].

32 Antonio Gallardo, S.A. *Spanish Patent* 278,780 [*Chem. Abstr.*, ***1964***, *60*: P2779a].

33 (a) Kaken Chem. Co. Ltd., *Fr. Patent* 1,360,648 [*Chem. Abstr.*, ***1964***, *61*: P11895f]; (b) Sakai, S.; Miyaji, Y.; Furutani, H.; Kobata, M.; Hachisuka, T.; Nakayama, A.; Takada, S.; Hayashi, T. *Jpn.* 12,264 ('62) [*Chem. Abstr.*, ***1963***, *59*: P9805e].

34 Kurono, M.; Shigeoka, S.; Sakai, Y.; Itoh, H. *Ger.Offen.* 2,323,043 [*Chem. Abstr.*, ***1974***, *80*: P37450z].

35 Denki Kagaku Kogyo, K.K. *Jpn. Kokai Tokkyo Koho* JP 57,183,749 [*Chem. Abstr.*, ***1983***, *98*: P106811d].

36 Kolb, H.C.; Bennani, Y.L.; Sharpless, K.B. *Tetrahedron Asymmetry*, ***1993***, *4*, 133.

37 Fujimoto, Y.; Koshimoto, S. *Jpn.* 71 08,682 [*Chm. Abstr.*, ***1971***, *75*: 36680j].

38 Denki Kagaku Kogyo K.K. *Jpn. Kokai Tokkyo Koho* 81 68,649 [*Chem. Abstr.*, ***1981***, *95*: P169788k].

39 Casati, P.; Carmeno, M.; Benedetti, A.; Fuganti, C.; Grasselli, P. *Eur. Pat. Appl.* EP 169,614 [*Chem. Abstr.*, ***1986***, *105*: P173041d].

40 Kurono, M.; Shigeoka, S.; Miyamoto, S.; Imaki, K. *Jpn. Kokai* 76 39,634 [*Chem. Abstr.*, ***1976***, *85*: P62690p].

41 D'Alo', F.; Masserini, A. *Farmaco Ed. Sci.*, ***1964***, *19*, 30.

42 Pinza, M.;Pifferi, G. *J. Pharm. Sci*, ***1978***, *67*, 120.

43 Wieland, H. *Ber. dtsch. Chem. Ges.*, ***1910***, *43*, 3362.

44 Halling, K.; Thomsen, I.; Torssell, K.B.G. *Liebigs Ann.Chem.*, ***1989***, 985.

45 Doebner, O. *Ber.*, ***1900***, *33*, 2140.

46 (a) Balenović, K.; Jambrešić, I.; Urbas, B. *J. Org. Chem.*, ***1954***, *19*, 1589; (b) Italseber, Sp.A. *Fr. Patent* M4329 [*Chem. Abstr.*, ***1967***, *67*: P100415r].

47 Kaneko, T.; Yoshida, R. *Bull Chem. Soc., Jpn.*, ***1962***, *35*, 1153.

48 Zhou, B.; Gopalan, A.S.; Van Middlesworth, F.; Shieh, W.-R.; Sih, C.J. *J. Am. Chem. Soc.*, ***1983***, *105*, 5925.

49 (a) Mohr, P.; Rösslein, L.; Tamm, C. *Helv. Chim. Acta*, ***1987***, *70*, 142; (b) Idem, *Tetrahedron Lett.*, ***1989***, *30*, 2513.

50 Gopalan, A.S.; Sih, C.J. *Tetrahedron Lett.*, ***1984***, *25*, 5235.

51 (a) Curtius, T. *J. Prakt. Chem.*, ***1894***, *50*, 275; (b) Smith, P.A.S. *Org. React.*, ***1946***, *3*, 337; (c) Saunders, J.H.; Slocombe, R.J. *Chem. Rev.*, ***1948***, *43*, 203; (d) Mundy, B.P.; Ellerd, M.G. *Name Reactions and Reagents in Organic Synthesis*, Wiley, New York, ***1988***, pp. 60-61

52 Lu, Y.; Miet, C.; Kunesch, N.; Poisson, J. *Tetrahedron Assymmetry*, ***1990***, *1*, 707.

53 Ogura, K.; Yamashita, M.; Suzuki, M.; Furukawa, S.; Tsuchihashi, G. *Bull. Chem. Soc., Jpn.*, ***1984***, *57*, 1637.

54 Aubé, J.; Wang, Y.; Ghosh, S.; Langhans, K.L. *Synth. Commun.*, ***1991***, *21*, 693.

55 Bongini, A.; Cardillo, G.; Orena, M.; Porzi, G.; Sandri, S. *Tetrahedron*, ***1987***,*43*, 4377.

56 Braun, M.; Waldmüller, D. *Synthesis*, ***1989***, 856.

57 Bock, K.; Lundt, I.; Pedersen, C. *Acta Chem. Scand., Ser.B*, ***1983***, *B37*, 341.

58 Jung, M.E.; Shaw, T.J. *J. Am. Chem. Soc.*, ***1980***, *102*, 6304.

59 Häusler, J.; Schmidt, U. *Liebigs Ann. Chem.*, ***1979***, 1881.

60 Häusler, J. *Monatsh. Chem.*, ***1987***, *118*, 865.

61 (a) Mori, K.; Takigawa, T.; Matsuo, T. *Tetrahedron*, ***1979***, *35*, 933; (b) Saito, S.; Hasegawa, T.; Inaba. M.; Nishida, R.; Fujii, T.; Nomizu, S.; Moriwake, T.

Chem. Lett., ***1984***, 1389; (c) Tanaka, A.; Yamashita, K. *Synthesis*, ***1987***, 570.

62 Larchevêque, M.; Henrot, S. *Tetrahedron*, ***1990***, *46*, 4277.

63 Stucky, G.; Seebach, D. *Chem. Ber.*, ***1989***, *122*, 2365.

64 Pellegata, R.; Dosi, I.; Villa, M.; Lesma, G.; Palmisano, G. *Tetrahedron*, ***1985***, *41*, 5607.

65 Kasai, N.; Sakaguchi, K.; Takano, S.; Ogasawara, K. *Kokai Tokkyo Koho* J P 63,316,759 [*Chem. Abstr.*, ***1989***,*111*: P57076u].

66 (a) Takano, S.; Yanase, M.; Sekiguchi, Y.; Ogasawara, K. *Tetrahedron Lett.*, ***1987***, *28*, 1783; (b) Takano, S.; Ogasawara, K.; Sekiguchi, Y.; Kasai, N.; Sakaguchi, K. *Jpn. Kokai Tokkyo Koho* JP 63,174,957 [*Chem. Abstr.*, ***1989***, *110*: P24308x]; (c) Idem, *Jpn. Kokai Tokkyo Koho* JP 63,174,960 [*Chem. Abstr.*, ***1989***, *110*: P23339w].

67 Carlsen, P.H.J.; Katsuki, T.; Martin, V.S.; Sharpless, K.B. *J. Org. Chem.*, ***1981***, *46*, 3936.

68 Rossiter, B.E.; Sharpless, K.B. *J. Org. Chem.*, ***1984***, *49*, 3707.

69 Bubov, Yu.N.; Lavrinovich, L.I.; Zykov, A.Yu.; Ignatenko, A.V. *Mendeleev Commun.*, ***1992***, 86 [*Chem. Abstr.*, ***1992***, *117*: 234489c].

70 Bose, S.; Gurjar, M.K. *Synth. Commun.*, ***1989***, *19*, 3313.

71 (a) Rinehart, K.L. Jr.; Gloer, J.B.; Hughes, R.G. Jr.; Renis, H.E.; McGovren, J.P.; Swynenberg, E.B.; Stringfellow, D.A.; Kuentzel, S.L.; Li, L.H. *Science*, ***1981***, *212*, 933; (b) Rinehart, K.L. Jr.; Gloer, J.B.; Cook, J.C. Jr. *J. Am. Chem. Soc.*, ***1981***, *103*,1857; (c) Umezawa, H.; Aoyagi, T.; Morishima, H.; Masuzaki, M.; Hamada, M.; Takeuchi, T. *J. Antibiotics*, ***1970***, *23*, 259.

72 (a) Boger, J.; Lohr, N.S.; Ulm, E.H.; Poe, M.; Blaine, E.H.; Fanelli, G.M.; Lin, T.-Y.; Payne, L.S.; Schorn, T.W.; LaMont, B.I.; Vassil, T.C.; Stabilito, I.I.; Veber, D.F. *Nature*, ***1983***, *303*, 81; (b) Rich, D.H. *J. Med. Chem.*, ***1985***, *28*, 263.

73 Rich, D.H.; Sun, E.T.; Boparai, A.S. *J. Org. Chem.*, ***1978***, *43*, 3624.

74 Rittle, K.E.; Homnick, C.F.; Ponticello, G.S.; Evans, B.E. *J. Org. Chem.*, ***1982***, *47*, 3016.

75 Roberts, D.A.; Bradbury, R.H.; Brown, D.; Faull, A.; Griffiths, D.; Major, J.S.; Oldham, A.A.; Pearce, R.J.; Ratcliffe, A.H.; Revill, J.; Waterson, D. *J. Med. Chem.*, ***1990***, *33*, 2326.

76 Wuts, P.G.M.; Putt, S.R. *Synthesis*, ***1989***, 951.

77 (a) Mukaiyama, T. *Org. React.*, ***1982***, *28*, 203; (b) Mukaiyama, T. *Angew. Chem. Int. Ed., Engl.*, ***1977***, *16*, 817.

78 (a) Ohta, T.; Shiokawa, S.; Sakamoto, R.; Nozoe, S. *Tetrahedron Lett.*, ***1990***, *31*, 7329; (b) Takemoto, Y.; Matsumoto, T.; Ito, Y.; Terashima, S. *Chem. Pharm. Bull.*, ***1991***, *39*, 2425.

79 (a) Rague, B.; Fehrentz, J.-A.; Guegan, R.; Chapleur, Y.; Castro, B. *Bull. Soc. Chim. Fr.*, ***1983***, II-230; (b) Castro, B.; Guegan, R. *Fr. Demande* FR 2,531,076 [*Chem. Abstr.*, ***1984***, *101*: P23943c].

80 (a) Reformatsky, S. *Ber.*, ***1887***, *20*, 1210; (b) Rathke, M.W. *Org. React.*, ***1975***, *22*, 423; (c) Diaper, D.G.M.; Kuksis, A. *Chem. Rev.*, ***1959***, *59*, 89.

81 Liu, W.-S.; Smith, S.C.; Glover, G.I. *J. Med. Chem.*, ***1979***, *22*, 577.

82 Liu, W.-S.; Glover, G.I. *J. Org. Chem.*, ***1978***, *43*, 754.

83 Atsumi, T.; Yamamoto, H. *Jpn. Kokai* 78 05,114 [*Chem. Abstr.*, ***1978***, *88*: P190098a].

84 Schnabel, E. *Liebigs Ann. Chem.*, ***1967***, *702*, 188.

85 Steulmann, R.; Klostermeyer, H. *J. L. Ann. Chem.*, ***1975***, 2245.

86 (a) Mailbaum, J.; Rich, D.H. *J. Org. Chem.*, ***1988***, *53*, 869; (b) Kirihata, M.; Tokumori, H.; Ichimoto, I. *Nippon Nogei Kagaku Kaishi*, ***1978***, *52*, 135 [*Chem. Abstr.*, ***1978***, *89*: 146371f]; (c) Morishina, H.; Takita, T.; Umezawa, H. *J. Antibiot.*, ***1973***, *26*, 115; (d) Harris, B.D.; Joullié, M.M. *Tetrahedron*, ***1988***, *44*, 3489; (e) Reetz, M.T.; Drewes, M.W.; Matthews, B.R.; Lennick, K. *J. Chem. Soc., Chem. Commun.*, ***1989***, 1474.

87 Nishi, T.; Kitamura, M.; Ohkuma, T.; Noyori, R. *Tetrahedron Lett.*, ***1988***, *29*, 6327.

88 Nishi, T.; Saito, F.; Nagahori, H.; Kataoka, M.; Morisawa, Y.; Yabe, Y.; Sakurai, M.; Higashida, S.; Shoji, M.; Matsushita, Y.; Iijima, Y.; Ohizumi, K.; Koike, H. *Chem. Pharm. Bull.*, ***1990***, *38*, 103.

89 Schuda, P.F.; Greenlee, W.J.; Chakravarty, P.K.; Eskola, P. *J. Org. Chem.*, ***1988***, *53*, 873.

90 Andrew, R.G.; Conrow, R.E.; Elliott, J.D.; Johnson, W.S.; Ramezani, S. *Tetrahedron Lett.*, ***1987***, *28*, 6535.

91 Sakaitni, M.; Ohfune, Y. *J. Am. Chem. Soc.*, ***1990***, *112*, 1150.

92 Nakano, M.; Atsuumi, S.; Koike, Y.; Tanaka, S.; Funabashi, H.; Hashimoto, J.; Ohkubo, M.; Morishima, H. *Bull. Chem. Soc., Jpn.*, ***1990***, *63*, 2224.

93 Katsuki, T.; Yamaguchi, M. *Bull. Chem. Soc., Jpn.*, ***1976***, *49*, 3287.

94 (a) Dieckmann, W. *Ber.*, ***1894***, *27*, 102, 965; (b) Idem. *Ibid*, ***1900***, *33*, 2670; (c) Dieckmann, W.; Groenveld, A. *Ibid*, ***1900***, *33*, 595; (d) Hauser, C.R.; Hudson, B.E. *Org. React.*, ***1942***, 1 266 (see p. 274); (e) Schaefer, J.P.; Bloomfield, J.J. *Ibid.*, ***1967***, *15*, 1.

95 Klutchko, S.; O'Brien, P.; Hodges, J.C. *Synth. Commun.*, ***1989***, *19*, 2573.

96 Woo, P.W.K. *Tetrahedron Lett.*, ***1985***, *26*, 2973.

97 see (a) Van Horn, D.E.; Masamune, S. *Tetrahedron Lett.*, ***1979***, 2229; (b) Evans, D.A.; Nelson, J.V.; Vogel, E.; Taber, T.R. *J. Am. Chem. Soc.*, ***1981***, *103*, 3099.

98 Kogen, H.; Nishi, T. *J. Chem. Soc., Chem. Commun.*, ***1987***, 311.

99 Misiti, D.; Zappia, G. *Tetrahedron Lett.*, ***1990***, *31*, 7359.

100 Murrary, D.H.; Prokop, J. *J. Pharm. Sci.*, ***1965***, *54*, 1468.

101 (a) Kinoshita, M.; Hagiwara, A.; Aburaki, S. *Bull. Chem. Soc., Jpn.*, ***1975***, *48*, 570; (b) Kinoshita, M.; Aburaki, S.; Hagiwara, A.; Imai, J. *J. Antibiot.*, ***1973***, *26*, 249.

102 Yanagisawa, H.; Kanazaki, T.; Nishi, T. *Chem. Lett.*, ***1989***, 687.

103 Jouin, P.; Castro, B.; Poncet, J.; Nisato, D. *Rept. Proc. Eur. Pept. Symp., 19th*, ***1986***, (Pub. 1987), 659-62 [*Chem. Abstr.*, ***1988***, *108*: 38381m].

104 Koot, W.-J.; Van Ginkel, R.; Kranenburg, M.; Hiemstra, H.; Louwrier, S.; Moolenaar, M.J.; Speckamp, W.N. *Tetrahedron Lett.*, ***1991***, *32*, 401.

105 Kunieda, T.; Higuchi, T.; Abe, Y.; Hirobe, M. *Tetrahedron*, ***1983***, *39*, 3253.

106 Kunieda, T.; Ishizuka, T.; Higuchi, T.; Hirobe, M. *J. Org. Chem.*, ***1988***, *53*, 3381.

107 Shinozaki, K.; Mizuno, K.; Oda, H.; Masaki, Y. *Chem. Lett.*, ***1992***, 2265.

108 Kano, S.; Yuasa, Y.; Yokomatsu, T.; Shibuya, S. *J. Org. Chem.*, ***1988***, *53*, 3865.

109 Halling, K.; Torssell, K.B.G.; Hazell, R.G. *Acta Chem. Scand.*,***1991***, *45*, 736.

110 (a) Katsuki, T.; Sharpless, K.B. *J. Am. Chem. Soc.*, ***1980***, *102*, 5974; (b) Rossiter, B.E.; Katsuki, T.; Sharpless, K.B. *J. Am. Chem. Soc.*, ***1981***, *103*, 464; (c) Martin, V.S.; Woodard, S.S.; Katsuki, T.; Yamada, Y.; Ikeda, M.; Sharpless, K.B. *Ibid.*, ***1981***, *103*, 6237.

111 Mitsunobu, O. *Synthesis*, ***1981***, 1.

112 Saïah, M.; Bessodes, M.; Antonakis, K. *Tetrahedron Asymmetry*, ***1991***, *2*, 111.

113 (a) Mulzer, J.; Angermann, A. *Tetrahedron Lett.*, ***1983***, *24*, 2843; (b) Mulzer, J.; Angermann, A.; Münch, W. *Liebigs Ann. Chem.*, ***1986***, 825.

114 Mulzer, J.; Büttelmann, B.; Münch, W. *Liebigs Ann. Chem.*, ***1988***, 445.

115 (a) Foye, W.O. *Principles of Medicinal Chemistry, 3rd Ed.*, Lea and Febiger, Philadelphia, ***1989***, pp. 66-75; (b) Langmuir, I. *J. Am. Chem. Soc.*, ***1919***,*41*, 868; (c) Idem. *Ibid*, ***1919***, *41*, 1543; (d) Burger, A. in *Medicinal Chemistry*, Burger, A (Ed.), 3rd Ed., New York, Wiley, ***1970***, p. 74.

116 Boger, J.; Payne, L.S.; Perlow, D.S.; Lohr, N.S.; Poe, M.; Blaine, E.H.; Ulm, E.H.; Schorn, T.W.; LaMont, B.I.; Lin, T.-Y.; Kawai, M.; Rich, D.H.; Veber, D.F. *J. Med. Chem.*, ***1985***, *28*, 1779.

117 Burgess, K.; Cassidy, J.; Henderson, I. *J. Org. Chem.*, ***1991***, *56*, 2050.

118 (a) Rinehart, K.L.; Sakai, R.; Kishore, V.; Sullins, D.W. *J. Org. Chem.*, ***1992***, *57*, 3007; (b) Rinehart, K.L.; Kishore, V.; Bible, K.C.; Sakai, R.; Sullins, D.W.; Li, K.-M. *J. Nat. Prod.*, ***1988***, *51*, 1.

119 Jouin, P.; Poncet, J. Dufour, M.-N.; Maugras, I.; Pantaloni, A.; Castro, B. *Tetrahedron Lett.*, ***1988***, *29*, 2661.

120 Schmidt, U.; Kroner, M.; Griesser, H. *Synthesis*, ***1989***, 832.

121 Rinehart, K.L. Jr.; Gloer, J.B.; Cook, J.C. Jr.; Mizsak, S.A.; Scahill, T.A. *J. Am. Chem. Soc.*, ***1981***, *103*, 1857.

122 Coppola, G.M.; Shuster, H.F. *Asymmetric Synthesis*, Wiley, New York, ***1987***, pp. 237-242.

123 (a) Naito, T.; Nakagawa, S. *Brit.* 1,466,001 [*Chem. Abstr.*, ***1977***, *87*: P102649g]; (b) Idem, *Jpn. Kokai* 74 24,914 [*Chem. Abstr.*, ***1974***, *81*: P4264j]; (c) Idem, *US* 3,823,187 [*Chem.Abstr.*, ***1974***, *81*: P91929w].

124 Woo, P.W.K.; Dion, H.W.; Bartz, O.R. *Tetrahedron Lett.*, ***1971***, 2617.

125 Kawaguchi, H.; Naito, T.; Nakagawa, S.; Fujisawa, K. *J. Antibiot.*, ***1972***, *25*, 695.

126 Herdeis, C.; Lütsch, K. *Tetrahedron Asymmetry*, ***1993***, *4*, 121.

127 Horiuchi, Y.; Akita, E.; Ito, T. *Agr. Biol. Chem.*, ***1976***, *40*, 1649.

128 (a) Liwschitz, Y.; Singerman, A.; Luwish, M. *Israel J. Chem.*, ***1963***, *1*, 441; (b) also see Umemura, E.; Tsuchiya, T.; Umezawa, S. *J. Antibiot.*, ***1988***, *41*, 530.

129 Martin, H.; Herrmann, R.; Ugi, I. *Z. Naturforsch. B, Chem. Sci.*, ***1987***, *42*, 1588.

130 Beatty, M.F.; Jenings-White, C.; Avery, M.A. *J. Chem. Soc., Chem. Commun.*, ***1991***, 351

131 Rinehart, K.L.; Harada, K.; Namikoshi, M.; Chen, C.; Harvis, C.A.; Munro, M.H.G.; Blunt, J.W.; Mulligan, P.E.; Beasley, V.R.; Dahlem, A.M.; Carmichael, W.W. *J. Am. Chem. Soc.*, ***1988***, *110*, 8557.

132 Namikoshi, M.; Rinehart, K.L.; Dahlem, A.M.; Beasley, V.R.; Carmichael, W.W. *Tetrahedron Lett.*, ***1989***, *30*, 4349.

133 DiPardo, R.M.; Bock, M.G. *Tetrahedron Lett.*, ***1983***, *24*, 4805.

134 (a) Ikekawa, T.; Iwami, F.; Hiranaka, H.; Umezawa, H. *J. Antibiot. (Tokyo)*, ***1964***, *17A*, 194; (b) Umezawa, H.; Maeda, K.; Takeuchi, T.; Okami, Y. *Ibid.*, ***1966***, *19A*, 200; (c) Umezawa, H. Suhara, Y.; Takita, T.; Maeda, K. *Ibid.*, ***1966***, *19A*, 210.

135 (a) McGowan, D.A.; Jordis, U.; Minster, D.K.; Hecht, S.M. *J. Am. Chem. Soc.*, ***1977***, *99*, 8078; (b) Arai, H.; Hagmann, W.K.; Suguna, H.; Hecht, S.M. *Ibid.*, ***1980***, *102*, 6631; (c) Levin, M.D.; Subrahamanian, K.; Katz, H.; Smith, M.B.; Burlett, D.J.; Hecht, S.M. *Ibid.*, ***1980***, *102*, 1452; (d) Hecht, S.M.; Rupprecht, K.M.; Jacobs, P.M. *Ibid.*, ***1979***, *101*, 3982; (e) Ohgi, T.; Hecht, S.M. *J. Org. Chem.*, ***1981***, *46*, 1232; (f) Pozsgay, V.; Ohgi, T.; Hecht, S.M. *Ibid.*, ***1981***, *46*, 3761; (g) Aoyagi, Y.; Suguna, H.; Murugesan, N.; Ehrenfeld, G.M.; Chang, L-H.; Ohgi, T.; Shekhani, M.S.; Kirkup, M.P.; Hecht, S.M. *J. Am. Chem. Soc.*, ***1982***, *104*, 5237; (h) Aoyagi, Y.; Katano, K.; Suguna, H.; Primeau, J.; Chang, L.-H.; Hecht, S.M. *Ibid.*, ***1982***, *104*, 5537.

136 Yoshioka, T.; Hara, T.; Takita, T.; Umezawa, H. *J. Antibiot.*, ***1974***, *27*, 356.

137 Correa, A.; Denis, J.N.; Greene, A.E.; Grierson, D.S. *PCT Int.Appl.* WO 91,17,976 [*Chem. Abstr.*, ***1992***, *116*: P129620v].

138 Vorlander, D.; Knotzsch, A. *Ann. Chem.*, ***1897***, *294*, 317 (see p. 319).

139 Suhara, Y.; Sasaki, F.; Koyama, G.; Maeda, K.; Umezawa, H.; Ohno, M. *J. Am. Chem. Soc.*, ***1972***, *94*, 6501.

140 (a) Wolff, L. *Ann.*, ***1912***, *394*, 23 (see p. 25); (b) Kirmse, W. *Carbene Chemistry, 2nd Ed.*, Academic Press, New York, ***1971***, pp. 475-492; (c) Reference 51d, pp. 232-233.

141 Johnson, R.L.; Verschoor, K. *J. Med. Chem.*, ***1983***, *26*, 1457.

142 Herold, P.; Duthaler, R.; Rihs, G.; Angst, C. *J. Org. Chem.*, ***1989***, *54*, 1178.

143 Wolfenden, R. *Nature*, ***1969***, *223*, 704.

144 Nishizawa, R.; Saino, T.; Takita, T.; Suda, H.; Aoyagi, T.; Umezawa, H. *J. Med. Chem.*, ***1977***, *20*, 510.

145 Umezawa, H.; Aoyagi, T.; Suda, H.; Hamada, M.; Takesuchi, T. *J. Antibiot.*, ***1976***, *29*, 97.

146 Luly, J.; Dellaria, J.F.; Plattner, J.J.; Soderquist, J.L.; Yi, N. *J. Org. Chem.*, ***1987*** *52*, 1487.

147 (a) Askin, D.; Wallace, M.A.; Vacca, J.P.; Reamer, R.A.; Volante, R.P.; Shinkai, I. *J. Org. Chem.*, ***1992***, *57*, 2771; (b) also see Askin, D.; Volante, R.P.; Ryan, K.M.; Reamer, R.A.; Shinkai, I. *Tetrahedron Lett.*, ***1988***, *34*, 4245.

148 Saito, Y.; Hashimoto, M.; Seki, H.; Kamiya, T. *Tetrahedron Lett.*, ***1970***, 4863.

149 (a) Kopelevich, V.M.; Bulanova, L.N.; Gunar, V.I. *Tetrahedron Lett.*, ***1979***, 3893; (b) Bulanova, L.N. *Khim. Biokhim. Funkts. Primen. Pantotenovoi Kisloty Mater. Grodn. Simp., 4th*, ***1977***, 24 [*Chem. Abstr.*, ***1979***, *91*: 21033g].

150 Musich, J.A.; Rapoport, H. *J. Am. Chem. Soc.*, ***1978***, *100*, 4865.

151 Tolstikov, A.G.; Yamilov, R.Kh.; Spirikhin, L.V.; Khalilov, L.M.; Tolstikov, G.A. *Bioorg. Khim.*, ***1991***, *17*, 988 [*Chem. Abstr.*, ***1991***, *115*: 279633v].

152 (a) Horner, L.; Hoffmann, H.; Wippel, J.H.; Klahre, G. *Ber.*, ***1959***, *92*, 2499; (b) Wadsworth, W.S. Jr.; Emmons, W.D. *J. Am. Chem. Soc.*, ***1961***, *83*, 1733; (c) Boutagy, J.; Thomas, R. *Chem. Rev.*, ***1974***, *74*, 87.

153 Sakai, N.; Ohfune,Y. *Tetrahedron Lett.*, ***1990***, *31*, 4151.

154 Garner, P.; Park, J.M.; Rotello, V. *Tetrahedron Lett.*, ***1985***, *26*, 3299.

155 (a) Shiumada, N.; Uekusa, M.; Denda, T.; Ishii, Y.; Iizuka, T.; Sato, Y.; Hatori, T.; Fukui, M.; Sudo, M. *J. Antibiot., Ser. A.*, ***1955***, *8*, 67; (b) Terawaki, A.; Greenberg, J . *Nature,**1966***, *209*, 481; (c) Lown, J.W.; Majumdar, K.C. *Can. J. Biochem.*, ***1977***, *55*, 630; (d) Onda, M.; Konda, Y.; Omura, S.; Hata, T. *Chem. Pharm. Bull.*, ***1971***, *19*, 2013.

156 Yoshimura, J.; Hara, K.; Yamaura, M.; Mikami, F.; Hashimoto, H. *Bull. Chem. Soc., Jpn.*, ***1982***, *55*, 933.

157 Shibuya, M. *Tetrahedron Lett.*, ***1983***, *24*, 1175.

158 Mitchell, D.L. *Can. J. Chem.*, ***1963***, *41*, 214.

159 Kamiya, T.; Saito, Y.; Hashimoto, M.; Seki, H. *Tetrahedron*, ***1972***, *28*, 899.

160 Yoo, S.; Lee, S.-H.; Kim, N.-J. *Tetrahedron Lett.*, ***1988***, *29*,2195.

161 Mancuso, A.J.; Swern., D. *Synthesis*, ***1981***, 165.

7

Aminocyclic and Heterocyclic Amino Acids

This chapter will discuss cycloalkanes that have an amino group and a carboxyl group on a ring, but these groups are not on the same carbon. An example is 2-aminocyclo-butane-1-carboxylic acid (*7.1*). A few examples will also be presented

7.1

7.2

7.3

7.4

where the ring has an attached aminoalkyl group [R_2N-$(CH_2)_n$-] or RO_2C-$(CH_2)_n$- [a carboxy-alkyl group] rather than having these groups attached directly to the ring. There are two fundamental structures for the aminocyclobutanoic acids discussed in

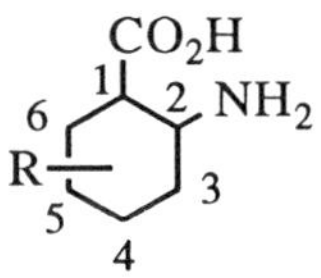
7.5

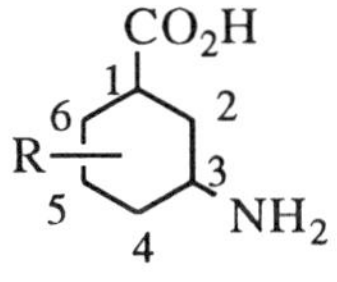
7.6

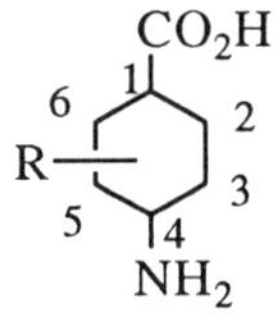
7.7

this chapter, *7.1* and *7.2*. The cyclopentane carboxylic acid derivatives discussed in this chapter have an amino group at C_2 or C_3 relative to the carbon bearing the carboxyl group, giving *7.3* or *7.4*. Six-membered rings lead to 1,2-, 1,3-, and 1,4-amino-cyclohexane carboxylic acid derivatives. The 1,2-derivatives are represented by *7.5*; the 1,3-derivatives by *7.6*; and, the 1,4-derivatives are represented by *7.7*. Derivatives will be presented that have an amino group at the C_2, C_3, or C_4 position relative to the carbon bearing the carboxyl carbon.

7.1. From Cyclic Precursors

One method to generate cyclic amino acids is to begin with a cyclic precursor. Refunctionalization of the existing groups lead to the amino acid.

7.1.A. Functional Group Interchange to Form Amines

Diethyl malonate was converted to cyclobutane derivative *7.8* by classical condensation chemistry.[1] Saponification of the ester moieties was followed by thermal decarboxylation to give *7.9*. Esterification, deprotection of the alcohol and conversion to a tosylate allowed its displacement by ammonia to give methyl 3-amino-cyclobutane carboxylate (*7.10*, as a 33:67 *cis:trans* mixture) but the yield was remarkably poor.[1]

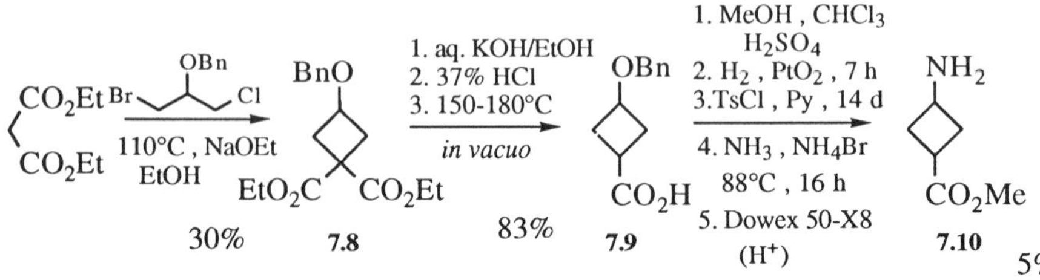

A better method used *trans*-cyclobutane dicarboxylate (*7.11*) and selective hydrolysis of one ester liberated an acid that could be reduced with borane to give *7.12*.[2] Conversion of that alcohol to a tosylate was followed by reaction with potassium cyanide to give nitrile ***7.13***. The ester was converted to an N-amino amide and then to an amine, allowing hydrolysis of the nitrile to give ***7.14***.[2]

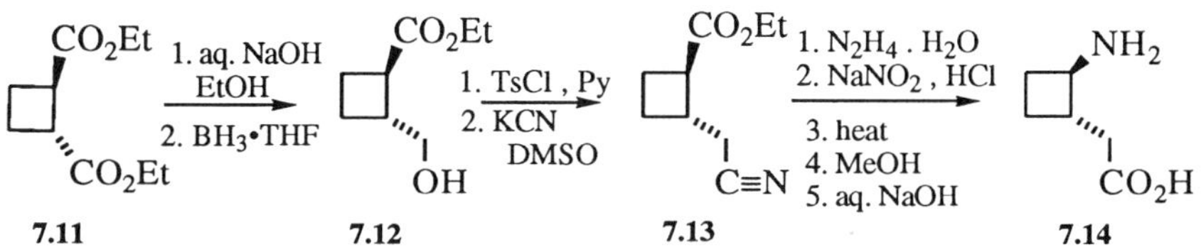

A different approach converted the ketone moiety in *cis*-pinononic acid [***7.15***, derived by permanganate oxidation of (-)-2-hydroxy pinocamphone] to an oxime (***7.16***).[3] Raney nickel reduction and Beckmann rearrangement[4] of the oxime gave ***7.17***. Subsequent treatment with hydrazine gave 3-amino-2,2-dimethyl-cyclobutane carboxylic acid (***7.18***).

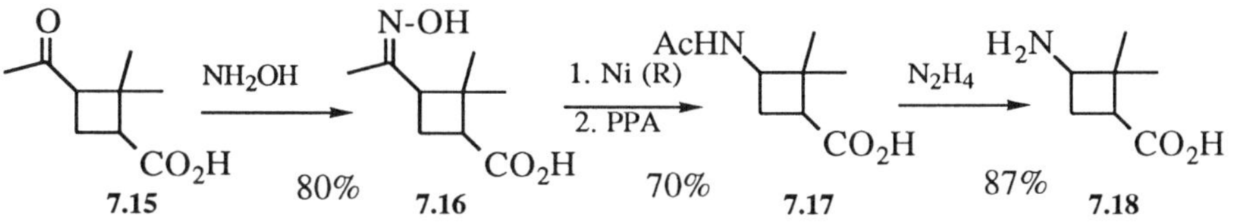

Another refunctionalization reaction converted an allylic acetate to a 38:62 mixture of allylic azides (***7.20*** and ***7.21***) via the palladium catalyzed reaction of ***7.19***

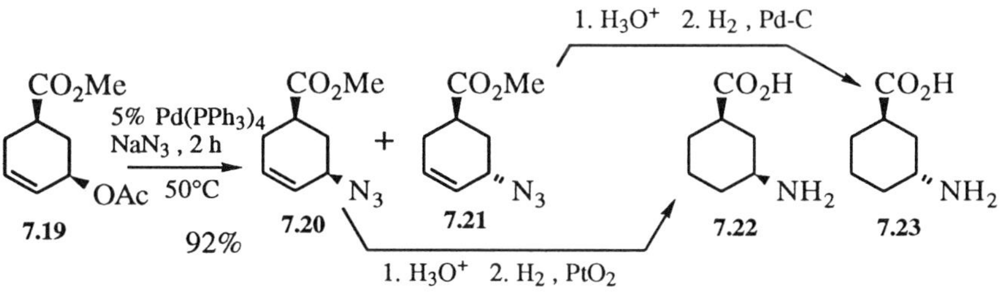

with sodium azide.[5] These azido-esters were separated and independently hydrogenated[6] to give *cis*-3-aminocyclohexane-1-carboxylic acid (***7.22***, quantitative) or *trans*-3-aminocyclohexane-1-carboxylic acid (***7.23***, 99% yield). The use of *bis*-

diphenylphosphinopropane (dppp) rather than triphenylphosphine to generate the palladium (0) species reversed the diastereoselectivity and led to an 84:16 mixture of ***7.22:7.23***, but in only 38% yield.[5]

Dicarboxylic acid derivatives are another useful source of substituted cyclic amino acids since one acid moiety can be converted to an amine moiety. Conversion

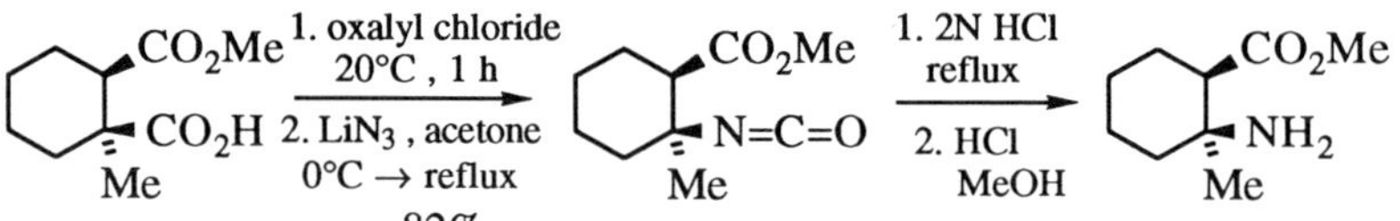

7.24 **7.25** **7.26**

of the mono-ester of dicarboxylic acid ***7.24***[7] to isocyanate ***7.25*** via a Curtius rearrangement[8] was followed by acid hydrolysis to give methyl 2-amino-2-methylcyclohexanecarboxylate, ***7.26*** .[9] Dicarboxylic acid ***7.27*** was also converted to ***7.28*** (3-amino-5-*t*-butylcyclohexanecarboxylic acid) by a modification of the "azide-incorporating" reaction.[10]

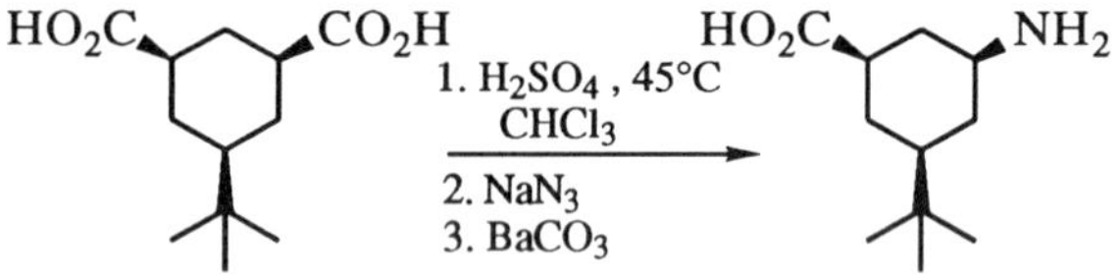

7.27 **7.28**

There are a handful of examples where amino acids with ring sizes of seven members and larger have been prepared using the carboxyl refunctionalization reaction just shown. Reaction of anhydride ***7.29***[11] with ammonia, for example, gave amide-acid ***7.30***. Hofmann rearrangement[12] of the amide led to *cis*-2-aminocycloheptane-1-carboxylic acid, ***7.31***.[13] An alternative synthesis conjugate addition of ammonia to cycloheptenecarboxylic acid, ***7.32***.[14] This reaction gave *trans*-2-aminocycloheptane-1-carboxylic acid, ***7.33***.

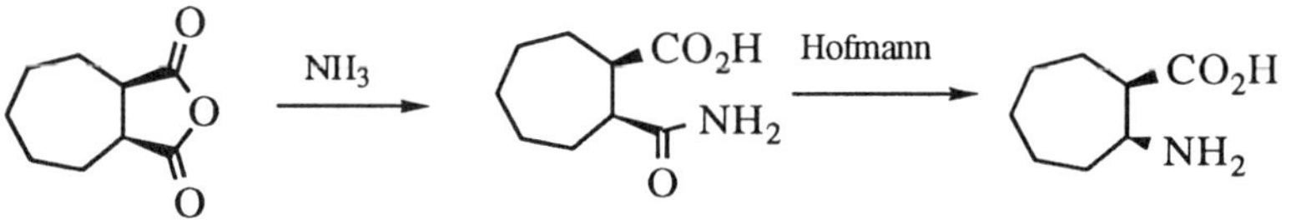

7.29 7.30 7.31

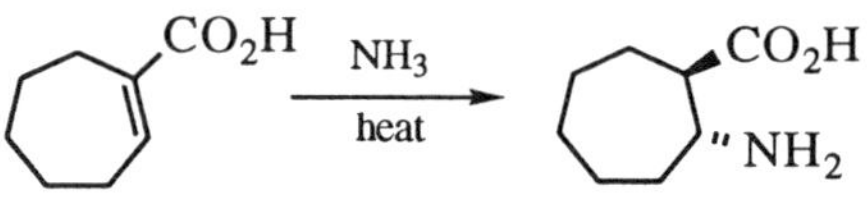

7.32 7.33

A lactone can be refunctionalized to an amino acid as well as a carboxylic acid. Reaction of ***7.34*** with potassium phthalimide opened the lactone ring to give a phthalimido-acid. Removal of the phthalimide group gave 2-methylamino-1-phenyl-cyclopropane-1-carboxylic acid, ***7.35***.[15] Amino acid ***7.35*** was tested as a potential antidepressant, and several aryl analogs were prepared: the 4-chlorophenyl, the 4-methylphenyl, and the 4-methoxyphenyl derivatives.

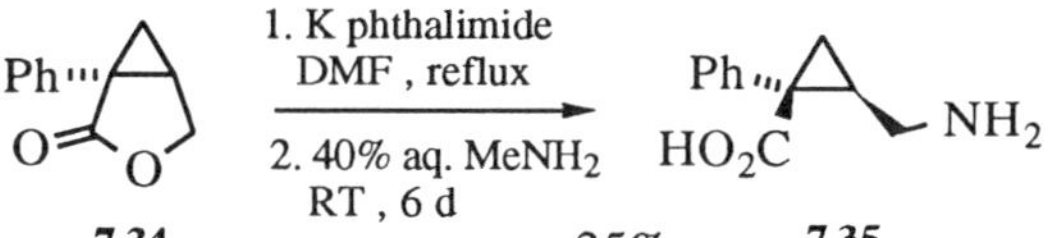

7.34 35% 7.35

An important source of cyclic dicarboxylic acids involves oxidative cleavage of bicyclic alkenes. *cis*-Cyclopentane dicarboxylic acid was prepared by ozonolysis of norbornene and then converted to its dimethyl ester (***7.36***).[16] Kinetic resolution of

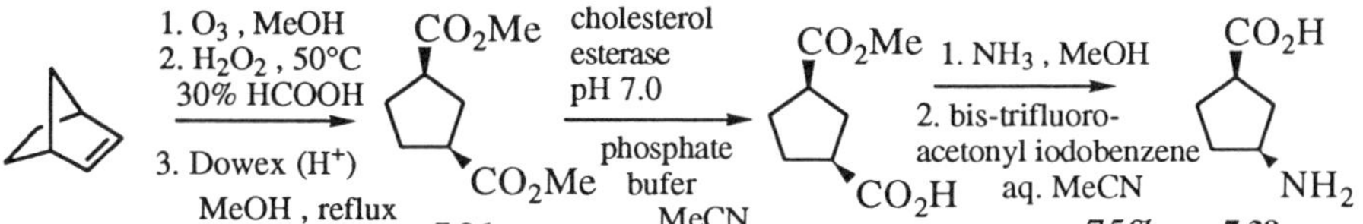

85% 7.36 7.37 7.38

this racemic diester with cholesterol esterase led to separation and isolation of

7.37. Subsequent conversion of the acid moiety to an amide was followed by rearrangement to 1S,3R-3-aminocyclopentane-1-carboxylic acid (*7.38*).[16] An alternative synthesis converted norbornanol to dicarboxylic acid ***7.39*** by oxidative cleavage with permanganate.[17] This was then converted to ***7.38***.[18] Dicarboxylic acid ***7.39*** was also converted to the *trans-* acid (***7.40***) by literature procedures[19] and the same reaction sequence as before was used to prepare *trans*-3-aminocyclopentane-1-carboxylic acid, ***7.27***.[18,20]

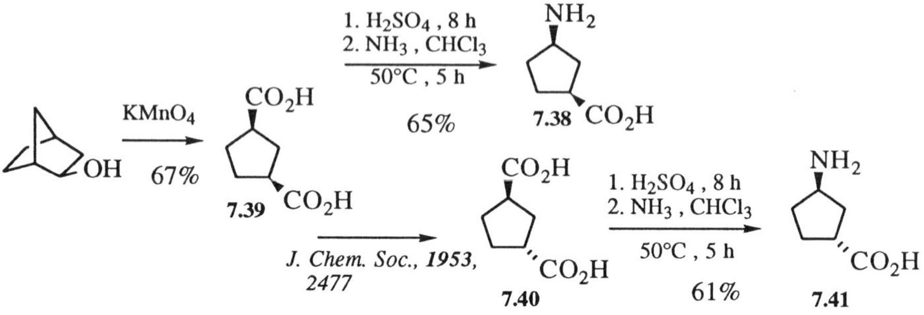

Alkenes are important vehicles for incorporating functionality into a ring. An example is the Michael addition of ammonia to 1-cyclopentene-1-carboxylic acid (***7.42***) to give 2-aminocyclopentane-1-carboxylic acid (***7.43***), as a mixture of *cis-* and *trans*-isomers.[21] The yield in this particular case was rather low, however. Hydrogenation of cycloalkene carboxylic acid derivatives is a convenient route to the

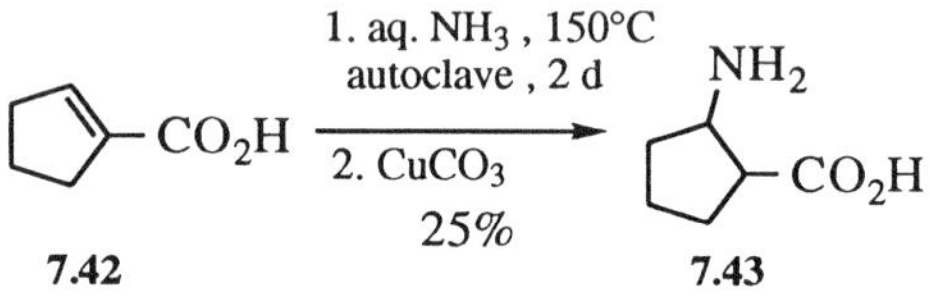

analogous saturated amino acids. Reaction of *trans-* methyl 3-amino-1-cyclopent-4-enoate (***7.44***) with hydrogen (palladium-on-carbon catalyst), for example, gave *trans*-3-aminocyclopentane-1-carboxylic acid, ***7.41***.[22]

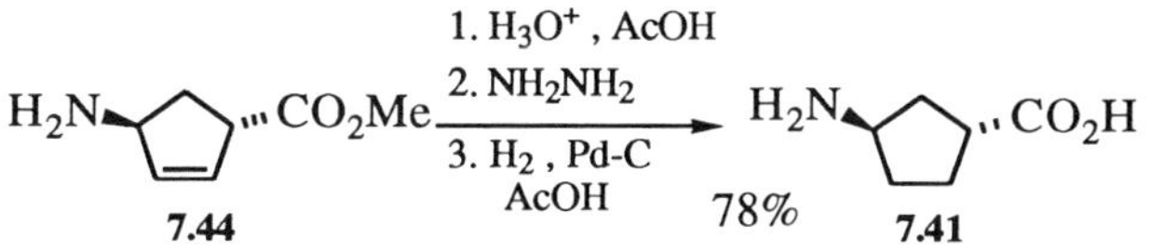

De Meijere showed that reactions of cyclopropylidene derivatives lead to aminocyclopropane carboxylic acids via conjugate addition of an amine surrogate. Reac-

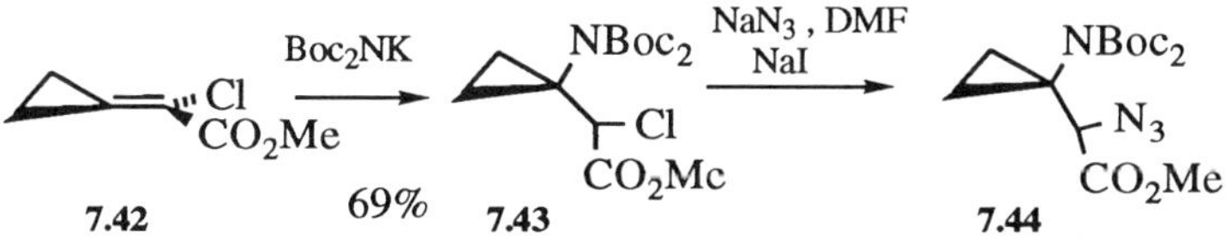

tion of ***7.42*** with nucleophiles such as $KN(Boc)_2$, for example, gave methyl 2-chloro-2-(1-(di-Bocaminocyclopropyl)ethanoate, ***7.43***.[23] Subsequent treatment with azide gave methyl 2-azido-2-[1-(di-Boc-amino)cyclopropyl]ethanoate, ***7.44***.

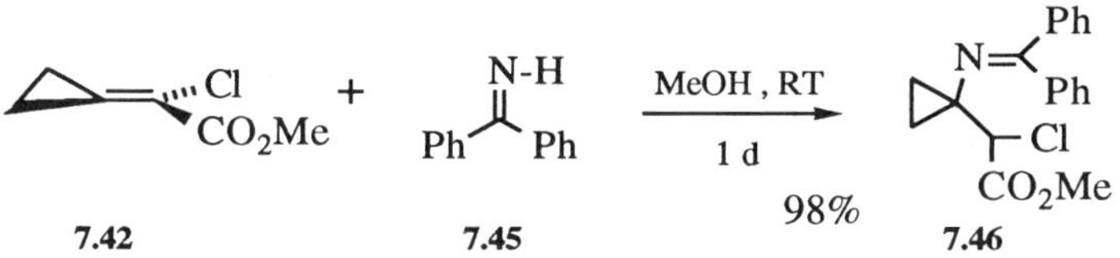

Another application involved reaction of ***7.42*** with imine ***7.45*** to give ***7.46***.[24] As with ***7.43***, ***7.46*** can react with nucleophiles such as azide or alkyl thiolates. A

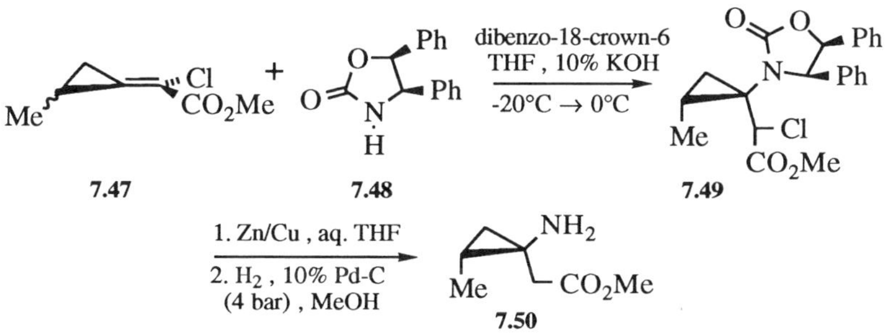

variation in this reaction generated optically active amino acids. Reaction of ***7.47*** with chiral, non-racemic ***7.48***, for example, gave ***7.49***.[25] Reduction of the chlorine moiety with zinc/copper and catalytic hydrogenation to remove the oxazolidine protecting group gave methyl (1E-amino-2-methylcyclopropyl) ethanoate, ***7.50***.

A different approach using alkene derivatives involved allylic substitution reactions. Ethyl cyclohex-3-en-1-carboxylate (***7.51***) reacted with N-bromosuccinimide to give ***7.52***.[26] This allowed reaction with potassium phthalimide to give ***7.53*** in an overall yield of 41%. Catalytic hydrogenation and treatment with methylamine led to *cis*-3-aminocyclohexane-1-carboxylic acid, ***7.22***.[26]

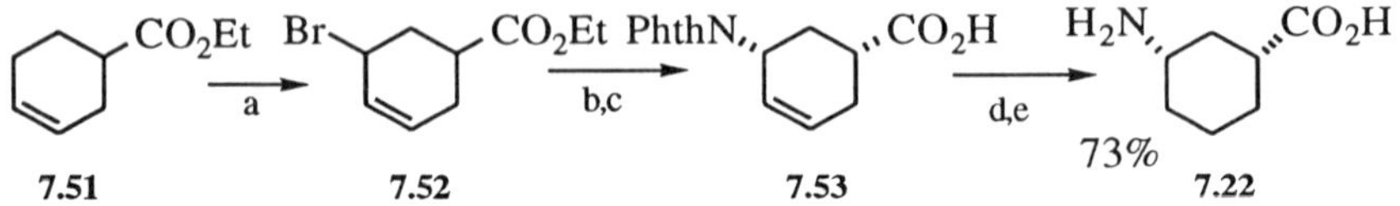

(a) NBS (b) K phthalimide (c) aq. H^+ (d) H_2 , Pd (e) $MeNH_2$

1-Cyclopentene-1-carbonitrile (***7.54***) also reacted with N-bromosuccinimide to give an allylic bromide, ***7.55***.[27] Subsequent reaction with ammonia displaced the bromine and acid hydrolysis of the nitrile led to 3-aminocyclopentene-1-carboxylic acid, ***7.56***. Cyclohexenyl acid derivatives have also been used. Bromination of

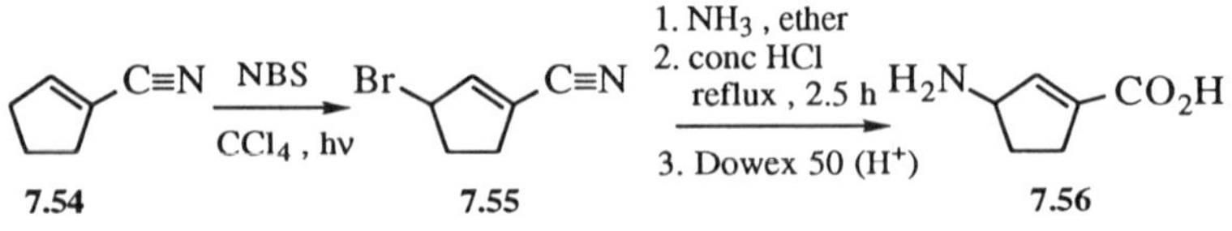

7.57 gave ***7.58***, which reacted with ammonia to give ethyl 4-amino-1-phenyl-2-cyclohexene-1-carboxylate (***7.59***) or with amines to give N-alkyl amines.[28] Even

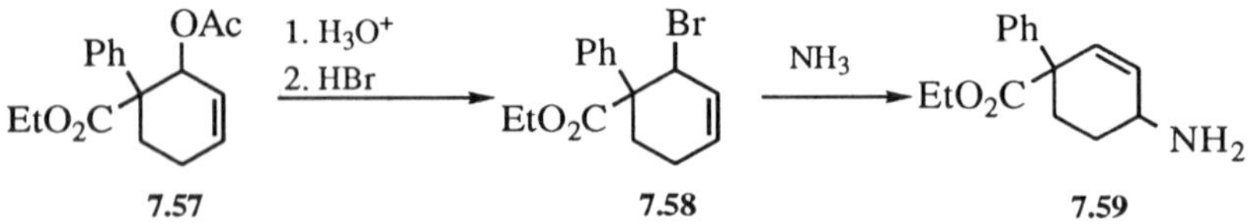

sterically hindered allylic bromides react, but often by an S_N2' mechanism.[29] An example is the reaction of ***7.60*** with diethylamine to give ethyl 4-(N,N-diethyl-amino)-1-phenyl-2-cyclohexene-1-carboxylate (***7.61***).[30]

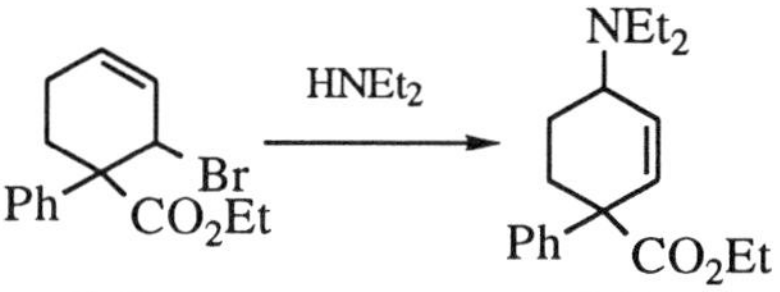

7.60 **7.61**

A "simple" reaction of ammonia and an alkyl halide can sometimes give unusual results. Treatment of ***7.62*** with ammonia, followed by treatment with acid, led to a product with the amine group distal to the carboxyl group. The alkene moiety also "migrated" into conjugation with the acid (derived from hydrolysis of the cyano group). In this reaction, ***7.62*** was converted to 4-amino-1-cyclohexene carboxylic acid (***7.63***).[31]

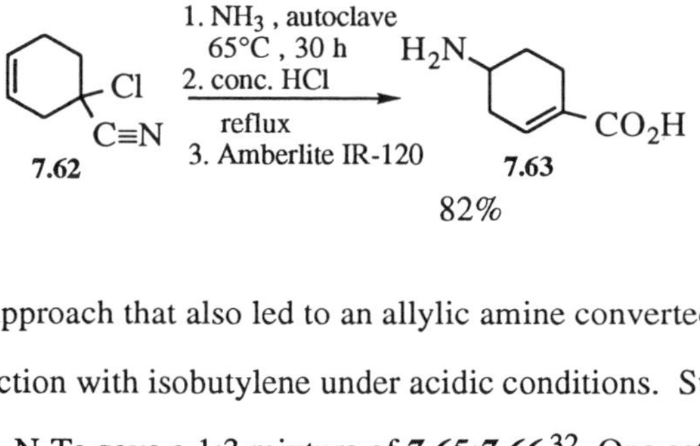

7.62 **7.63**

82%

An alternative approach that also led to an allylic amine converted ***7.64*** to the *t*-butyl ester by reaction with isobutylene under acidic conditions. Subsequent treatment with Ts-N=S=N-Ts gave a 1:3 mixture of ***7.65:7.66***.[32] One ester (***7.66***) was

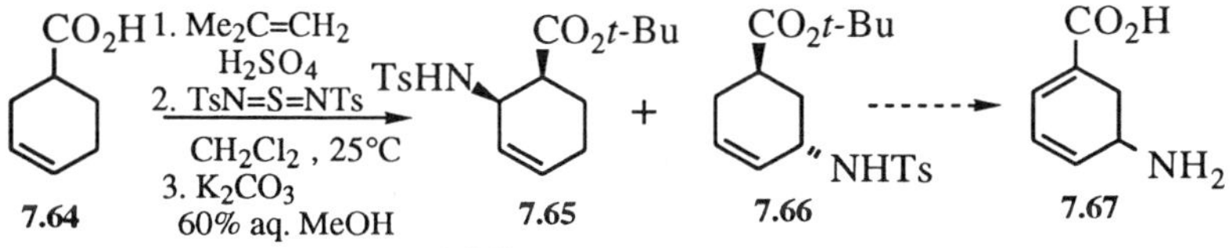

7.64 **7.65** **7.66** **7.67**

70%

isolated and then converted to gabaculine (5-aminocyclohexa-1,3-dienyl-1-carboxylic acid, ***7.67***). Gabaculine is reported to be an inhibitor[33] of GABA-T (GABA transaminase; see chapter six, section 6.1).[34] An alternative route converted 3-cyclohexene-1-carboxylic acid (***7.64***) to ***7.68***[35] and palladium catalyzed amination gave ***7.69***.[36] This alkylated derivative was subsequently converted to gabaculine.

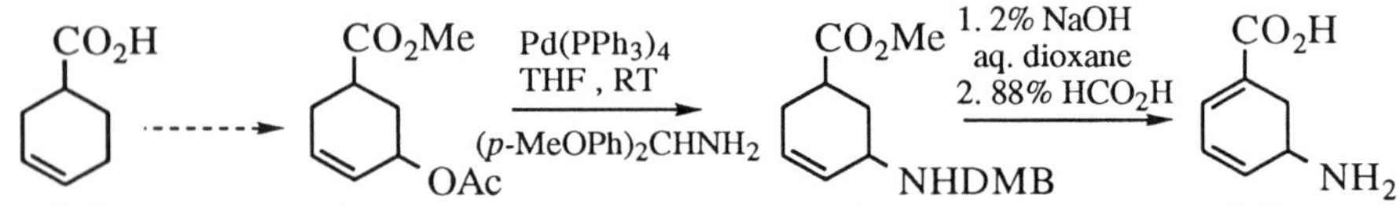

7.64 **7.68** **7.69** **7.67**

A different strategy used the isocyanate moiety in ***7.70***[37] as an amine surrogate. Reaction with HCl gave 2-amino-4-cyclohexene-1-carboxylic acid, ***7.71***.[38]

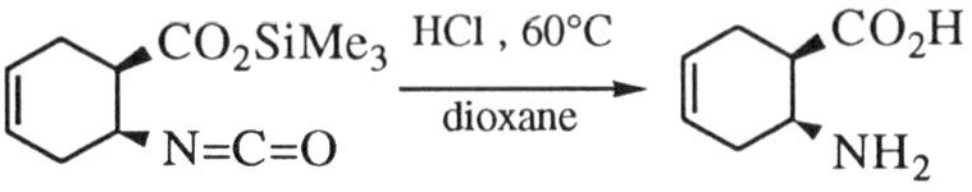

7.70 **7.71**

The conjugate addition of a nitrogen surrogate can lead to a different product if the terminus of the alkene moiety is somewhat hindered. When ***7.72*** reacted with sodium azide, an S_N2' reaction led to ***7.73***, which was reduced to methyl 2-amino-5-methylcyclohexane-1-carboxylate (***7.74***), which was shown to be a microbicide.[39]

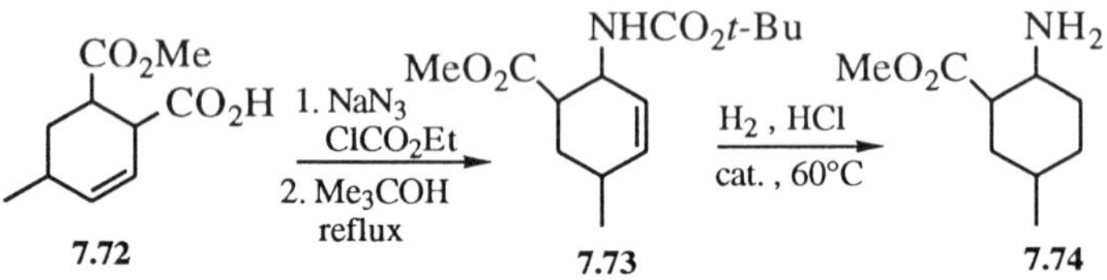

7.72 **7.73** **7.74**

7.1.B. Lactam Based Syntheses

Lactams are cyclic molecules that can be used as synthetic precursors to cyclic amino acids, usually by refunctionalization reactions. One example is the conversion of the alkene moiety in bicyclic lactam ***7.75*** (2-azabicyclo[2.2.1]hept-5-en-3-

one)[40] to the diol (*7.76*).[41] Acid hydrolysis of the lactam gave 3-amino-4,5-dihydroxy-cyclopentane-1-carboxylic acid, *7.77*.

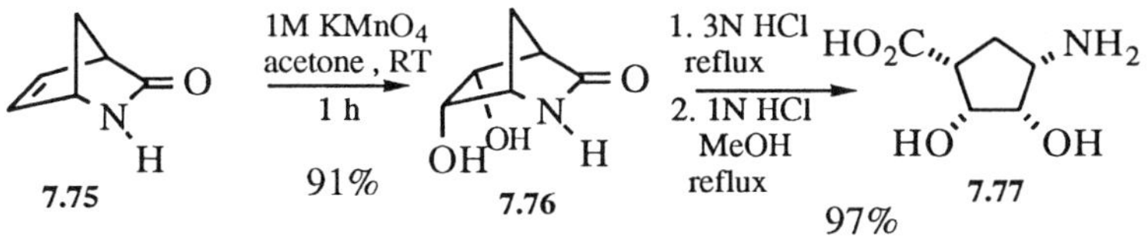

Lactam *7.78* was prepared from the corresponding succinimide derivative and then converted to *7.79*, allowing cyclization to *7.80* (as a 95:5 *endo:exo* mixture).[42]

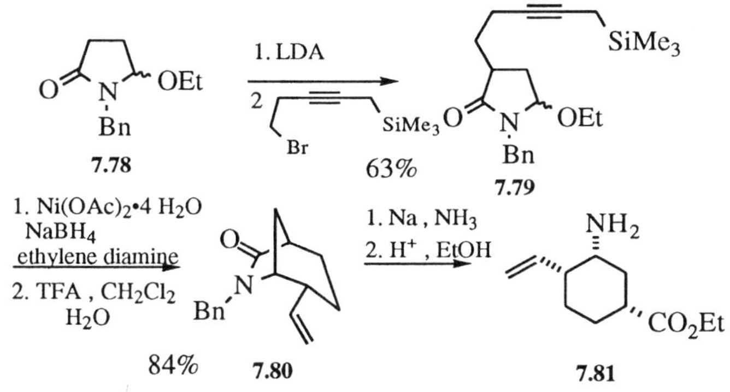

Dissolving metal reduction cleaved the N-benzyl group and acid hydrolysis opened the lactam ring to give ethyl 3-amino-4-ethenylcyclohexane-1-carboxylate, ***7.81***.[42]

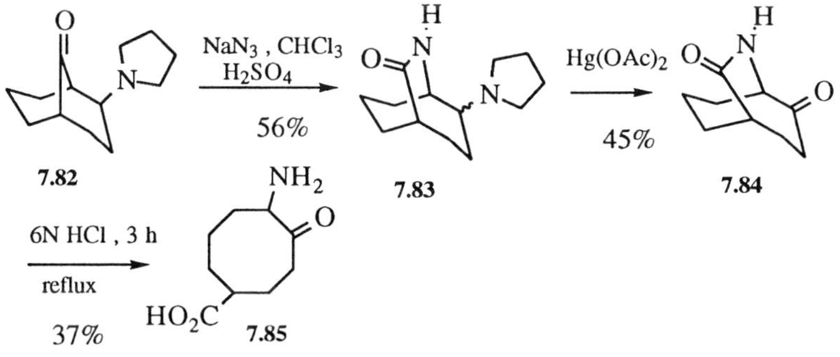

Eight-membered ring derivatives are also accessible. Reaction of ***7.82***[43] with sodium azide in sulfuric acid gave lactam ***7.83***(48% *exo*: 8% *endo*).[44] Subsequent reaction with mercuric acetate converted the amino group to a ketone moiety (see ***7.84***). This bicyclic lactam was hydrolyzed to give 5-amino-4-oxocyclooctane-1-carboxylic acid, ***7.85***.[44]

7.1.C. β-Lactam Intermediates

As in other chapters of this book, particularly chapter one, β-lactams are important precursors to non-α-amino acids. In this section, bicyclic β-lactam derivatives will be used to prepare cyclic amino acids. In a typical example, reaction of cyclopentene with chlorosulfonyl isocyanate (CSI) gave a β-lactam and treatment with basic potassium iodide cleaved the sulfonyl group to give ***7.86***.[45,46] Acid hydrolysis gave *cis*-2-aminocyclopentane-1-carboxylic acid, ***7.87***. Another method used

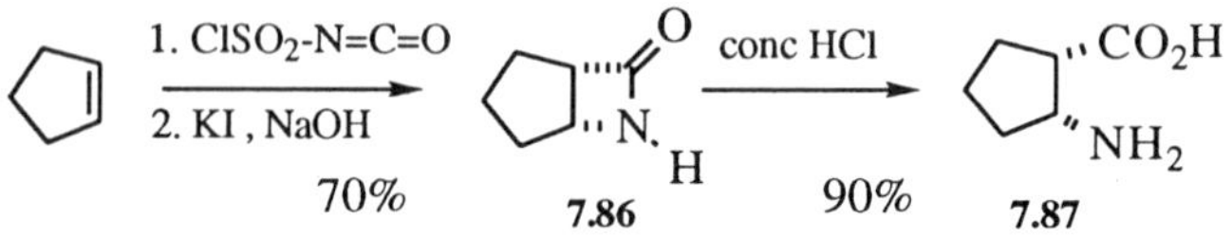

bicyclic lactam ***7.88***, which was racemic and obviously related to ***7.86***.[47] Incubation with *Rhodococcus equi* (NCIB 40213), also known as ENZA-1[48] led to kinetic resolution, allowing the isolation of (+)-***7.88***.[47] Catalytic hydrogenation of the double bond in this chiral, non-racemic bicyclic lactam was followed by hydrolytic ring opening to give (-)-***7.87*** [*cis*-2-aminocyclopentane-1-carboyxlic acid; also known as (-)-cispentacin], an anti-fungal antibiotic.[49]

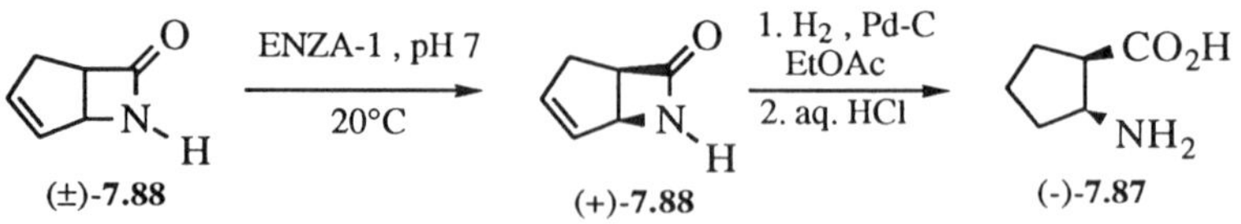

Aminocyclohexanecarboxylic acids can also be prepared by this general method.

Condensation of cyclohexene with N-benzyl isocyanate gave ***7.89***.[50] Acid hydrolysis of the lactam ring led to 2-aminocyclohexane-1-carboxylic acid, ***7.90***.

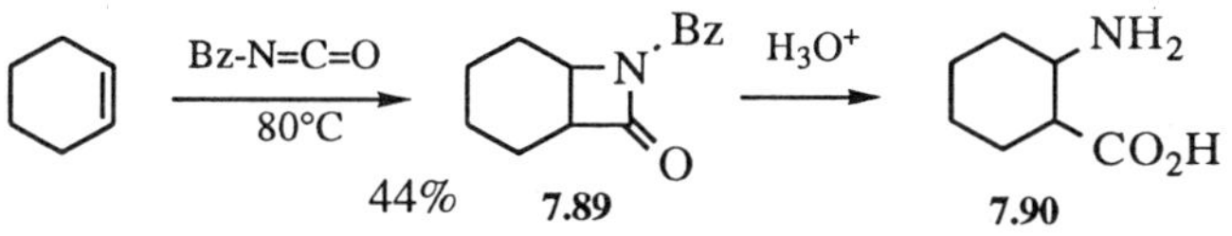

A ten-membered ring derivative was prepared by the reaction of allene ***7.91*** with N-chlorosulfonyl isocyanate to give β-lactam ***7.92***.[51] Removal of the sulfonyl group and acid hydrolysis gave 10-amino-1-cyclodecene-1-carboxylic acid, ***7.93***. Similar

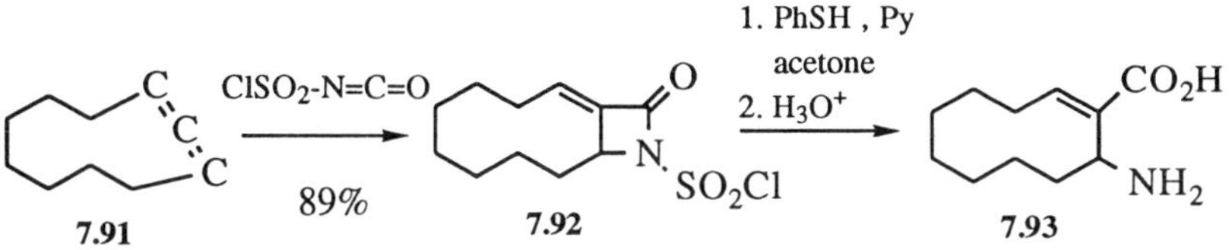

reaction of allene ***7.94*** with N-chlorosulfonyl isocyanate gave ***7.95*** .[51] Deprotection and hydrolysis gave 2-(1-amino-1-cyclohexyl)but-2E-enoic acid, ***7.96***. Although ***7.96*** is not formally categorized with the cycloalkylamino acids in this chapter, it is included to show the scope of the methods and the structural types available through synthesis.

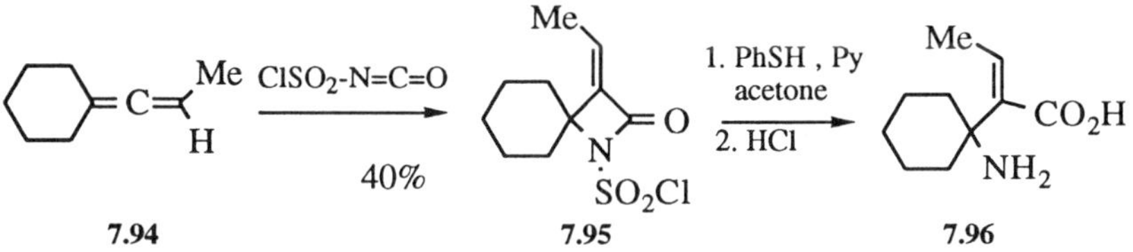

Another "isocyanate strategy" used a cyclohexadiene derivative (***7.97***) and generated a mixture of ***7.98*** and ***7.99*** in a 1:1 ratio.[33] These interesting iodoiso-cyanates had been previously reported.[52] Subsequent reaction with DABCO gave ***7.100*** which was converted to gabaculine, ***7.67*** (*vide supra*).[33]

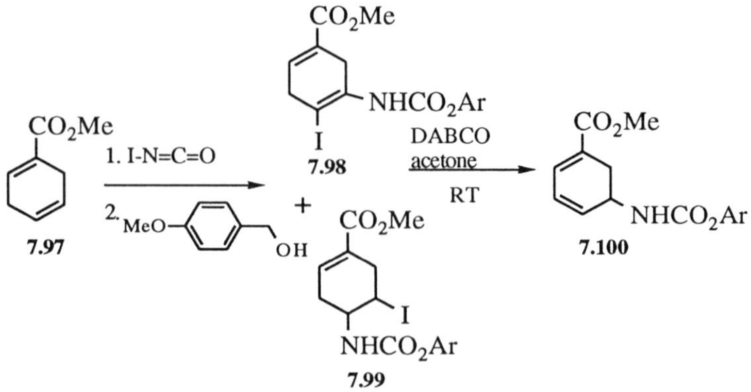

7.1.D. From Aromatic Precursors

Perhaps the most common synthetic route to aminocyclohexane carboxylic acids is reduction of nitro- and amino-substituted benzoic acid derivatives. Reduction of *ortho*-substituted acids give ***7.106***; reduction of *meta*-substituted acids give ***7.104***; and reduction of *para*-substituted acids give ***7.57102***

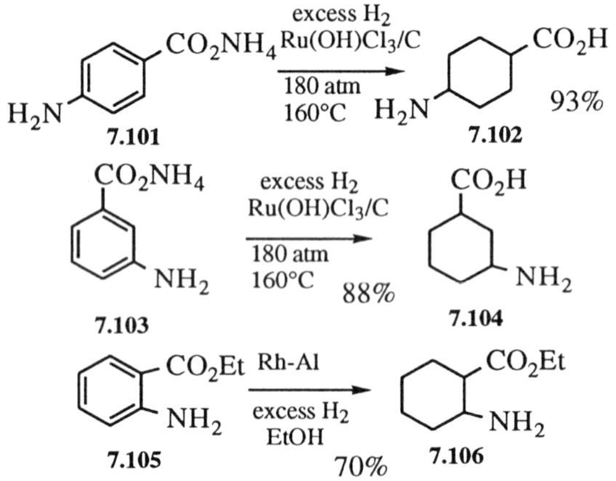

Catalytic hydrogenation of *p*-amino ammonium salt ***7.101*** gave 4-aminocyclohexane carboxylic acid (***7.102***), as a 72:28 *cis:trans* mixture.[53] Similarly, reduction of the *meta*-derivative (***7.103***) gave 3-amino-1-cyclohexene carboxylate, ***7.105***.[53] A rhodium-aluminum catalyst was used to reduce the *ortho*-amino-ester (***7.105***) to ethyl 2-aminocyclohexane carboxylate (***7.106***).[54] Other aromatic systems were reduced in a similar manner using a nickel-rhenium catalyst to prepare ***7.102*** and ***7.104***;[55] a ruthenium-on-carbon-ruthenium oxide catalyst was used for ***7.102*** and ***7.104***;[56] a rhenium-aluminum oxide catalyst for ***7.102***[57] and for ***7.104***;[58] a platinum oxide catalyst for ***7.102*** and ***7.104***;[59] a ruthenium oxide catalyst for ***7.102***,[60] and also for ***7.104***;[61] and a ruthenium catalyst for ***7.102***.[62]

Substituted aromatic acids provide a route to other cyclohexyl derivatives. An example is the catalytic hydrogenation of the benzene ring in ***7.107*** to give 4-amino-methyl-1-cyclohexene carboxylic acid, ***7.108***.[53] Partial reduction of a benzene ring

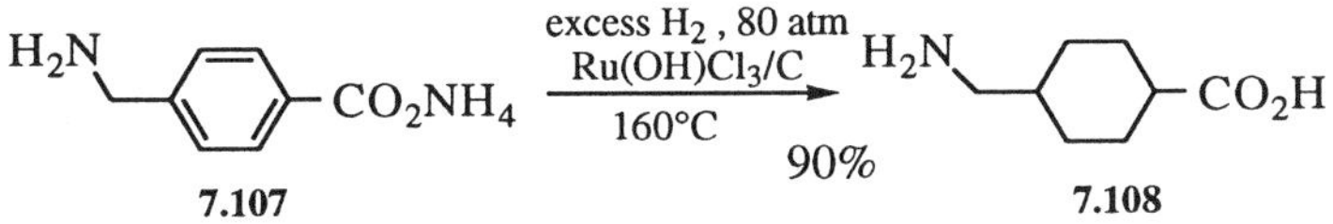

7.107 **7.108**

is also possible, allowing isolation of a cyclohexene carboxylic acid. An example is the conversion of ***7.105*** to ethyl 2-amino-1-cyclohexene carboxylate (***7.109***).[54a]

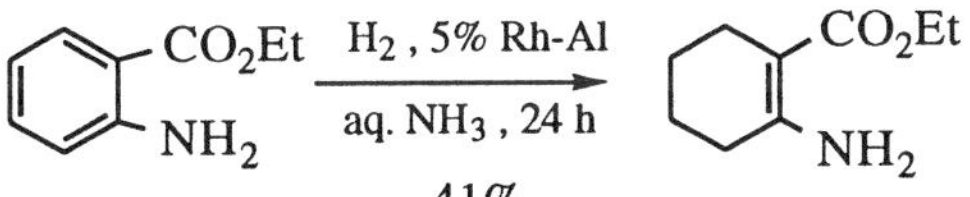

7.105 **7.109**

An alternative method of reducing the aromatic ring relies on metals (sodium, lithium calcium, etc.), in a so-called dissolving metal reduction. An example is the

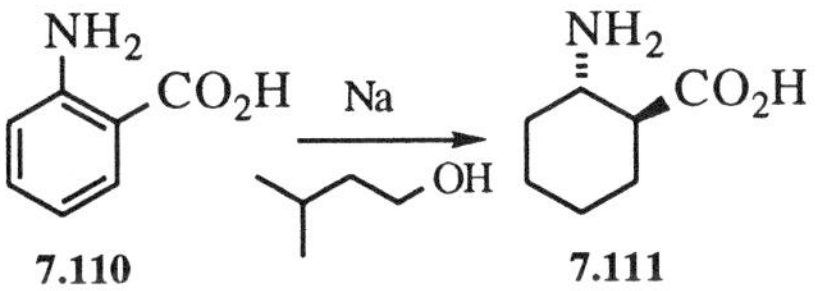

7.110 **7.111**

reaction of ***7.110*** with sodium in isoamyl alcohol to give ***7.111***.[63] Sodium in ammonia was used for the same transformation (as early as 1894), but with less *cis/trans* selectivity.[64]

A related method initially reduced ***7.112*** to ***7.113*** (63% yield) and required separated the enantiomers by resolution techniques. This gave *R-**7.113***.[26] Conversion to the oxime and its reduction with sodium in ammonia gave 1R,3S-*7.22* (*cis*-3-aminocyclohexane-1-carboxylic acid), but the yield was very poor.

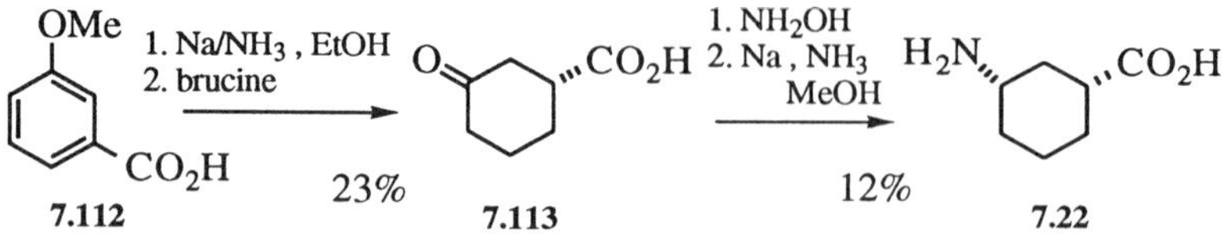

7.1.E. From Cyclic Ketones

As with acyclic β-keto esters, treatment of a cyclic β-keto ester with ammonia or

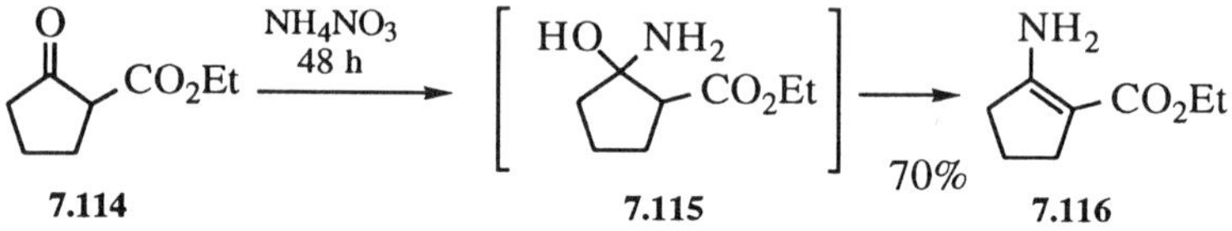

an amine leads to cyclic alkenyl amino acids. The reaction of 2-carboethoxycyclopentanone (***7.114***) with ammonium nitrate gave ***7.115*** as an intermediate, but the final product is ethyl 2-amino-1-cyclopentene carboxylate (***7.116***).[65] Dieckmann postulated the intermediacy of ***7.115*** in this reaction as early as 1901.[66] This reaction and minor variations have been reported many times.[67] Variations include

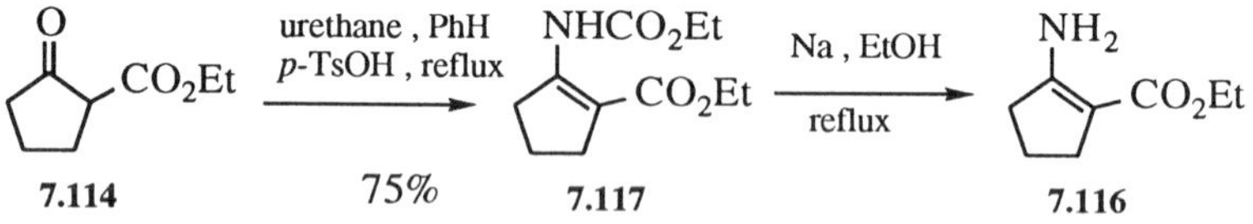

the reaction of ***7.114*** with urethane to give ***7.117***, and this gave ethyl 2-amino-1-cyclopentene-1-carboxylate (***7.116***) after reaction with sodium in ethanol.[68] Amino-ester ***7.116*** was also formed by reaction of ***7.114*** with phosphazines such as $Ph_3P{=}NSiMe_3$[69] in 91% yield.[70] Once the alkenyl amino acid is in hand, catalytic hydrogenation or another reduction technique gives the saturated derivative. The reaction with ammonia can be modified to give the saturated derivative directly. Treatment of ***7.118*** with ammonia and Raney nickel, for example, led to ***7.43*** (2-aminocyclopentane-1-carboxylic acid) via reductive amination. This amino acid was isolated as a mixture of *cis*- and *trans*-isomers.[18]

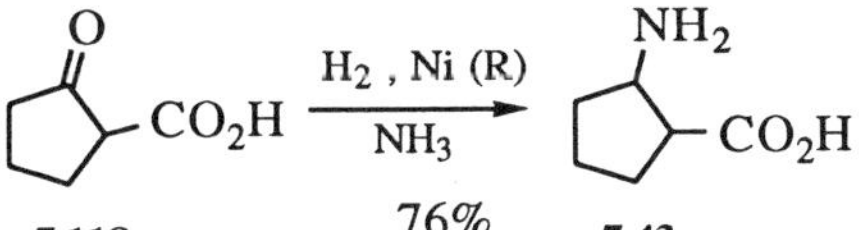

7.118 **7.43**

This method can also be used for the preparation of cyclohexene carboxylic acid derivatives. Reaction of ***7.119*** with ammonia gave ethyl 2-amino-1-cyclohexene carboxylate, ***7.109***.[66] The conversion of ***7.119*** to ***7.109*** has been reported several

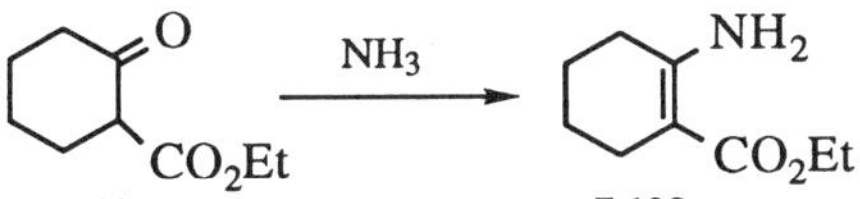

7.119 **7.109**

times by using ammonia[71,67a,1,72] or amines (see below).[67a] Analogous to the conversion of ***7.114*** to ***7.117***, ***7.109*** can be prepared both by reaction of ***7.119*** with urethane,[67] and by reaction with $Ph_3P{=}N{-}SiMe_3$.[69]

Amines also react with cyclic with β-keto-esters. Keto-ester ***7.114*** reacted with propylamine, for example, to give ethyl 2-(N-propylamino)-1-cyclopentene carboxylate, ***7.120***.[73] This work was based on the earlier reported reaction with ***7.114*** with

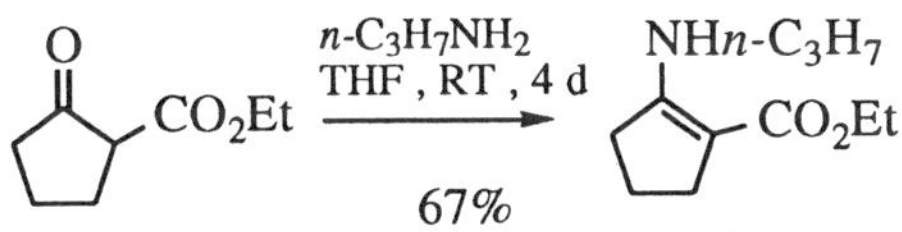

7.114 **7.120**

benzylamine to give the N-benzyl derivative of ***7.120***.[74] In some cases the reaction of an amine with a β-keto-ester (such as ***7.114***) can be "stopped" at the hydroxy-amine (see ***7.121***),[72] before elimination gives the alkenyl amino acid (see chapter 1, section 1.1.C). It is more normal, however, that the alkenyl amino acid is the final product.

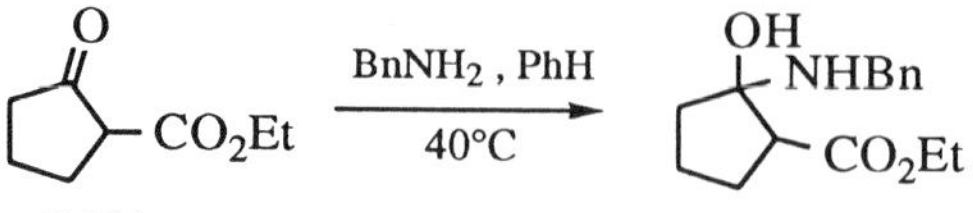

7.114 **7.121**

Just as with the ammonia reaction, cycloalkenyl amino acids can be reduced to their saturated derivatives. Reaction of ***7.119*** with methylamine, for example, gave ***7.122***, which was converted to ***7.123*** via catalytic hydrogenation.[75]

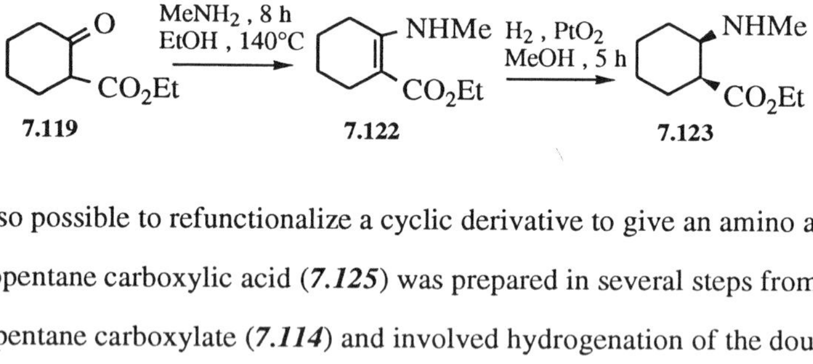

7.119 **7.122** **7.123**

It is also possible to refunctionalize a cyclic derivative to give an amino acid. 1,2-Cyclopentane carboxylic acid (***7.125***) was prepared in several steps from ethyl 2-oxocyclopentane carboxylate (***7.114***) and involved hydrogenation of the double bond in an intermediate product, ***7.124***.[67] An early version of this transformation simply

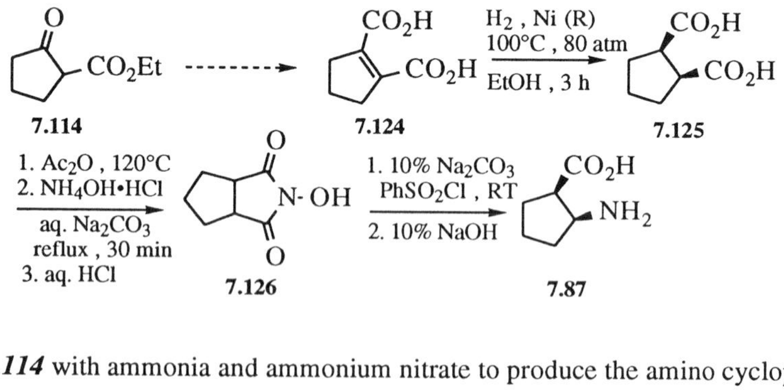

7.126 **7.87**

treated ***7.114*** with ammonia and ammonium nitrate to produce the amino cyclopentene carboxylic acid,[74] which could then be hydrogenated. Dicarboxylic acid ***7.125*** was, in the reaction shown, converted to an anhydride, and then reacted with ammonium hydroxide to give an imide (***7.126***). A subsequent Lossen rearrangement[76] led to 2-aminocyclopentane carboxylic acid (***7.87***).[74] An alternative preparation of ***7.125*** from ***7.114*** has also been reported.[45]

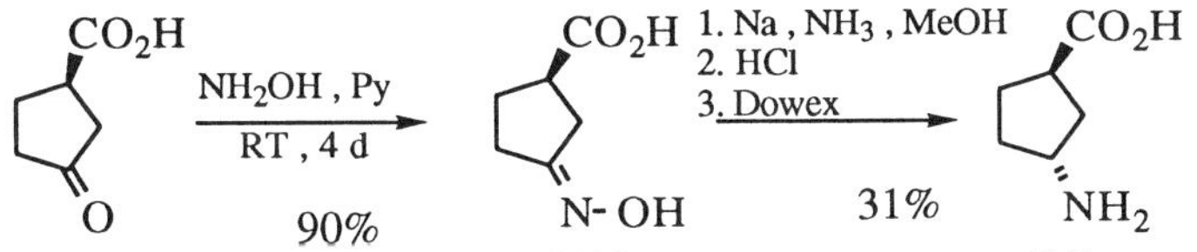

7.127 **7.128** **7.41**

The ketone moiety of a cyclic keto-ester need not be in conjugation with the carboxyl moiety. This allows other methods for introduction of the amine group. Conversion of R-***7.127*** (resolved from the racemic material with brucine) to its oxime (***7.128***) was accomplished by treatment with hydroxylamine and pyridine.[20]

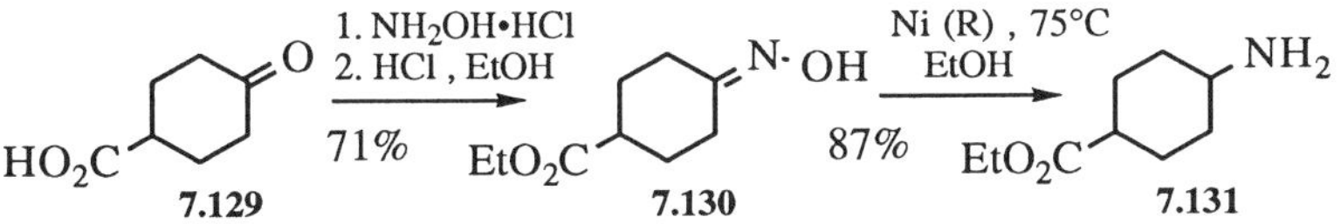

7.129 **7.130** **7.131**

Dissolving metal reduction of the oxime and hydrolysis gave 1R,3R-3-aminocyclopentane-1-carboxylic acid, ***7.41***. An identical approach was used to prepare cyclohexane derivatives. Reaction of ***7.129***[62] with hydroxylamine[76b,77] and subsequent

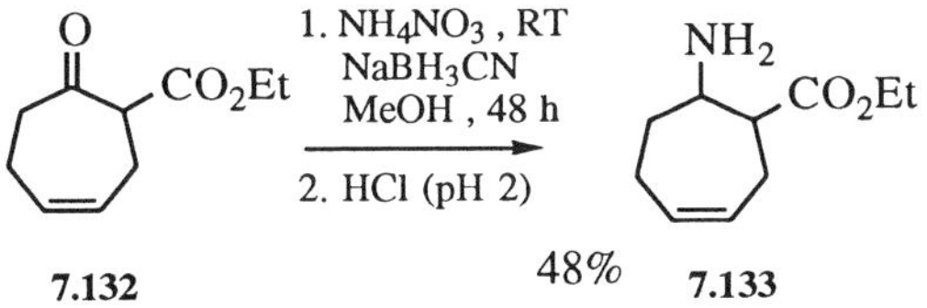

7.132 **7.133**

esterification gave ***7.130***.[76,78] Reduction of the oxime with Raney nickel gave ethyl 4-aminocyclohexane-1-carboxylate, ***7.131***. The conversion of the ketone moiety in

7.129 to an amine via reductive amination gave ***7.130*** (ethyl 2-aminocyclo-hept-5-ene-1-carboxylate) is another example of this type of conversion.[79]

More sophisticated derivatives can be prepared from the appropriate keto-ester. Reaction of ***7.134*** with phenethylamine gives a protected aminocyclopentane derivative. Both ***7.134a*** and ***7.134b*** were converted to the corresponding amine (***7.135a*** or ***7.135b***). Deprotection by catalytic hydrogenation led to ***7.136a*** [ethyl 2-(2-amino-1-cyclopentyl)ethanoate; 4% overall yield] or ***7.136b*** [ethyl 2-(2-amino-1-cyclopentyl)propanoate; 38% overall yield].[80] These amino acids are unusual for the categories specified in this chapter, but are shown as examples of structural variations possible in this type of reaction.

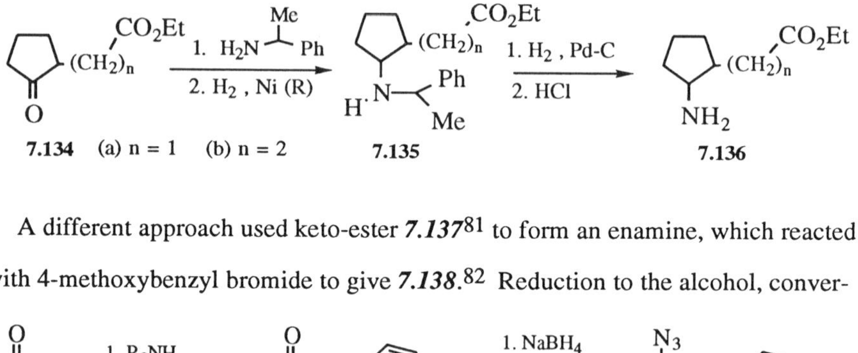

7.134 (a) n = 1 (b) n = 2 **7.135** **7.136**

A different approach used keto-ester ***7.137***[81] to form an enamine, which reacted with 4-methoxybenzyl bromide to give ***7.138***.[82] Reduction to the alcohol, conver-

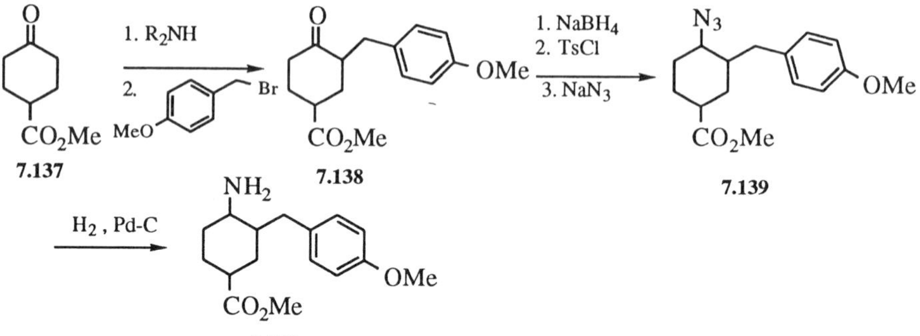

7.137 **7.138** **7.139** **7.140**

sion to the tosylate, and reaction with sodium azide gave ***7.139***. Subsequent hydrogenation gave methyl 4-amino-3-(4-methylphenylmethyl)-1-cyclohexenecarboxylate, ***7.140***. In this synthesis, an enamine generated by reaction of ***7.137*** with a dialkylamine, was a reactive intermediate. Another variation in refunctionalization of a ketone moiety was the conversion of ***7.141***[83] to enamine ***7.142***, which was the final product[84] Although this is not formally an amino acid, reduction of enamines to give the amine is possible (although elimination to the alkene is often a problem). As seen in the previous example, enamines can also be reactive intermediates. This example is also included as an example of an 'unusual' cyclohexene amino acid.

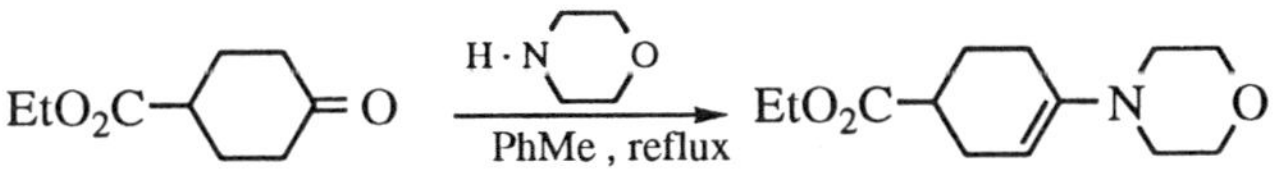

7.141 **7.142**

In another variation, keto-ester ***7.14*** reacted with ethyl α-cyanoacetate to give ***7.143***. Subsequent reaction with ammonium hydroxide led to ***7.116*** in good yield.[85] Amines were also reacted with ***7.143*** to give N-alkyl derivatives of ***7.116***.

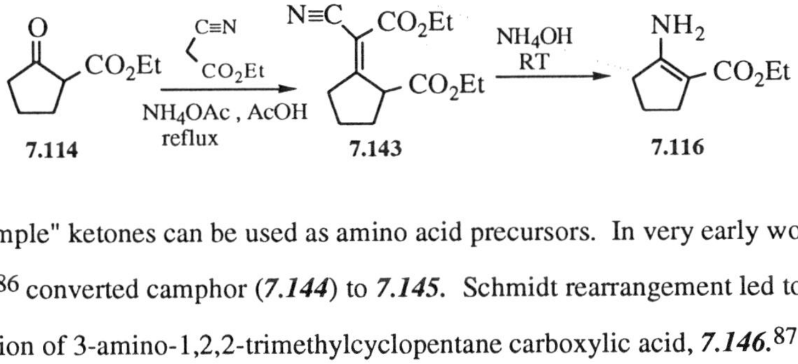

7.114 **7.143** **7.116**

"Simple" ketones can be used as amino acid precursors. In very early work, Noyes[86] converted camphor (***7.144***) to ***7.145***. Schmidt rearrangement led to the formation of 3-amino-1,2,2-trimethylcyclopentane carboxylic acid, ***7.146***.[87] A

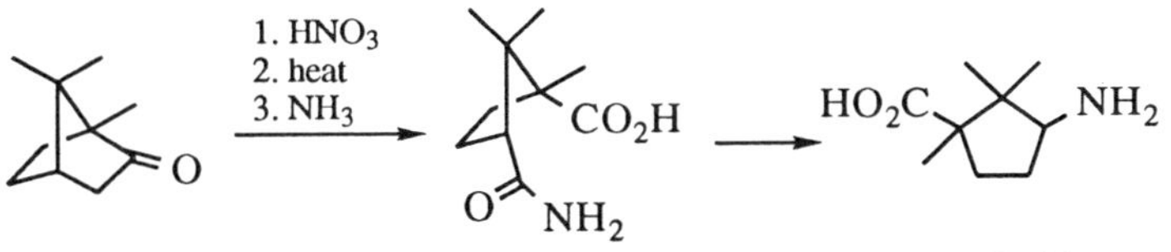

7.144 **7.145** **7.146**

related route converted ***7.145*** to its methyl ester, ***7.147***. Reaction with lead tetraacetate and tin tetrachloride in *t*-butyl alcohol gave ***7.148***.[88] Treatment of ***7.148*** with ethanolic HCl gave methyl 3-amino-2,2,3-trimethylcyclopentane-1-carboxylate, ***7.149*** (the methyl ester of ***7.146***).

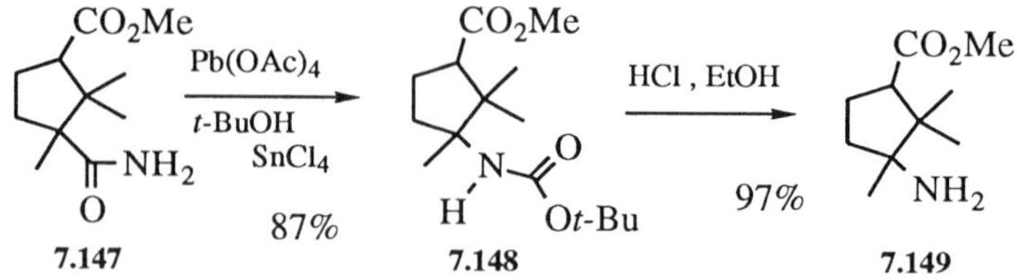

Another approach using ketones converted cyclohexanone to 1-cyano-1-cyclohexene (***7.150***).[89] Conjugate addition of ammonia and esterification led to *trans*-ethyl 2-aminocyclohexane-1-carboxylate, ***7.151***.

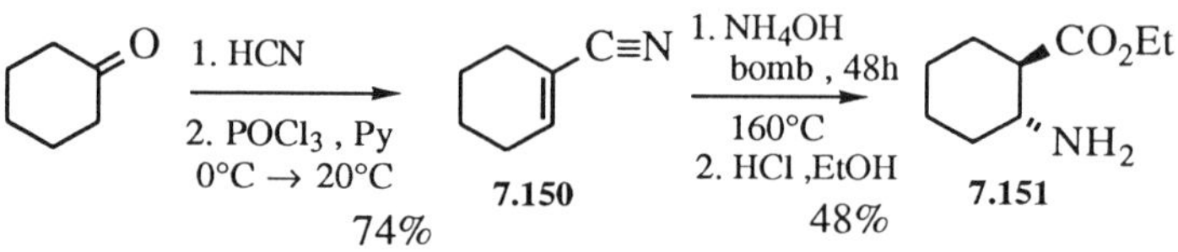

Ketone ***7.152***[90] (derived from 1,3-cyclohexanedione) served as an amino acid precursor by reaction with 3-methoxyphenylmagnesium bromide (***7.153***), to give

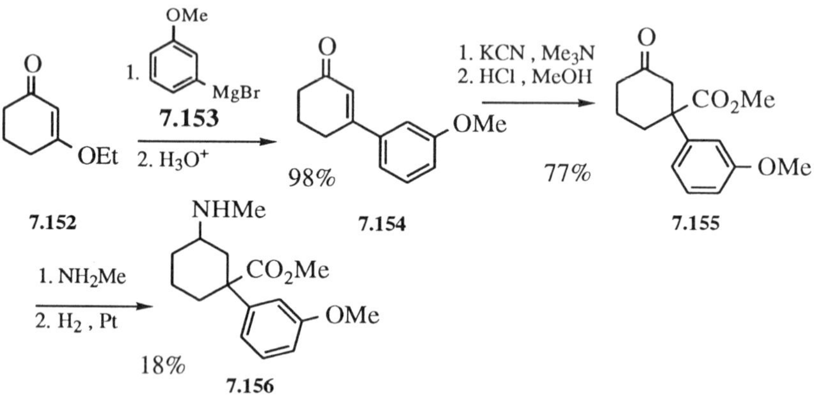

7.154 via conjugate addition and elimination of ethoxide.[91] A second conjugate

addition of potassium cyanide followed by acid hydrolysis gave ***7.155***. Reaction of the ketone moiety with methylamine to give the imine and catalytic hydrogenation of the imine moiety gave 3-(N-methylamino)-1-(3-methoxyphenyl)-1-cyclohexene carboxylate (***7.156***), but in very poor yield. Both the 2-methoxy- and the 4-methoxy derivative, as well as the simple phenyl derivative, were prepared by this method. In some cases, amino-esters such as ***7.156*** cyclized to a bicyclic lactam.[91]

7.1.F. Miscellaneous Syntheses

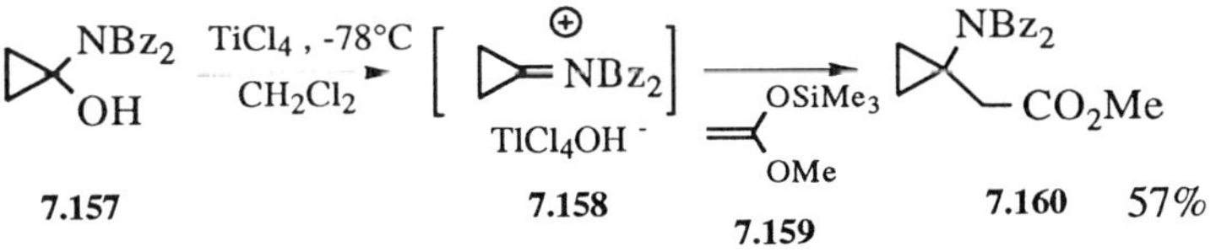

An interesting approach reacted a cyclopropylcarbinolamine such as ***7.157*** with titanium tetrachloride to generate iminium salts such as ***7.158***.[92] When this salt was generated in the presence of a silyloxy ethylene derivative such as ***7.159***, an amino-ester was formed. This Mukaiyama reaction[93] (see chapter four, section 4.2) generated 2-(1-N,N-dibenzoylaminocyclopropyl)ethanoate, ***7.160***.[92]

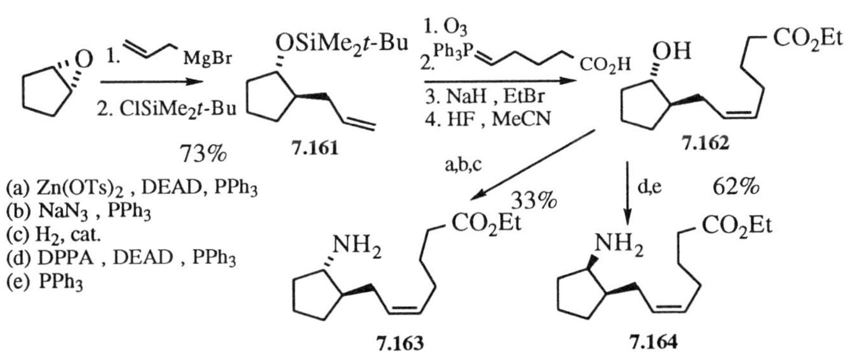

Cyclopentene oxide reacted with allylmagnesium bromide and subsequent protection of the resultant alcohol gave ***7.161***.[94] Ozonolysis, Wittig olefination,[95]

and deprotection gave *7.162*, which was converted to *trans*- or *cis*-ethyl 7-(2-amino-1-cyclopentyl)hept-5Z-enoate (*7.163* or *7.164*, respectively).[94]

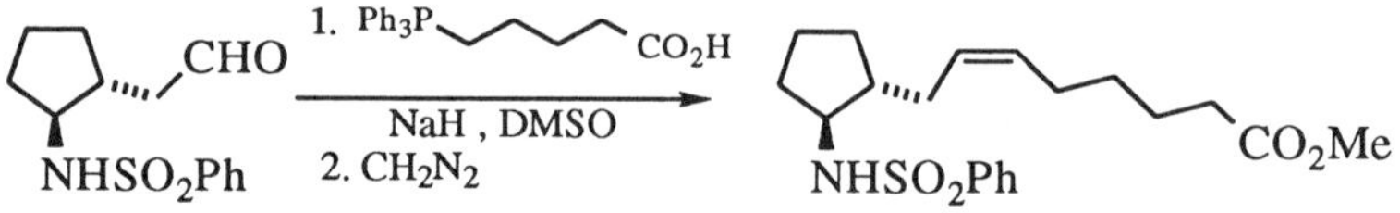

7.165 **7.166**

The Wittig methodology applied to the synthesis of *7.162* can be used in a variety of situations. Aldehyde *7.165*, for example, was converted to *7.166* [methyl 8-(2-N-benzenesulfonylamino-1-cyclopentyl)oct-6-enoate] via a Wittig reaction and diazomethane esterification.[96] The corresponding acid of *7.166* was converted to its sodium carboxylate, and was shown to inhibit arachidonic acid induced blood platelet aggregation in rabbits.[96] A quite different approach converted anhydride *7.167* to *7.168*.[97] Note the similarity to cleavage of bicyclic derivatives in section 7.1.B (see *7.75* and *7.78*).

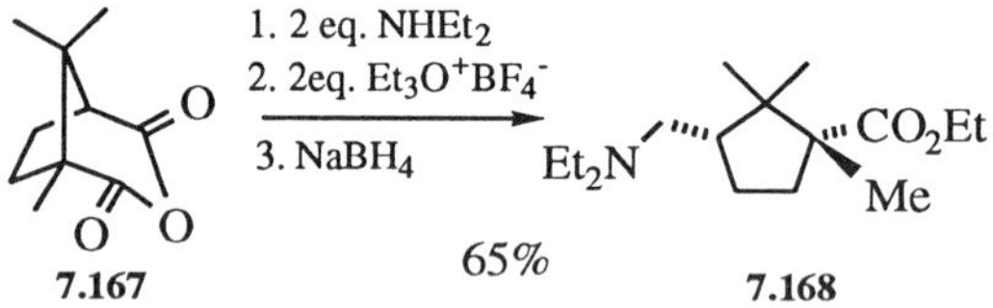

7.167 **7.168**

A highly specialized synthesis converted (-)-shikimic acid (*7.169*) to *7.170*.[98] Reaction with azide and catalytic hydrogenation gave 5-aminoshikimate (*7.171*, methyl 5-amino-3,4-dihydroxy-1-cyclohexene carboxylate) in 21% overall yield. Amino acid *7.171* is an inhibitor of EPSP synthetase, and can be used as an herbicide.

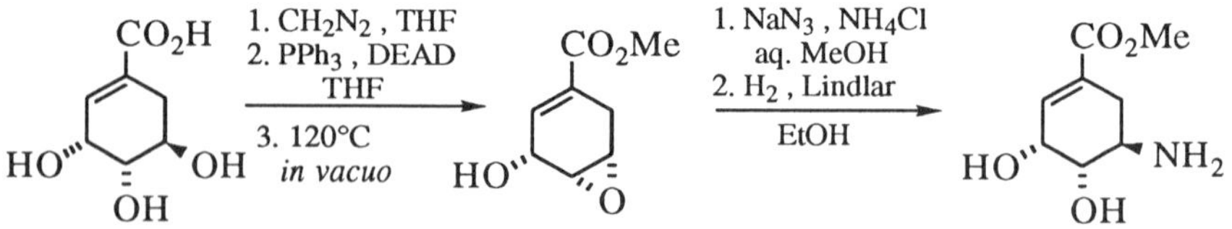

7.169 **7.170** **7.171**

7.2. Diels-Alder Strategies

The Diels-Alder reaction is one of the more powerful reactions in all of organic chemistry.[99] A Diels-Alder strategy can be used to prepare cyclic amino acids. Dienes can react with tosyl cyanide to generate lactams. Cyclopentadiene reacted with *p*-toluenesulfonyl cyanide (N-tosyl cyanide; TsCN; $MeC_6H_4SO_2CN$), for example, to give azabicycloheptene derivative, ***7.172***.[100] Treatment with aqueous

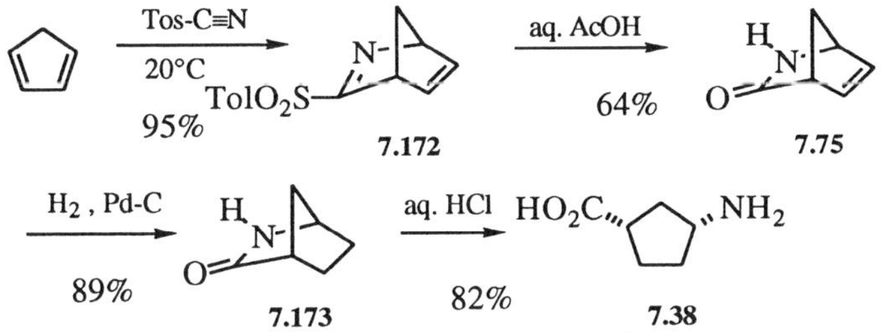

acetic acid converted the sulfonyl-imine moiety into lactam ***7.75***. Subsequent catalytic hydrogenation of the double bond gave ***7.173***. Acid hydrolysis gave *cis*-3-aminocyclopentane-1-carboxylic acid, ***7.38***. The same basic strategy was used in the Diels-Alder reaction of cyclopentadiene with TsCN[101] to give lactam ***7.75***, after

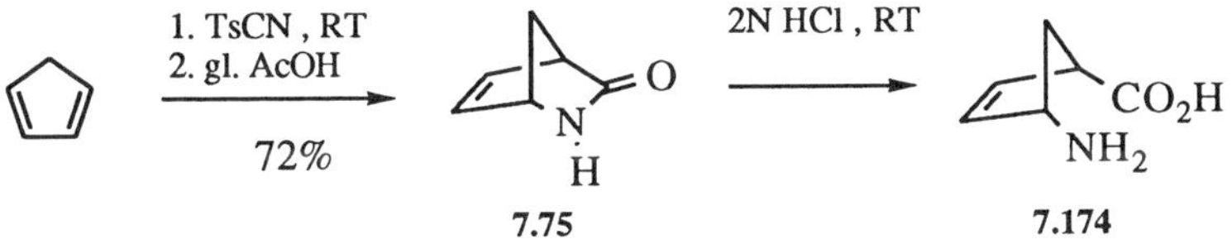

treatment with glacial acetic acid.[41,102] Reaction with HCl gave *cis*-4-amino-2-cyclopentene carboxylic acid, ***7.174***,[102,103,27,22] in good yield. Treatment of ***7.75*** with a lactamase (ENZA-1,[48] see section 7.1.C) selectively hydrolyzed racemic ***7.75*** to give (-)-***7.74***, leaving (+)-***7.75*** untouched.[104]

An alternative route to bicyclic anhydrides uses a Diels-Alder reaction[99] of 1,3-butadiene and maleic anhydride to give ***7.175***.[105] This was converted to ***7.176*** by treatment with ammonia and then catalytic hydrogenation.[106] Treatment with sodium hydroxide and bromine (NaOBr; the Hofmann reaction)[12] gave *cis*-2-amino-cyclohexane-1-carboxylic acid, ***7.177***.[106,63]

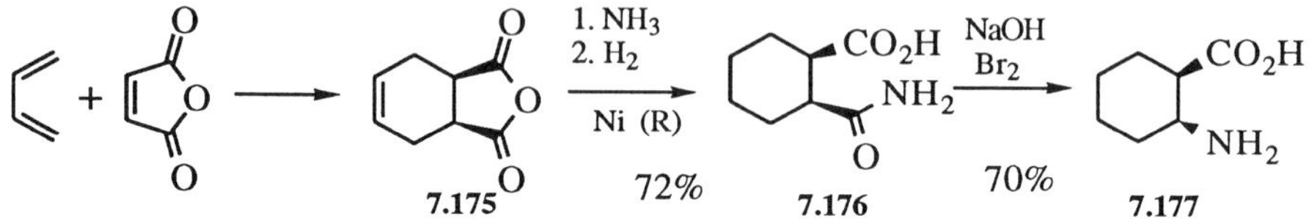

Tetrahydrophthalic anhydride (***7.175***) was converted directly to *cis*-5-amino-3-cyclohexene carboxylic acid, ***7.71*** by reaction with ammonia and then NaOCl.[107,14] If ***7.175*** is first opened, isomerized, and converted to the *trans*-diester (***7.178***), formation of the *trans*-anhydride is possible (***7.179***). This anhydride was then converted to *trans*-5-amino-3-cyclohexene carboxylic acid, ***7.180***.[72] A related route converted ***7.175*** to racemic *cis*-methyl 2-carboxylcyclohex-4-enoate, which was resolved with cinchonidine.[108] The resolved (+) or (-)-half-esters were then converted to (-)-***7.71*** or (+)-***7.71***.

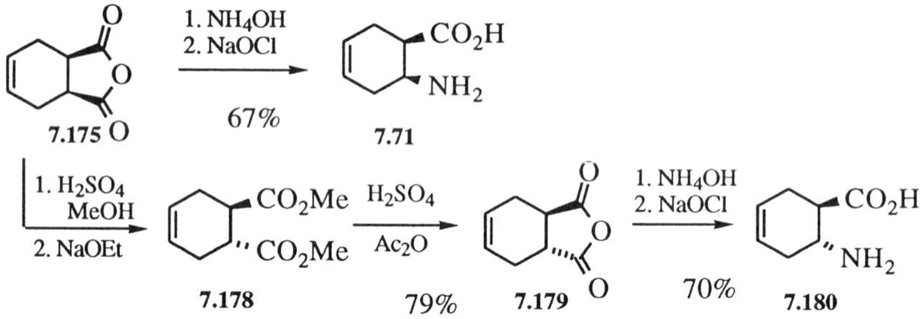

Derivatives of mono- and dicarboxylic acids are another important source of *ortho*- and *meta*-aminocyclohexane carboxylic acids, but not the *para*-derivatives. Anhydride ***7.181*** was converted to the amide, then to ***7.177*** (*cis*-2-aminocyclohex-ane-1-carboxylic acid) via a Hofmann reaction,[12] and finally to *cis*- ethyl 2-aminocyclohexane-1-carboxylate (*7.182*).[109] The amino acid (*7.177*) is easily isomerized with heating to give the diastereomeric ***7.111.*** The anhydride (***7.181***) was prepared from cyclohexane dicarboxylic acid and this is an often used route to *trans*-2-aminocyclohexane-1-carboxylic acid (***7.82***) and ***7.11.***[110]

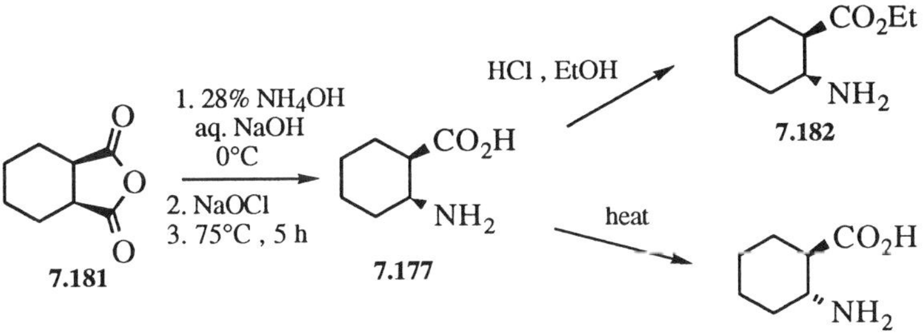

7.181 7.177 7.182 7.111

An "updated" version of this sequence is the reaction of the butadiene surrogate 3-sulfolene with maleic anhydride to give ***7.175***. Subsequent hydrogenation of the double bond gave anhydride ***7.181***, and this was then treated with trimethylsilyl azide to give *cis*-2-aminocyclohexane-1-carboxylic acid, ***7.177.***[111] The transformation of ***7.187*** to ***7.177*** probably proceeds via a transient isocyanate, ***7.183***. Formation of an isocyanate such as this and its conversion to the amino acid (***7.183*** to ***7.177***) had been reported in earlier work.[38]

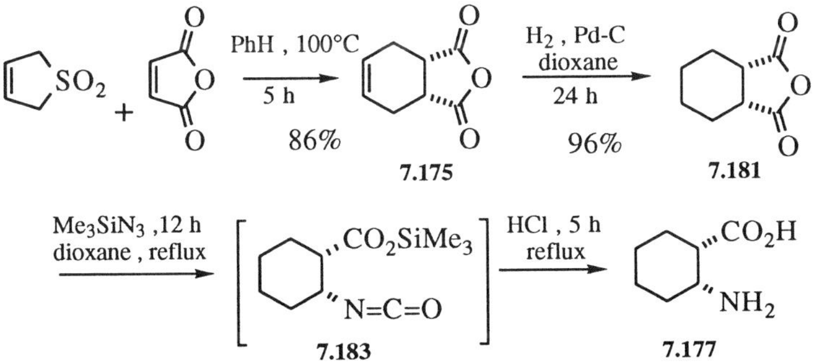

7.175 7.181 7.183 7.177

Another method converted the anhydride to an imide rather than a lactam. In this example, anhydride ***7.181*** was converted to ***7.184***[112] and then reacted with benzene-

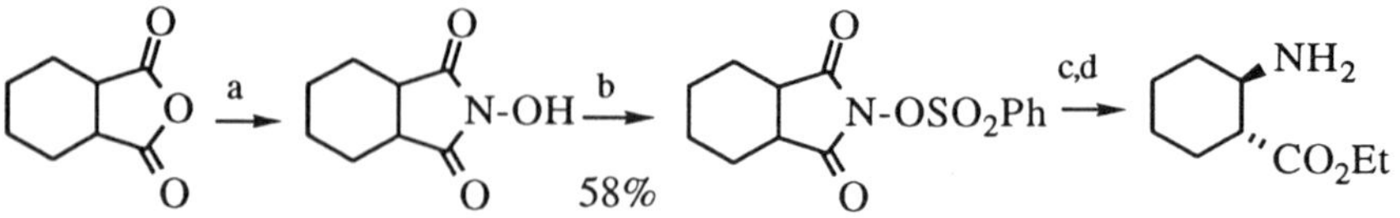

7.181 **7.184** **7.185** **7.151**

(a) NH_4OH (b) $PhSO_2Cl$, aq.Na_2CO_3 (c) aq. NaOH , steam bath
(d) conc HCl , 200°C , sealed tube

sulfonyl chloride to give ***7.185***.[113] Treatment with base led to a Lossen rearrangement,[76] which gave *trans*- ethyl 2-aminocyclohexane-1-carboxylate (***7.151***) after hydrolysis of the N-sulfonyl group.[101]

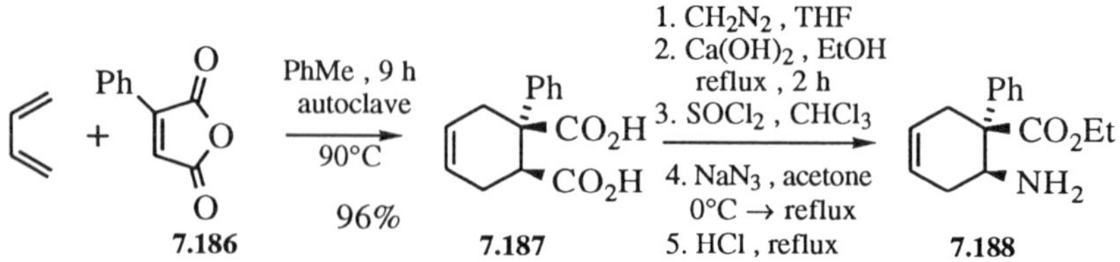

7.186 **7.187** **7.188**

Other anhydrides can be used in the Diels-Alder reaction, as in the reaction of 1,3-butadiene with ***7.186***[114] to give ***7.187***.[115] Esterification was followed by

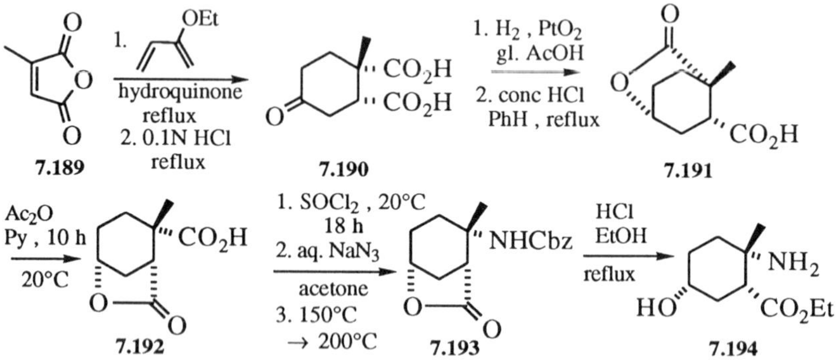

7.189 **7.190** **7.191** **7.192** **7.193** **7.194**

conversion to ethyl 6-amino-1-phenyl-3-cyclohexene-1-carboxylate (***7.188***). A closely related strategy used citraconic anhydride (***7.189***) in a reaction with 2-ethoxy-1,3-butadiene. The anhydride moiety in the resultant cycloadduct was hydrolyzed to give ***7.190***.[116] The ketone moiety was reduced to an alcohol, allowing formation of lactone ***7.191***. Treatment with acetic anhydride led to a new anhydride (***7.192***) and the acid moiety was converted to a Cbz amino moiety (in ***7.193***). Treatment of the lactone with HCl in ethanol led to ethyl 2-amino-5-hydroxy-2-methylcyclohexane-1-carboxylate, ***7.194***. It is noted that in different work, diester ***7.195*** was resolved to give ***7.196*** using pig liver esterase.[117] Subsequently, ***7.196*** was converted to ***7.71*** (*cis*-2-amino-4-cyclohexene-1-carboxylic acid) in six steps.[118]

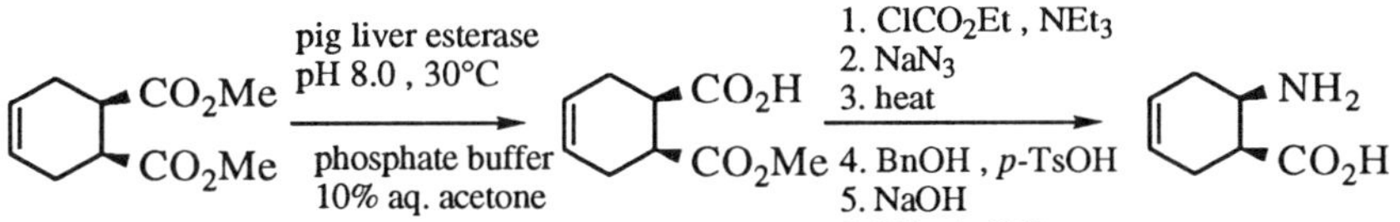

7.195 **7.196** **7.71**

A different approach cleaved the alkene moiety in ***7.197*** via permanganate oxidation (***7.197*** was an 80:20 *endo:exo* mixture) to give ***7.198***, along with a lesser amount of ***7.199*** (from the *exo*-acetate, ***7.197***).[119] Treatment of ***7.198*** with acetic anhydride gave the anhydride and subsequent reaction with ammonia gave ***7.200***. Beckmann rearrangement[4] and elimination gave 4-amino-1-cyclopentene carboxylic acid (***7.201***).

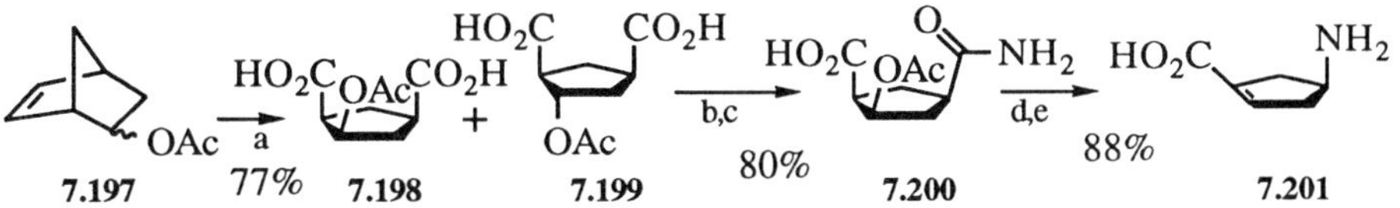

7.197 **7.198** **7.199** **7.200** **7.201**

(a) $KMnO_4$ (b) Ac_2O (c) NH_3 (d) Br_2 , NaOH (e) H_3O^+

Heteroatom-substituted dienes can be used to prepare amino acids. Azadiene ***7.202***[120] reacted with ***7.203***,[121] for example, to give ***7.204*** [methyl 3-(N-Boc amino)-2-nitrocyclohex-4-en-1-carboxylate].[122] Treatment with DBU gave a 70% yield of ***7.205***, which was converted to isogabaculine (***7.206***, 3-amino-1,4-cyclohexadiene-1-carboxylic acid). A virtually identical approach involved the reaction

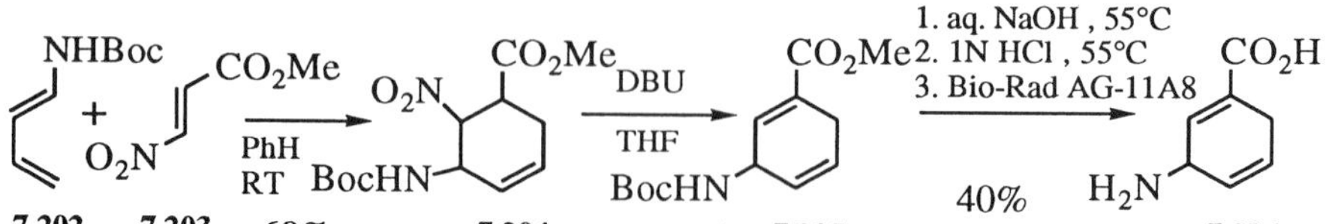

7.202 **7.203** 68% **7.204** 70% **7.205** **7.206**

of amino butadiene ***7.207*** (rather than the carbamate) and ***7.208*** to give a 1:3 mixture of ***7.209:7.210***.[123] Amino-ester ***7.50*** is known as tilidine [ethyl 2-(N,N-diethylamino)-1-phenyl-3-cyclohexenecarboxylate], a clinically used analgesic.[123a,d] Similarly, carbamate ***7.211*** was also used in a reaction with ***7.208***, and gave ethyl 2-(N-Cbz amino)-1-phenyl-3-cyclohexene-1-carboxylate (***7.212***), along with 20% of the isomeric ***7.213***.

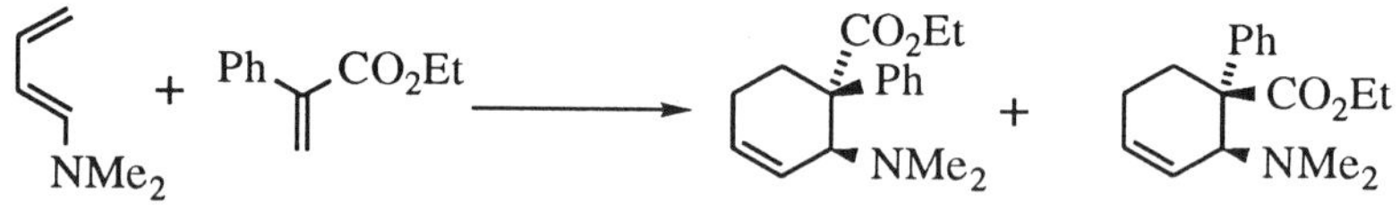

7.207 **7.208** **7.209** **7.210**

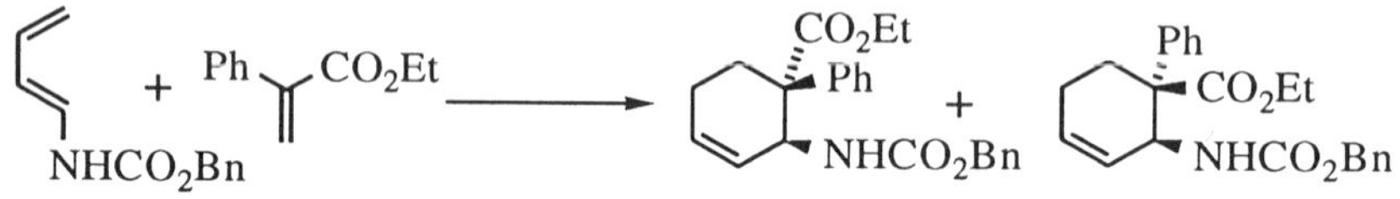

7.211 **7.208** **7.212** (71%) **7.213** (20%)

Finally, the Diels-Alder reaction of dihydropyridine ***7.214*** and methyl acrylate gave cycloadduct ***7.215***.[124] Subsequent treatment with base led to ***7.216***. Osmylation gave a mixture of diastereomers of methyl 3,4-dihydroxy-5-(N-Boc aminomethyl)-1-cyclohexene-1-carboxylate, ***7.217*** and ***7.218***.

7.3. From Acyclic Precursors

The Diels-Alder is an excellent example of forming rings via reaction of two acyclic precursors. Other cyclization reactions generate rings and will be discussed in this section.

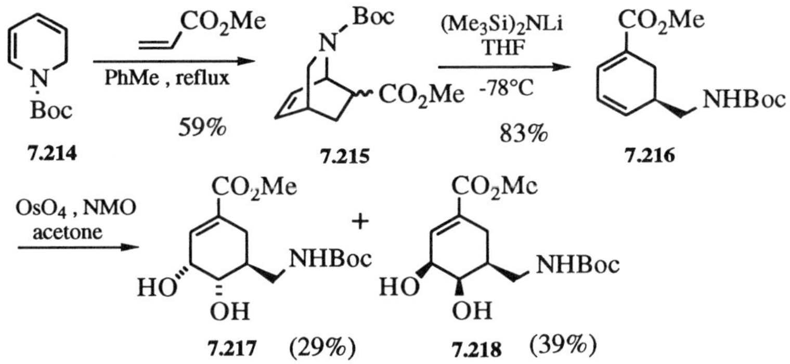

7.3.A. Via Thermal [2+2]-Cycloaddition Reactions

A particularly good route to four-membered rings is via a [2+2]-cycloaddition. With a suitably functionalized precursor, this is a good method for preparing amino-cyclobutane amino acids. Reaction of enamine ***7.219*** (derived from isobutyraldehyde)

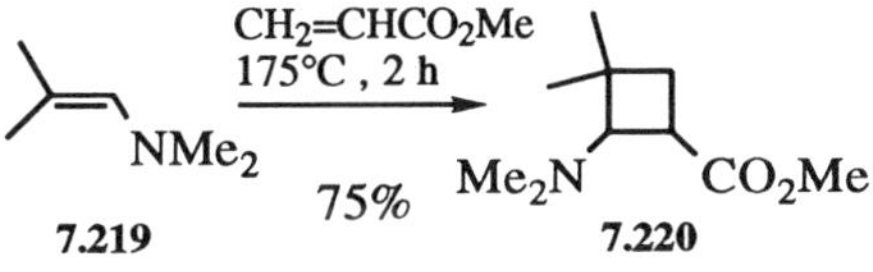

and methyl acrylate at 175°C, for example, gave methyl 2-(N,N-dimethylamino)-3,3-dimethylcyclobutane-1-carboxylate, ***7.220.***[125] Various N,N-dialkyl derivatives were prepared by using different enamines of isobutyraldehyde; in addition to ***7.220***, the 3-ethyl, 3-phenyl-, and 3,3-diethyl- derivatives were prepared.[125a] There is also an example in which ***7.221*** was converted to methyl 2-piperidino[3,5]spirononane-1-carboxylate, ***7.222.***[125a] This example is shown because it generates an interesting spirocyclic amino acid.

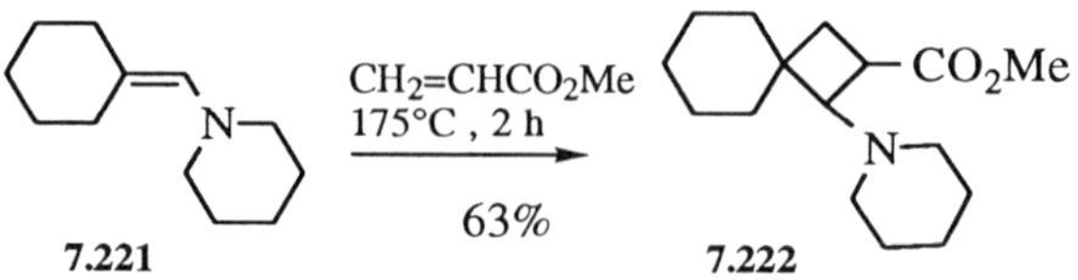

7.221 7.222

7.3.B. Other Acyclic Precursors

Cyclopropane derivatives can be prepared by several methods. Michael addition of the enolate of ethyl chloroacetate to ethyl acrylate generated the cyclopropane ring in ***7.223*** via addition to form a carbanion and internal expulsion of the chlorine moiety[126] Manipulation of functional groups allowed selective reduction to ***7.224*** and conversion to ***7.225*** (as a mixture of *cis*- and *trans*-isomers). Rearrangement and hydrolysis led to *cis*-2-(2-amino-1-cyclopropyl)ethanoic acid, ***7.226.*** The analogous cyclobutane derivatives were also prepared by a similar route.

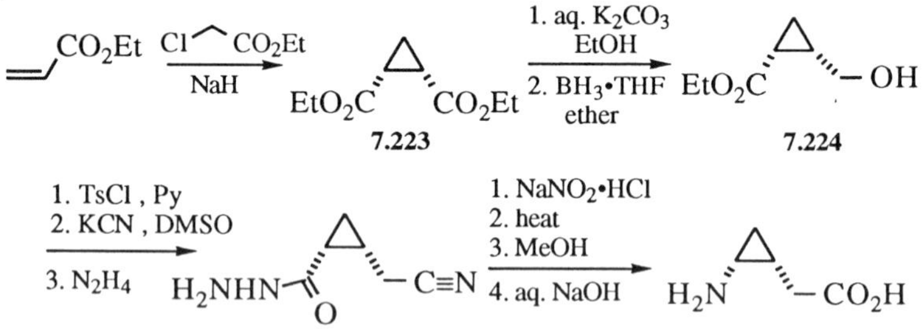

A different route involved rhodium catalyzed addition of the carbene derived from methyl α-diazoacetate to ***7.227*** to give ***7.228*** (as a 34:66 *cis:trans* mixture).[127]

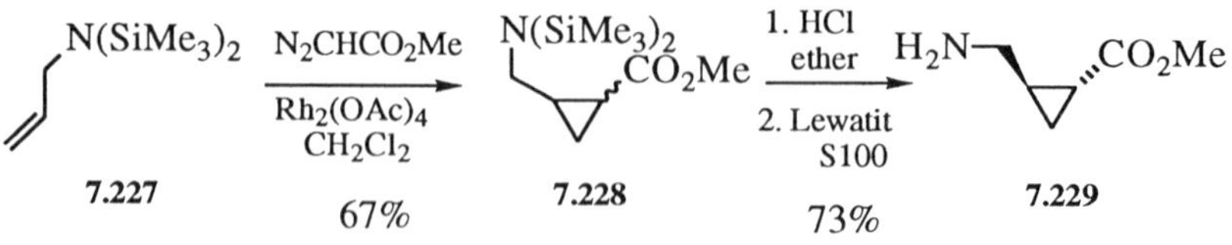

7.227 7.228 7.229

Reaction with HCl gave the amine and separation of diastereomers on Lewatit S-100 gave *trans*-methyl-2-aminomethyl-1-cyclopropanecarboxylate, ***7.229***.

A statine mimic (see chapter six, section 6.3) was prepared by conversion of aldehyde ***7.230*** to ***7.231*** (as a 5:1 mixture of diastereomers). Protection as the oxazolidine gave ***7.232*** for an overall yield of 50%.[128] Addition of the carbene derived from *t*-butyl diazo-acetate gave ***7.233*** and deprotection gave a mixture of ***7.234*** and ***7.235*** [*t*-butyl 2-(2-amino-1-hydroxy-3-cyclohexylpropyl)-3,3-dimethyl-cyclopropane-1-carboxylates].[128]

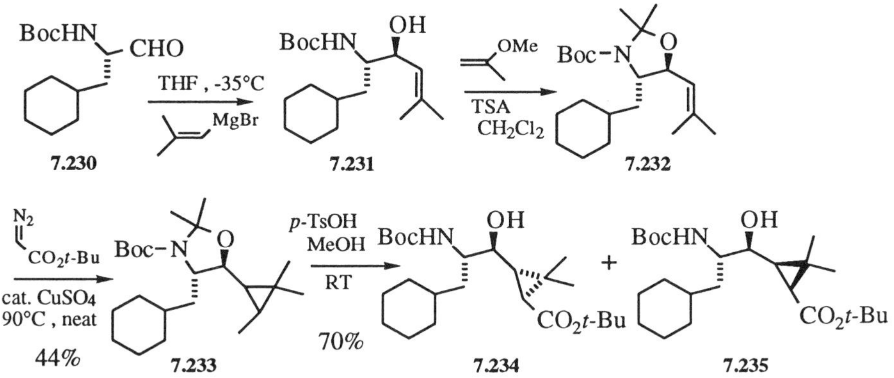

7.4. Cyclic Heteroatom Amino Acids

The compounds in this section are similar to the others in chapter seven, but a nitrogen, oxygen or a sulfur is incorporated in the ring. Since most cases involve a nitrogen, however, the final products are usually substituted aziridines, azetidines, pyrrolidines, piperidines, or hexahydroazepines with a carboxyl group pendant to the ring. Many synthetic strategies are unique to this class of amino acids.

7.4.A. Aziridine Carboxylic Acids

Treatment of alkyne ***7.236*** with sodium azide gave ***7.237***, and heating that azide in toluene led to formation of ***7.238***.[129] Reduction of the azirine moiety in ***7.238*** with sodium borohydride gave ethyl 3-heptylaziridine-2-carboxylate, ***7.239***. Alternatively, hydrolysis with aqueous sodium bicarbonate led to ethyl 3-heptyl-3-hydroxyaziridine-2-carboxylate, ***7.240***. A n-C_5H_{11} derivative was also prepared from the appropriate alkyne.

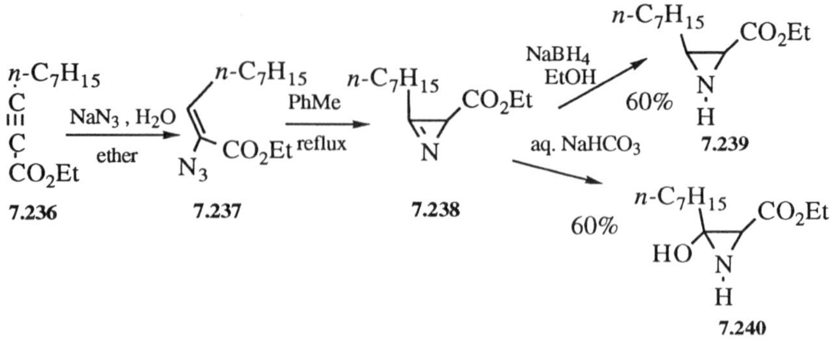

A more elaborate approach produced a chiral, non-racemic amino acid that was used in the synthesis of a statine derivative (see chapter six, section 6.3). Azide ***7.242*** was prepared from ***7.241***, and reaction with propanedithiol opened the ring to

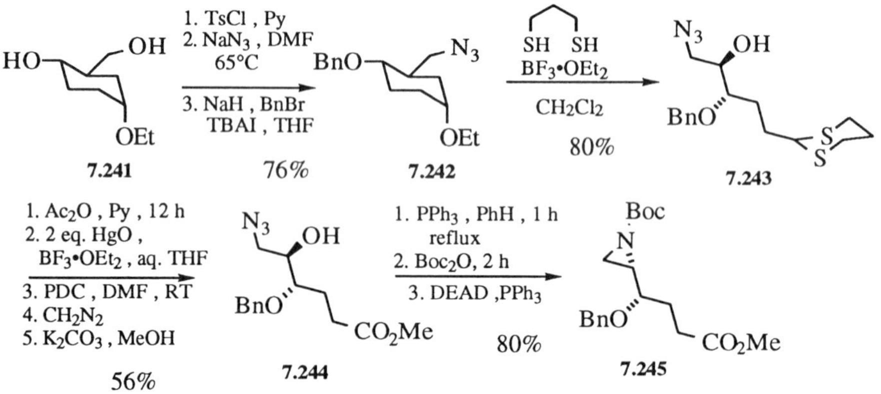

the protected aldehyde, ***7.243***.[130] Conversion of the dithiane to the aldehyde, oxidation with pyridinium dichromate in DMF, and esterification of the resulting acid moiety gave ***7.244***. The final sequence reduced the azide and generated the aziridine ring in *7.245* [methyl 4-(N-Boc-2-aziridino)-4-benzyloxybutanoate] via an internal Mitsunobu reaction.[131]

7.4.B. Azetidine Carboxylic Acids

Only one example was found where this class of amino acids was produced. Oxidation of *7.246* gave a ketone and treatment with boron trifluoride led to the conjugated ketone, *7.247*.[132] Conjugate addition of azide, reduction to the amine and cyclization gave diastereomeric methyl 9-(3-hexyl-2-aziridino)nonanoic acid (*7.248* and *7.249*). The final step in that sequence proceeded in only 37% yield. If the alkenyl moiety was converted to an epoxide moiety, aziridine carboxylic acids were prepared in good yield, via the azide.[132]

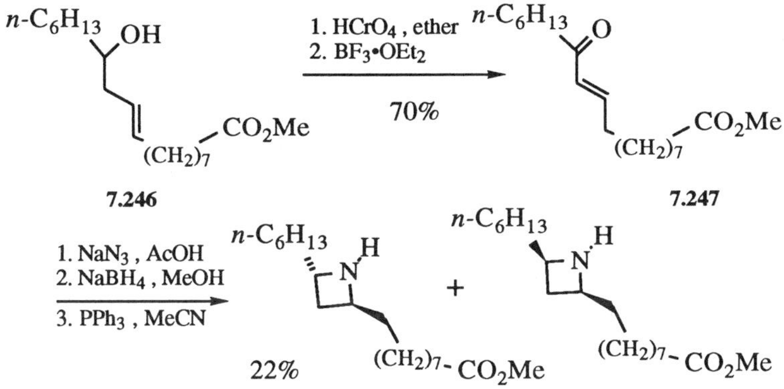

7.246 7.247 7.248 7.249

7.4.C. Pyrrolidine Carboxylic Acids

Several syntheses have appeared in which pyrrolidine carboxylic acids other than proline have been prepared. Refunctionalization of proline is an also excellent route to "non-proline" amino acids, however, because of the ready availability of proline and its derivatives. Proline (*7.250*) was converted to aldehyde *7.251* by standard procedures, and Wittig olefination[95] led to *7.252*.[133] In an interesting transformation, the alkynyl moiety in ***7.252*** was converted to an acid moiety, giving 3-(2-N-Boc pyrrolidino)-prop-2E-enoic acid (***7.253***). Alternatively, condensation of ***7.251***

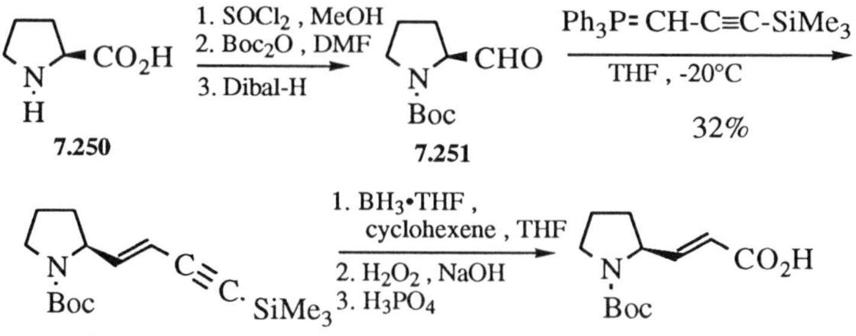

7.250 **7.251** **7.252** **7.253**

with ***7.254*** gave ***7.255*** [ethyl 2-methyl-3-hydroxy-3-(2-N-Boc pyrrolidino)propanoate], directly.[134] This was converted to a pyrrolidine amino acid[135] component of the anti-cancer agent, dolastatin 10.[136]

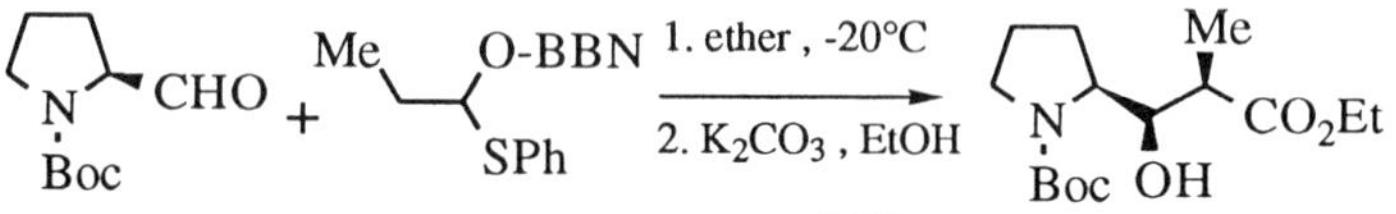

7.251 **7.254** **7.255**

Conjugated alkynyl esters are quite useful precursors when the five-membered ring is formed via an internal Michael addition. Reaction of ***7.256*** with methyl 6-aminohexanoate, for example, generated a transient acyclic derivative (***7.257***), which

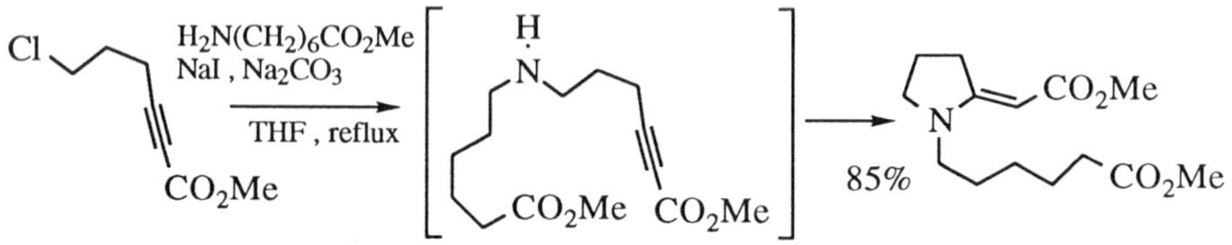

7.256 **7.257** **7.258**

cyclized to give methyl 6-(2-carbomethoxymethylene-1-pyrrolidino)hexanoate, ***7.258***.[137] Several other N-substituted derivatives were also prepared. A related synthesis converted azide ***7.259*** to amine ***7.260***, and this was cyclized to methyl 2-methyl-2-(2-pyrrolidino)ethanoate (***7.261***) as a 9:1 mixture of diastereomers.[138]

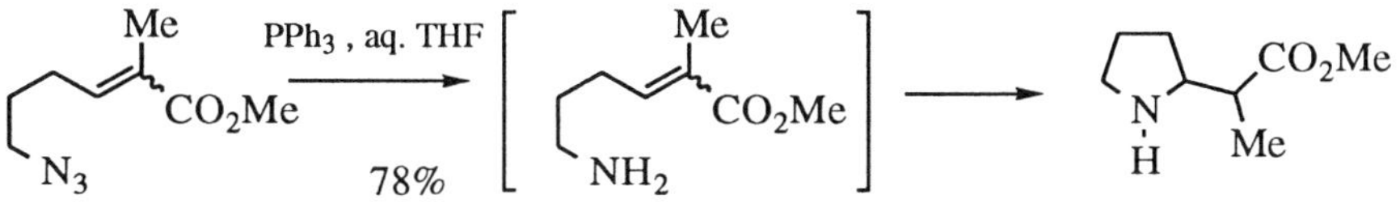

7.259 **7.260** **7.261**

A very different route used an unusual lactone (***7.262***) in a Mitsunobu[131] type transesterification reaction with methanol to give ***7.263***.[139] Reduction of the ester to an aldehyde with diisobutylaluminum hydride and Wittig olefination[95] gave methyl 3-(N-trityl-4-hydroxy-2-pyrrolidino)prop-2E-enoate, ***7.264***.

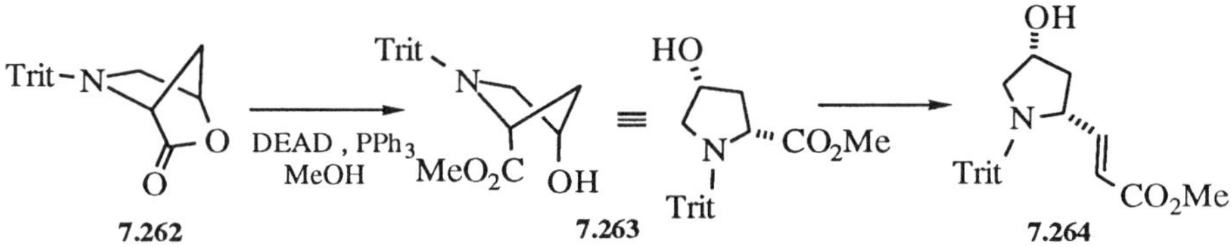

7.262 **7.263** **7.264**

Lactams can be used to prepare pyrrolidinone derivatives. The alkenyl group in β -lactam ***7.265***, for example, was oxidatively cleaved to ***7.266***, allowing ring expansion[140] to ***7.267***.[141] Several 2-aryl derivatives and N-aryl derivatives were

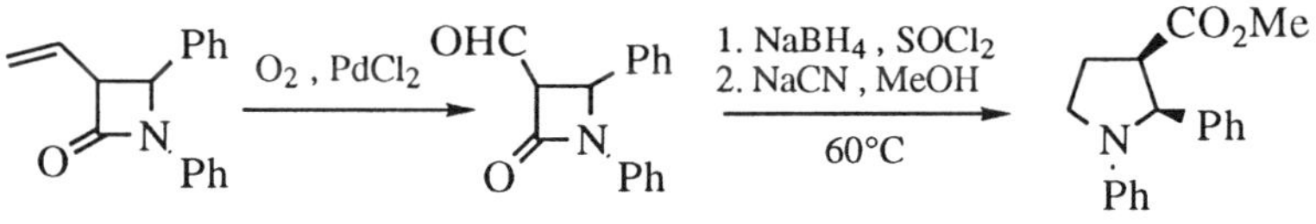

7.265 **7.266** **7.267**

also prepared by this method. Methoxypyrrolidine derivative ***7.268*** (derived from 2-pyrrolidinone) was condensed with ***7.269*** under Mukaiyama conditions[142] to give methyl 1-methyl-2-(N-carbomethoxy-2-pyrrolidino)propanoate, ***7.270*** in another synthesis of pyrrolidine derivatives.[143] It is noted that piperidine and hexahydroazepine analogs were prepared by this method.

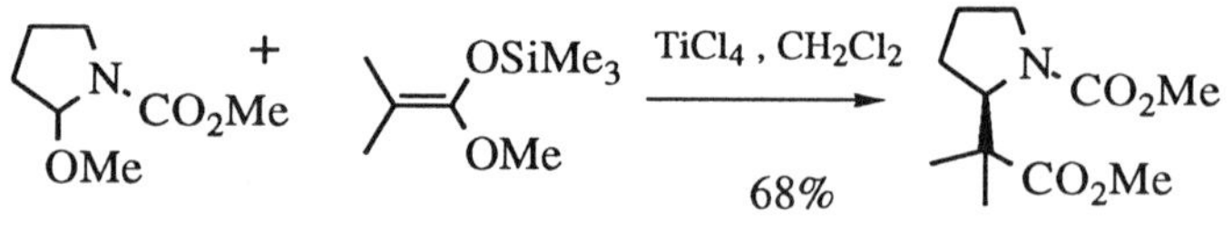

7.268 **7.269** **7.270**

In a different approach that utilized heterocyclic intermediates, amino alcohol *7.271* was protected as an oxazolidine (*7.272*), allowing manipulation of the

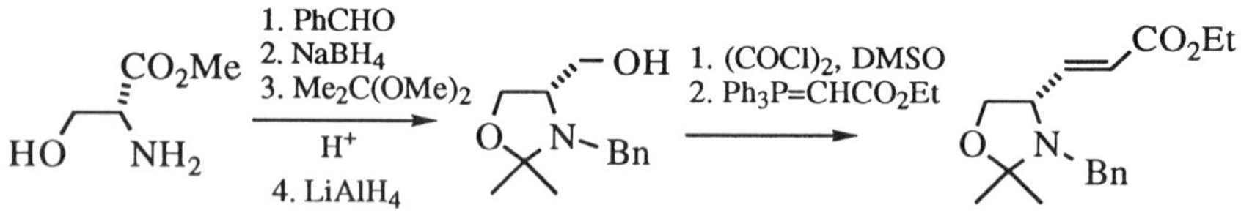

7.271 **7.272** **7.273**

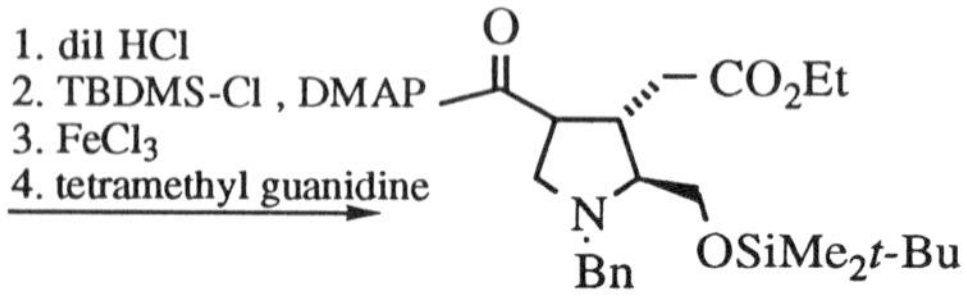

7.274

hydroxymethyl group to give the conjugated ester moiety in *7.273*.[144] Deprotection and cycloaddition led to ethyl 2-(N-benzyl-4-acetyl-2-dimethyl-*t*-butylsilyloxy-methyl-3-pyrrolidino)ethanote (*7.274*), in good yield.

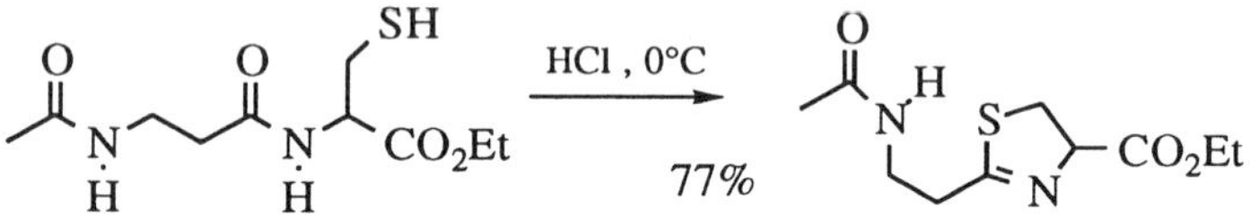

7.275 **7.276**

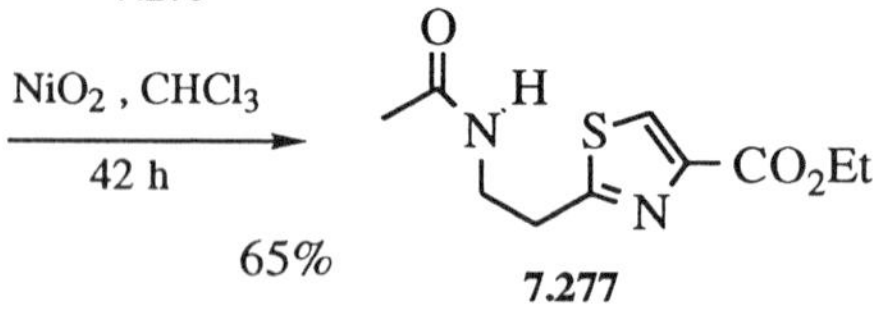

7.277

A related synthesis involved the preparation of a thiazole amino acid used in

Hecht's synthesis of bleomycin A_2.[145,146] This synthesis clearly does not involve a pyrrolidine moiety, but does contain a five-membered heterocyclic ring and is included here as an example of how the strategies presented in this section can be applied to other systems. Treatment of dipeptide ***7.275*** with HCl gave thiazoline ***7.276***.[146a] Subsequent oxidation with nickel oxide led to ethyl 2-(2-N-acetylamino-ethyl)-thiazole-5-carboxylate, ***7.227***.

7.4.D. Piperidine Carboxylic Acids

There are many examples of piperidine carboxylic acids, prepared by a variety of methods. Reduction of azide ***7.278*** led to an amine, and subsequent internal Michael addition to the conjugated ester gave methyl 2-(2-piperidino)ethanoate, ***7.279***.[138]

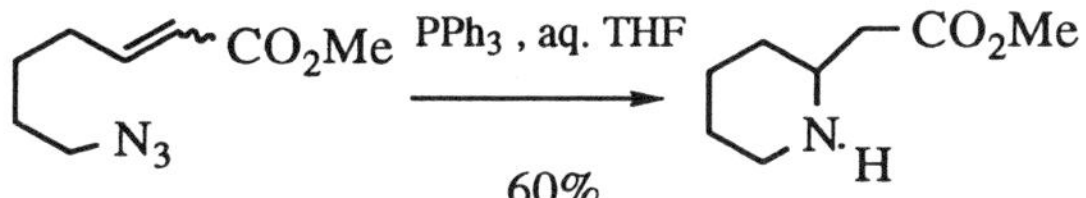

7.278 **7.279**

A more involved synthesis prepared ***7.280***. using a "double" Michael sequence, and then changed the N-protecting group to a carbamate (see ***7.281***).[147] Dieckmann

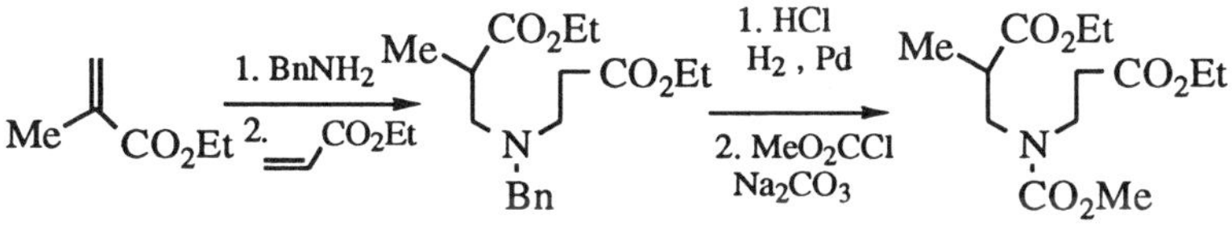

7.280 **7.281**

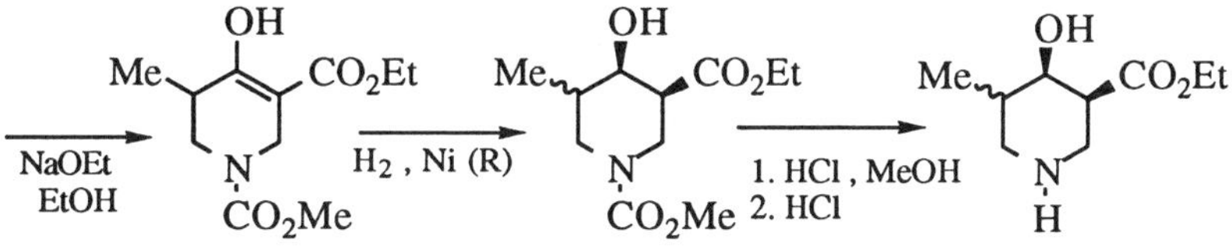

7.282 **7.283** **7.284**

cyclization[148] gave ***7.282***,[149] which was hydrogenated to give ***7.283***. Acid hydrolysis of the carbamate moiety gave ethyl 4-hydroxy-5-methylpiperidine-3-carboxylate, ***7.284***. A related route converted ***7.285*** to ***7.286***. Dieckmann cyclization and reduction gave ***7.287***.[150] Other 6-alkyl and 6-aryl derivatives were prepared by the method shown for the synthesis of the 2-ethyl derivative (***7.287***).

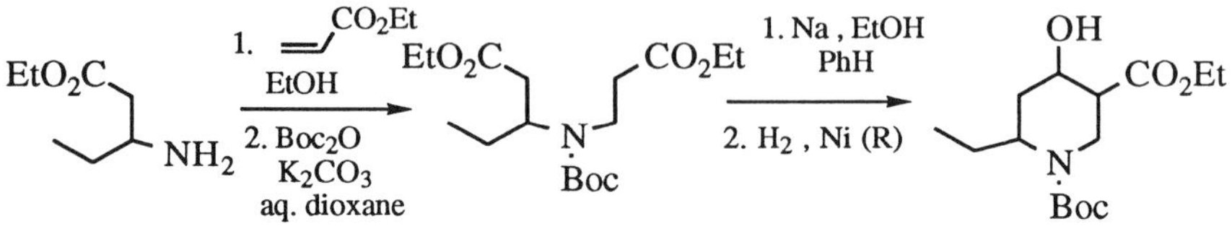

7.285 **7.286** **7.287**

Another synthetic route used hydropiperidone ***7.288*** as a precursor. Catalytic hydrogenation of the ketone moiety in ***7.288***, followed by transesterification gave methyl ester ***7.289***.[151]. Acid hydrolysis of the carbamate and ester moieties gave 3-hydroxypiperidine-4-carboxylic acid, ***7.290***.[152]

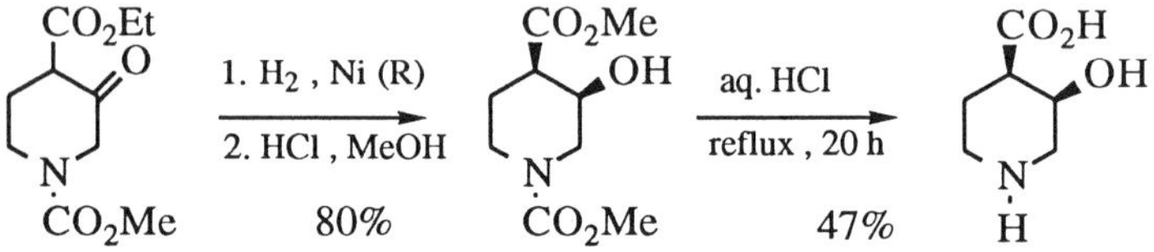

7.288 **7.289** **7.290**

An interesting synthesis proceeded by shrinking a seven-membered ring, specifically, α-bromocaprolactam (***7.291***) to 3,4,5,6-tetrahydropyridine-2-carboxylic acid, ***7.292***.[153] Both the five (5H 3,4-dihydropyrrole-2-carboxylic acid) and eight-membered ring (3,4,5,6,7,8-hexahydroazacine-2-carboxylic acid) analogs of ***7.292*** were prepared from α-bromopiperidone and α-bromoazacyclooctan-2-one, respectively.

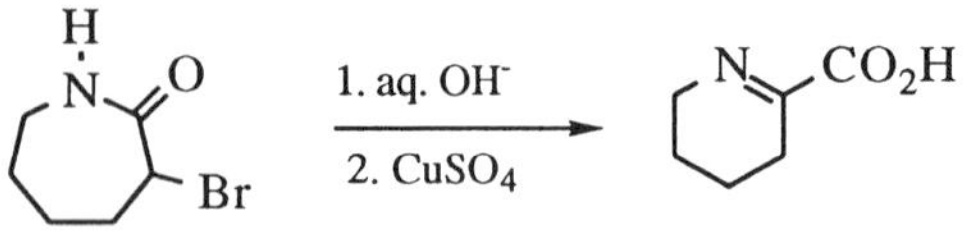

7.291 **7.292**

7.4.E. Hexahydroazepine Carboxylic Acids

The conversion of ***7.288*** to ***7.290*** can be modified to produce seven-membered rings. The six membered ring in amino-ketone ***7.293***, prepared from the N-benzyl derivative,[154] was expanded by reaction with diazoacetate and rearrangement to give the seven-membered ring amino ketone, ***7.294***.[154] Reduction, elimination and catalytic hydrogenation gave hexahydroazepine-4-carboxylic acid, ***7.295.*** The latter

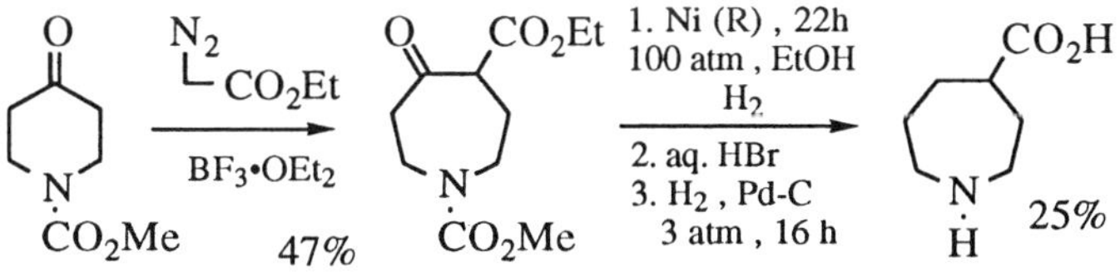

7.293 **7.294** **7.295**

stages of this synthesis were used in a closely related transformation in which a carbamoyl protected amino ketone (***7.296***)[155] was converted to hexahydroazepine-3-carboxylic acid (***7.297***).[155]

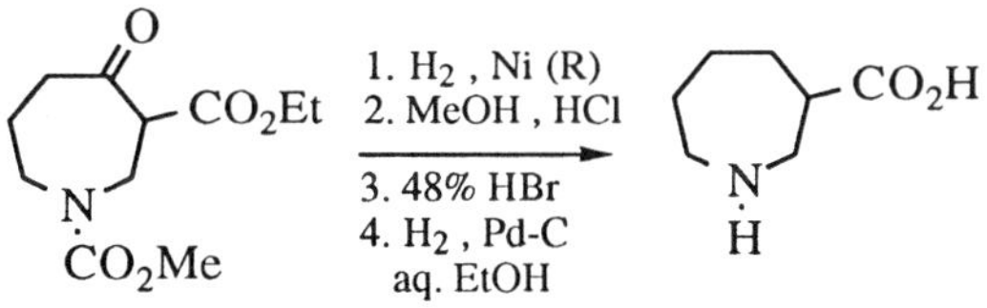

7.296 **7.297**

REFERENCES

1 Šafanda, J.; Sobotka, P. *Collect. Czech. Chem. Commun.*, ***1982***, *47*, 2440.

2 Kennewell, P.D.; Matharu, S.S.; Taylor, J.B.; Westwood, R.; Sammes, P.G. *J. Chem. Soc., Perkin Trans. I*, ***1982***, 2553, 2563.

3 Avotins, F.; Gilis, A.; Gudriniece, E.; Spince, B. *Latv. PSR Zinat. Akad. Vestis, Kim. Ser.*, ***1984***, 339 [*Chem. Abstr.*, ***1984***, *101*: 191182x].

4 (a) Beckmann, E. *Ber.*, ***1886***, *19*, 988; (b) Donaruma, L.G.; Heldt, W.Z. *Org. React.*, ***1960***, *11*, 1; (c) Mundy, B.P.; Ellerd, M.G. *Name Reactions and*

Reagents in Organic Synthesis, Wiley, New York, ***1988***, pp. 24-25.
5 Murahashi, S.; Taniguchi, Y.; Imada, Y.; Tanigawa, Y. *J. Org. Chem.* , ***1989***, *54*, 3292.
6 Inoue, G.; Oshima, N. *Jpn. Patent* 68 26,502 [*Chem. Abstr.*, ***1969*** *70*: P67756q].
7 Nazarov, I.N.; Kucherov, V.F. *Izvest. Akad. Nauk. SSSR, Ser. Khim.*, ***1954***, 63.
8 (a) Curtius, T. *J. Prakt. Chem.*, ***1894***, *50*, 275; (b) Smith, P.A.S. *Org. React.*, ***1946***, *3*, 337; (c) Saunders, J.H.; Slocombe, R.J. *Chem. Rev.*, ***1948***, *43*, 203.
9 Armarego, W.L.F. *J. Chem. Soc., C*, ***1971***, 1812.
10 Armitage, B.J.; Kenner, G.W.; Robinson, M.J.T. *Tetrahedron*, ***1964***, *20*, 723, 747.
11 Sicher, J.; Šipoš, F.; Jonáš, J. *Coll. Czech. Chem. Commun.*, ***1961***, *26*, 262.
12 (a) Hofmann, A.W. *Ber.*, ***1881***, *14*, 2725; (b) Wallis, E.S.; Lane, J.F. *Org. React.*, ***1949***, *3*, 267; (c) Mundy, B.P.; Ellerd, M.G. *Name Reactions and Reagents in Organic Synthesis*, Wiley, New York, ***1988***, pp. 112-113.
13 Bernáth, G.; Göndös, Gy.; Kovács, K.; Sohár, P. *Tetrahedron*, ***1973***, *29*, 981.
14 Bernáth, G.; Kovács, K.; Láng, K.L. *Acta Chim. Budapest*, ***1970***, *64*, 183 [*Chem. Abstr.*, ***1970*** *73*: 34856f].
15 Bonnaud, B.; Cousse, H.; Mouzin, G.; Briley, M.; Stenger, A.; Fauran, F.; Couzinier, J.-P. *J. Med. Chem.*, ***1987***, *30*, 318.
16 Chênevert, R.; Martin, R. *Tetrahedron Asymmetry*, ***1992***, *3*, 199.
17 Roberts, J.D.; Lee, C.C.; Saunders, W.H. *J. Am. Chem. Soc.*, ***1954***, *76*, 4501.
18 Berger, H.; Paul, H.; Hilgetag, G. *Chem. Ber.*, ***1968***, *101*, 1525.
19 Birch, S.F.; Dean, R.A. *J. Chem. Soc.*, ***1953***, 2477.
20 Also see Allan, R.D.; Johnston, G.A.R.; Twitchin, B. *Aust. J. Chem.*, ***1979***, *32*, 2517.
21 (a) Connors, T.A.; Ross, W.C.J. *J. Chem. Soc.*, ***1960***, 2119; (b) Allan, R.D.; Dickenson, H.W.; Fong, J. *Eur. J. Pharmacol.*, ***1986***, *122*, 339.
22 Allan, R.D.; Fong, J. *Aust. J. Chem.*, ***1986***, *39*, 855.
23 (a) Wessjohann, L.; Krass, N.; Yu, D.; de Meijere, A. *Chem. Ber.*, ***1992***, *125*, 867; (b) de Meijere in *New Aspects of Organic Chemistry II, Proceedings of the Fifth International Kyoto Conference on New Aspects of Organic Chemistry* (Yoshida, Z.; Ohshiro, Y., Eds), Kodansha/CVH, Tokyo/ Weinheim, ***1992***, pp. 181-213.

24 Wessjohann, L.; McGaffin, G.; de Meijere, A. *Synthesis*, ***1989***, 359.

25 Es-Sayed, M.; Gratkowski, C.; Krass, N.; Meyers, A.I.; de Meijere, A. *Tetrahedron Lett.*, ***1993***, *34*, 289.

26 Allan, R.D.; Johnston, G.A.R.; Twitchin, B. *Aust. J. Chem.*, ***1981***, *34*, 2231.

27 Allan, R.D.; Twitchin, B. *Aust. J. Chem.*, ***1980***, *33*, 599.

28 Satzinger, G.; Herrmann, M. *Ger. Offen.* 2,166,019 [*Chem. Abstr.*, ***1973***, *78*: P3822r].

29 (a) Magid, R.M. *Tetrahedron*, ***1980***, *36*, 1901; (b) Bordwell, F.G.; Pagani, G.A. *J. Am. Chem. Soc.*, ***1975***, *97*, 118; (c) Bordwell, F.G.; Mecca, T.G. *Ibid*, ***1975***, *97*, 123, 127; (d) Bordwell, F.G.; Wiley, P.F.; Mecca, T.T.G. *Ibid*, ***1975***, *97*, 132.

30 Satzinger, G.; Herrmann, M. *Ger. Offen.* 2,107,871 [*Chem. Abstr.*, ***1972***, *77*: P151688w].

31 Asahi, Chemical Industry Co., Ltd. *Jpn.* 100 ('67) [*Chem. Abstr.*, ***1967***, *66*: P55105f].

32 Singer, S.P.; Sharpless, K.B. *J. Org. Chem.*, ***1978***, *43*, 1448.

33 Kobayashi, K.; Miyazawa, S.; Terahara, A.; Mishima, H.; Kurihara, H. *Tetrahedron Lett.*, ***1976***, 537.

34 Roberts, E. *Biochem. Pharm.*, ***1974***, *23*, 2637.

35 (a) Kato, M.; Kageyama, M.; Tanaka, R.; Kuwahara, K.; Yoshikoshi, A. *J. Org. Chem.*, ***1975***, *40*, 1932; (b) Trost, B.M.; Timko, J.M.; Stanton, J.L. *J. Chem. Soc., Chem. Commun.*, ***1978***, 436.

36 Trost, B.M.; Keinan, E. *J. Org. Chem.*, ***1979***, *44*, 3451.

37 (a) Kricheldorf, H.R. *Chem. Ber.*, ***1972***, *105*, 3958; (b) Washburne, S.S.; Peterson, W.R. Jr.; Berman, D.A. *J. Org. Chem.*, ***1972***, *37*, 1738; (c) Kricheldorf, H.R. *Makromol. Chem.*, ***1973***, *173*, 13.

38 Kricheldorf, H.R. *J.L. Ann. Chem.*, ***1975***, 1387.

39 (a) Kunisch, F.; Babczinski, P.; Arlt, D.; Brandes, W.; Dehne, H.W.; Dutzmann, S.; Plempel, M. *Ger. Offen.* DE 4,033,415 [*Chem. Abstr.*, ***1992***, *117*: P69489t]; (b) Kunisch, F.; Babczinski, P.; Arlt, D.; Plempel, M. *Ger. Offen.*, DE 4,028,046 [*Chem. Abstr.*, ***1992***, *117*: P20486a].

40 Daluges, S.; Vince, R. *J. Org. Chem.*, ***1978***, *43*, 2311.

41 Kam, B.L.; Oppenheimer, N.J. *J. Org. Chem.*, ***1981***, *46*, 3268.

42 Klarer, W.J.; Hiemstra, H.; Speckamp, W.N. *Tetrahedron*, ***1988***, *44*,

6729.
43 Stork, G.; Landesman, H.K. *J. Am. Chem. Soc.*, ***1956***, *78*, 5129.
44 Plostnieks, J. *J. Org. Chem.*, ***1966***, *31*, 634.
45 Nativ, E.; Rona, P. *Isr. J. Chem.*, ***1972***, *10*, 55.
46 Konishi, M.; Nishio, M.; Saitoh, K.; Miyaki, T.; Oki, T.; Kawaguchi, H. *J. Antibiot.*, ***1989***, *42*, 1749.
47 (a) Berge, J.M.; Roberts, S.M.; Suschitzky, H.; Kemp, J.E.G. *J. Chem. Soc., S*, ***1978***, 255; (b) Evans, D.A.; Biller, S.A. *Tetrahedron Lett.*, ***1985***, *26*, 1907.
48 Evans, C.; McCague, R.; Roberts, S.M.; Sutherland, A.G.; Wisdom, R. *J. Chem. Soc., Perkin Trans. I*, ***1991***, 2276.
49 (a) Iwamoto, T.; Tsujii, E.; Ezaki, M.; Fujie, A.; Hashimoto, S.; Okuhara, M.; Kohsaka, M.; Imanaka, H.; Kawabata, K.; Inamoto, Y.; Sakane, K. *J. Antibiot.*, ***1990***, *43*, 1; (b) Kawabata, K.; Inamoto, Y.; Sakane, K.; Iwamoto, T.; Hashimoto, S. *Ibid*, ***1990***, *43*, 513.
50 Arbuzov, B.A.; Zobova, N.N. *Dokl. Akad. Nauk. SSSR*, ***1967***, *172*, 845 (Engl., p. 103).
51 Moriconi, E.J.; Kelly, J.F. *J. Org. Chem.*, ***1968***, *33*, 3036.
52 Hassner, A. *Accts. Chem. Res.*, ***1971***, *4*, 9.
53 Litvin. E.F.; Freidlin, L.Kh.; Gurskii, R.N.; Istratova, R.V.; Presnov, A.P. *Izv. Akad. Nauk. SSSR, Ser. Khim*, ***1979***, 2441 (Engl., p. 2258).
54 (a) Liska, K.J. *J. Pharm. Sci.*, ***1964***, *53*, 1427; (b) Smissman, E.E.; Steinman, M. *J. Med. Chem.*, ***1967***, *10*, 1054.
55 Palaima, A.I.; Poskiene, R.A.; Karpavicius, K.I.; Kil'disheva, O.V.; Knunyants, I.L. *Izv. Akad. Nauk. SSSR, Ser. Khim.*, ***1977***, 195 (Engl., p. 171).
56 Rhone-Poulenc *Fr. Patent* 1,473,246 [*Chem. Abstr.*, ***1968***, *68*: P12555b].
57 Vechkanov, G.N.; Zhizdyuk, B.I.; Malukh, V.A.; Chegolya, A.S. *Sin. Volokna*, ***1969***, 21 [*Chem. Abstr.*, ***1970***, *73*: 66100w].
58 Schneider, W.; Hoyer, J. *Archiv. Pharm. (Weinheim)*, ***1971***, *304*, 637.
59 Škarić, V.; Kovačević, M.; Škarić, D. *J. Chem. Soc., Perkin Trans. I*, ***1976***, 1199.
60 Freifelder, M.; Stone, G.R. *J. Org. Chem.*, ***1962***, *27*, 3568.
61 Ponomarev, A.A.; Ryzhenko, L.M.; Smirnova, N.S. *Zh. Org. Khim*, ***1969***, *5*, 75 (Engl, p. 73).
62 Schneider, W.; Hüttermann, A. *Archiv der Pharm. (Weinheim)*, ***1965***, *298*, 226.

63 Hünig, S.; Kahanek, H. *Chem. Ber.*, ***1953***, *86*, 518.
64 Einhorn, A.; Meyenberg, A. *Deutsch. Ges. Ber.*, ***1894***, *27*, 2466.
65 Prelog, V.; Szpilfogel, S. *Helv. Chim. Acta*, ***1945***, *28*, 1684.
66 Dieckmann, W. *J. L. Ann. Chem.*, ***1901***, *317*, 27.
67 (a) Durbeck, H.W.; Duttka, L.L. *Tetrahedron*, ***1973***, *29*, 4285; (b) Takaya, T.; Yoshimoto, H.; Imoto, E. *Bull. Chem. Soc., Jpn.*, ***1967***, 40, 2844.
68 Takaya, T.; Yoshimoto, H.; Imoto, E. *Bull. Chem. Soc., Jpn.*, ***1968***, *41*, 2176.
69 Birkofer, L.; Ritter, A.; Richter, P. *Chem. Ber.*, ***1963***, *96*, 2750.
70 Kloek, J.A.; Leschinsky, K.L. *J. Org. Chem.*, ***1978***, *43*, 1460.
71 Takaya, T.; Yoshimoto, H.; Imoto, E. *Bull. Chem. Soc., Jpn.*, ***1967***, 40, 2844.
72 Buckler, R.T.; Hartzler, H.E. *J. Med. Chem.*, ***1975***, *18*, 509.
73 Pennington, F.C.; Kehret, W.D. *J. Org. Chem.*, ***1967***, *32*, 2034.
74 Triebs, W.; Mayer, R.; Madejski, M. *Chem. Ber.*, ***1954***, *87*, 356.
75 Horii, Z.; Watanabe, T.; Ikeda, M.; Tamura, Y. *Yakugaku Zasshi*, ***1963***, *83*, 930 [*Chem. Abstr.*, ***1964***, *60*: 4106d].
76 (a) Lossen, W. *Ann.*, ***1872***, *161*, 347; (b) *Ibid*, ***1874***, *175*, 271, 313; (c) Popp, F.O.; McEwen, W.E. *Chem. Rev.*, ***1958***, *58*, 321 (see p. 374), and references cited therein.
77 Mills, W.H.; Bain, A.M. *J. Chem. Soc.*, ***1910***, *97*, 1866.
78 Schneider, W.; Hüttermann, A. *Archiv. der Pharm. (Weinheim)*, ***1965***, *298*, 226.
79 Borch, R.F.; Ho, B.C. *J. Org.Chem.*, ***1977***, *42*, 1225.
80 Omar, F.; Frahm, A.W. *Archiv. Pharm. (Weinheim)*, ***1989***, *322*, 461.
81 Stork, G.; Brizzolara, A.; Landesman, H.; Szmuszkovicz, J.; Terrell, R. *J. Am. Chem. Soc.*, ***1963***, *85*, 207.
82 Bégué, J.-P.; Fétizon, M. *C.R. Acad. Sci.*, ***1965***, *260*, 3425.
83 Musso, H.; Naumann, K.; Grychtol, K. *Chem. Ber.*, ***1967***, *100*, 3614.
84 Rosenmund, P.; Haase, W.H.; Bauer, J.; Frische, R. *Chem Ber.*, ***1973***, *106*, 1459.
85 Kasturi, T.R.; Srinivasan, A. *Tetrahedron*, ***1966***, *22*, 2575.
86 (a) Noyes, W.A. *J. Am. Chem. Soc.*, ***1894***, *16*, 500; (b) Königs, W.; Hörlin, J. *Ber. Deutsch. Chem. Ges.*, ***1893***, *26*, 811 (see p. 817).
87 (a) Faigle, H.; Karrer, P. *Helv. Chim. Acta.*, ***1962***, *45*, 73; (b) Gassman, P.G.; Cryberg, R.L. *J. Am. Chem. Soc.*, ***1969***, *91*, 2047.

88 Baumgarten, H.E.; Smith, H.L.; Staklis, A. *J. Org Chem.*, ***1975***, *40*, 3554.

89 Moriconi, E.J.; Mazzocchi, P.H. *J. Org. Chem.*, ***1966***, *31*, 1372.

90 Grewe, R.; Nolte, E.; Rotzoll, R.-H. *Chem. Ber.*, ***1956***, *89*, 600.

91 Takeda, M.; Inoue, H.; Noguchi, K.; Honma, Y.; Kawamori, M.; Tsukamoto, G.; Yamawaki, Y.; Saito, S. *Chem. Pharm. Bull.*, ***1976***, *24*, 1514.

92 Mertin, A.; Thiemann, T.; Hanss, I.; de Meijere, A. *Synlett*, ***1991***, 87.

93 (a) Mukaiyama, T. *Org. React.*, ***1982***, *28*, 203; (b) Mukaiyama, T. *Angew. Chem. Int. Ed., Engl.*, ***1977***, *16*, 817.

94 Ladouceur, G.; Mais, D.E.; Jakubowski, J.A.; Utterback, B.G.; Robertson, D.W. *Bioorg. & Medicinal Chem. Lett.*, ***1991***, *1*, 173.

95 (a) Wittig, G.; Rieber, M. *Ann.*, ***1949***, *562*, 187; (b) Wittig, G.; Geissler, G. *Ibid*, ***1953***, *580*, 44; (c) Wittig, G.; Schöllkopf, U. *Chem. Ber.*, ***1954***, *87*, 1318; (d) Gensler, W.J. *Chem. Rev.*, ***1957***, *57*, 191 (see p. 218).

96 Kamata, S.; Haga, N.; Tsuchima, T.; Kawada, K.; Kishi, M. *Jpn. Kokai Tokkyo Koho* JP 63,277,655 [*Chem. Abstr.*, ***1989***,*111*: P7134r].

97 Bell, K.H.; Fullick, G. *Synth. Commun.*, ***1987***, *17*, 1965.

98 Pansegrau, P.D.; Anderson, K.S.; Widlanski, T.; Ream, J.E.; Sammons, R.D.; Sikorski, J.A.; Knowles, J.R. *Tetrahedron Lett.*, ***1991***, *32*, 2589.

99 Desimoni, G.; Tacconi, G.; Barco, A.; Pollini, G.P. *Natural Product Synthesis through Pericyclic Reactions*, Amer. Chem. Soc., Washington, ***1983***.

100 Jagt, J.C.; van Leusen, A.M. *J. Org. Chem.*, ***1974***, *39*, 564.

101 Adriaens, P.; Meesschaert, B.; Janssen, G.; Dumon, L.; Eyssen, H. *Recuil. Trav. Chim., Pays-Bas*, ***1978***, *97*, 260.

102 also see Daluges, S.; Vince, R. *Ibid*, ***1978***, *43*, 2311.

103 Ikbal, M.; Cerceau, C.; LeGoffic, F.; Sicsic, S. *Eur. J. Med. Chem.*, ***1989***, *24*, 415 .

104 Evans, C.T.; Roberts, S.M. *Eur. Pat. Appl.* EP 424,064 [*Chem. Abstr.*, ***1991***, *115*: P49382h].

105 (a) Kaufmann, Sr.; Rosenkranz, G.; López, J. *J. Am. Chem. Soc.*, ***1946***, *68*, 2733; (b) Wallis, E.S.; Nagel, S.C. *Ibid*, ***1931***, *53*, 2787.

106 Plieninger, H.; Schneider, K. *Chem. Ber.*, ***1959***, *92*, 1594.

107 (a) Bernáth, G.; Stájer, G.; Szabó, A.E.; Fülöp, F.; Sohár, P. *Tetrahedron*, ***1985***, *41*, 1353; (b) Bernáth, G.; Kovács,K.; Láng, K.L. *Acta Chim. Acad. Sci. Hung.*, ***1970***, *65*, 347 [*Chem. Abstr.*, ***1971***, *74*: 41680t].

108 Tamura, N.; Natsugari, H.; Kawano, Y.; Matsushita, Y.; Yoshioka, K.; Ochiai, M. *Chem. Pharm. Bull*, ***1987***, *35*, 996.

109 Nohira, H.; Kurokawa, M. *Yuki Gosei Kagaku Kyokai Shi*, ***1976***, *34*, 55 [*Chem. Abstr.*, ***1977***, *86*: 16325w].

110 Armarego, W.L.F.; Kobayashi, T. *J. Chem. Soc., C*, ***1969***, 1635.

111 Goodridge, R.J.; Hambley, T.W.; Ridley, D.D. *Aust. J. Chem.*, ***1986***, *39*, 591.

112 Orndorff, W.R.; Pratt, D.S. *J. Am. Chem. Soc.*, ***1912***, *34*, 89.

113 Bauer, L.; Miarka, S.V. *J. Org. Chem.*, ***1959***, *24*,, 1293.

114 (a) Miller, L.E.; Staley, H.B.; Mann, D.J. *J. Am. Chem. Soc.*, ***1949***, *71*, 374; (b) Miller, L.E.; Mann, D.J. *Ibid*, ***1950***, *72*, 1484; (c) *Ibid*, ***1951***, *73*, 45.

115 Satzinger, G. *Liebigs Ann. Chem.*, ***1972***, *758*, 43.

116 (a) Bruck, P.R.; Clark, R.D.; Davidson, R.S.; Günther, W.H.H.; Littlewood, P.S.; Lythgoe, B. *J. Chem. Soc., C*, ***1967***, 2529; (b) Davidson, R.S.; Littlewood, P.S.; Medcalfe, T.; Waddington-Feather, S.M.; Williams, D.H.; Lythgoe, B. *Tetrahedron Lett.*, ***1963***, 1413.

117 Kobayashi, S.; Kamiyama, K.; Iimori, T.; Ohno, M. *Tetrahedron Lett.*, ***1984***, *24*, 2557.

118 Kurihara, M.; Kamiyama, K.; Kobayashi, S.; Ohno, M. *Tetrahedron Lett.*, ***1985***, *26*, 5831.

119 Huckin, S.N.; Weiler, L. *Can. J. Chem.*, ***1974***, *52*, 1343.

120 Overman, L.E.; Taylor, G.F.; Petty, C.B.; Jessup, P.J. *J. Org. Chem.*, ***1978***, *43*, 2164.

121 McMurry, J.E.; Musser, J.H.; Fleming, I.; Fortunak, J.; Nübling, C. *Org. Synth. Coll. Vol. 6*, ***1988***, 799.

122 Danishefsky, S.; Hershenson, F.M. *J. Org. Chem.*, ***1979***, *44*, 1180.

123 (a) Overman, L.E.; Petty, C.B.; Doedens, R.J. *J. Org. Chem.*, ***1979***, *44*, 4183; (b) Warner-Lambert Pharmaceuticals Co. *Brit. Patent* 1,120,186 [*Chem. Abstr.*, ***1968***, *69*: P96062v]; (c) Satzinger, G.; Herrmann, W.; Novak, R.M. *Ger. Offen.* 1,923,619 [*Chem Abstr.*, ***1970***, *72*: P78530d]; (d) Satzinger, G. *J. L. Ann. Chem.*, ***1969***, *728*, 64.

124 Campbell, M.M.; Mahon, M.F.; Sainsbury, M.; Searle, P.A.; Davies, G.M. *Tetrahedron*, ***1992***, *48*, 8751.

125 (a) Brannock, K.C.; Bell, A.; Burpitt, R.D.; Kelly, C.A. *J. Org. Chem.*, ***1964***, *29*, 801; (b) *Ibid*, ***1961***, *26*, 625; (c) Brannock (Eastman Kodak, Co.)

Ger. Patent 1,131,207 [*Chem. Abstr.*, ***1962***, *57*: P13642d]; (d) Ficini, J.; Krief, A. *Tetrahedron Lett.*, ***1970***, 885.

126 Kennewell, P.D.; Matharu, S.S.; Taylor, J.B.; Westwood, R.; Sammes, P.G. *J. Chem. Soc., Perkin Trans. I*, ***1982***, 2553, 2563.

127 Paulini, K.; Reißig, H.-U. *Liebigs Ann. Chem.*, ***1991***, 455.

128 Melnick, M.J.; Bisaha, S.N.; Gammill, R.B. *Tetrahedron Lett.*, ***1990***, *31*, 961.

129 Haddach, M.; Pastor, R.; Riess, J.G. *Tetrahedron Lett.*, ***1990***, *31*, 1989.

130 Chakraborty, T.K.; Gangakhedkar, K.K. *Tetrahedron Lett.*, ***1991***, *32*, 1897.

131 Mitsunobu, O. *Synthesis*, ***1981***, 1.

132 Lie Ken Jie, M.S.F.; Syed-Rahmatullah, M.S.K. *J. Am. Oil. Chem. Soc.*, ***1992***, *69*, 359.

133 Miles, N.J.; Sammes, P.G.; Kennewell, P.D.; Westwood, R. *J. Chem. Soc., Perkin Trans. I*, ***1985***, 2299.

134 Tomioka, K.; Kanai, M.; Koga, K. *Tetrahedron Lett.*, ***1991***, *32*, 2395.

135 (a) Hanson, G.J.; Baran, J.S.; Lindberg, T. *Tetrahedron Lett.*, ***1986***, *27*, 3577; (b) Hamada, Y.; Shiori, T. *Chem. Pharm. Bull.*, ***1982***, *30*, 1921.

136 (a) Pettit, G.R.; Kamano, Y.; Herald, C.L.; Tuinman, A.A.; Boettner, F.E.; Kizu, H.; Schmidt, J.M.; Baczynskyj, L.; Tomer, K.B.; Bontems, R.J. *J. Am. Chem. Soc.*, ***1987***, *109*, 6883; (b) Pettit, G.R.; Singh, S.B.; Hogan, F.; Lloyd-Williams, P.; Herald, D.L.; Burkett, D.D.; Clewlow, P.J. *Ibid*, ***1989***, *111*, 5463.

137 Wilson, C.A. III; Bryson, T.A. *J. Org. Chem.*, ***1975***, *40*, 800.

138 Knouzi, N.; Vaultier, M.; Toupet, L.; Carrie, R. *Tetrahedron Lett.*, ***1987***, *28*, 1757.

139 (a) Papaioannou, D.; Stavropoulos, G.; Sivas, M.; Barlos, K.; Francis, G.W.; Aksnes, D.W.; Maartmann-Moe, K. *Acta Chem. Scand.*, ***1991***, *45*, 99; (b) Papaioannou, D.; Stavropoulos,G.; Karagiannis, K.; Francis, G.W.; Brekke, T.; Aksnes, D.W. *Ibid*, ***1990***, *44*, 243.

140 Cavanga, F.; Linkies, A.; Pietsch, H.; Reuschling, D. *Angew. Chem. Int. Ed., Int., Engl.*, ***1980***, *19*, 129.

141 Bose, A.K.; Krishnan, L.; Wagle, D.R.; Manhas, M.S. *Tetrahedron Lett.*, ***1986***, *27*, 5955.

142 (a) Mukaiyama, T. *Org. React.*, ***1982***, *28*, 203; (b) Mukaiyama, T. *Angew. Chem. Int. Ed., Engl.*, ***1977***, *16*, 817.

143 Shono, T.; Tsubata, K.; Okinaga, N. *J. Org. Chem.*, ***1984***, *49*, 1056.

144 (a) Barco, A.; Benetti, S.; Spalluto, G.; Casolari, A.; Pollini, G.P.; Zanirato, V. *J. Org. Chem.*, ***1992***, *57*, 6279; (b) Garner, P.; Min Park, J. *Ibid*, ***1987***, *52*, 2361.

145 (a) Ikekawa, T.; Iwami, F.; Hiranaka, H.; Umezawa, H. *J. Antibiot. (Tokyo)*, ***1964***, *17A*, 194; (b) Umezawa, H.; Maeda, K.; Takeuchi, T.; Okami, Y. *Ibid.*, ***1966***, *19A*, 200; (c) Umezawa, H. Suhara, Y.; Takita, T.; Maeda, K. *Ibid.*, ***1966***, *19A*, 210.

146 (a) McGowan, D.A.; Jordis, U.; Minster, D.K.; Hecht, S.M. *J. Am. Chem. Soc.*, ***1977***, *99*, 8078; (b) Arai, H.; Hagmann, W.K.; Suguna, H.; Hecht, S.M. *Ibid.*, ***1980***, *102*, 6631; (c) Levin, M.D.; Subrahamanian, K.; Katz, H.; Smith, M.B.; Burlett, D.J.; Hecht, S.M. *Ibid.*, ***1980***, *102*, 1452; (d) Hecht, S.M.; Rupprecht, K.M.; Jacobs, P.M. *Ibid.*, ***1979***, *101*, 3982; (e) Ohgi, T.; Hecht, S.M. *J. Org. Chem.*, ***1981***, *46*, 1232; (f) Pozsgay, V.; Ohgi, T.; Hecht, S.M. *Ibid.*, ***1981***, *46*, 3761; (g) Aoyagi, Y.; Suguna, H.; Murugesan, N.; Ehrenfeld, G.M.; Chang, L-H.; Ohgi, T.; Shekhani, M.S.; Kirkup, M.P.; Hecht, S.M. *J. Am. Chem. Soc.*, ***1982***, *104*, 5237; (h) Aoyagi, Y.; Katano, K.; Suguna, H.; Primeau, J.; Chang, L.-H.; Hecht, S.M. *Ibid.*, ***1982***, *104*, 5537.

147 Krogsgaard-Larsen, P.; Thyssen, K.; Schaumburg, K. *Acta Chem. Scand.*, ***1978***, *B32*, 327.

148 (a) Dieckmann, W. *Ber.*, ***1894***, *27*, 102, 965; (b) *Ibid*, ***1900***, *33*,2670; (c) Dieckmann, W.; Groenveld, A. *Ibid*, ***1900***, *33*, 595; (d) Hauser, C.R.; Hudson, B.E. *Org. React.*, ***1942***, 1 266 (see p. 274); (e) Schaefer, J.P.; Bloomfield, J.J. *Ibid.*, ***1967***, *15*, 1.

149 Krogsgaard-Larsen, P.; Hjeds, H. *Acta Chem. Scand.*, ***1974***, *B28*, 533.

150 N'Goka, V.; Schlewer, G.; Linget, J.-M.; Chambon, J.-P.; Wermuth, C.-G. *J. Med. Chem.*, ***1991***, *34*, 2547.

151 Krogsgaard-Larsen, P. *Acta Chem. Scand.*, ***1977***, *B31*, 584.

152 Krogsgaard-Larsen, P.; Roldskov-Christiansen, T. *Eur. J. Med. - Chim. Ther.*, ***1979***, *14*, 157.

153 Botta, A. *Ger. Offen.* 2,359,990 [*Chem. Abstr.*, ***1975***, *83*: P97045t].

154 Morosawa, S. *Bull. Soc. Chem., Jpn.*, ***1958***, *31*, 418.

155 Krogsgaard-Larsen, P.; Hjeds, H. *Acta Chem. Scand.*, ***1976***, *B30*, 884.

Conclusions

The variety of structural features and the variety of synthetic approaches illustrated in this book shows the importance of non-α-amino acids. These amino acids are useful for their biological properties, as shown for many specific examples. They are also very important as synthetic intermediates that can be transformed into other important organic molecules. To those who deal with amino acids on a day-to-day basis, the value is obvious. The ability to use this chemistry in a variety of applications also makes this book useful to the organic chemical community in general. The synthesis of amino acids requires refunctionalization of other molecules and selective positioning of functional groups in the new molecule. Both of these concepts are critically important to organic synthesis and many of the syntheses presented here should be valuable in a general sense, as well as a strategic sense. The wide range of organic chemical reactions used for the synthesis of amino acids also suggests that these transformations are a microcosm of the mainstream of organic chemistry rather than simply one branch. The ability to use chiral templates to synthesize chiral, nonracemic amino acids offers important strategic lessons to other synthetic projects. The ability to use chiral, nonracemic amino acids as chiral templates opens the door to new applications.

I believe that this book represents an important compilation of synthetic methods for a class of organic molecules that are increasingly important in organic chemistry. The large number of synthetic applications is testament to their current and past value. There are many antibiotics, anti-cancer drugs, and drugs active against HIV that contain non-α-amino acids, or use them as synthetic intermediates. A discussion of these techniques is valuable to those working in those areas. I believe there is a more lasting value for this book. The techniques and strategies presented for the synthesis of past and current important molecules are also applicable to yet undiscovered antibiotics, anti-cancer drugs and anti-viral drugs. The assessment that the techniques in this book can be readily applied to future problems is, perhaps, the best conclusion for a book of this type.

Amino Acid Index

This index is arranged alphabetically, first by the parent carboxylic acid, and then with the positional number of the amino group or functionality. For derivatives, the substituents are listed under the carboxylic acid, amine: 4-amino-5-methyl, 3-phenyl-hexanoic acid is, for example, listed as:

hexanoic acid, 4-amino
    5-methyl, 3-phenyl   XXX

Author Index

Subject Index